Mobile Communications and Networks

Mobile Communications and Networks

Edited by
Kameron Smith

Larsen & Keller
www.larsen-keller.com

Mobile Communications and Networks
Edited by Kameron Smith
ISBN: 978-1-63549-188-3 (Hardback)

☰ Larsen & Keller

Published by Larsen and Keller Education,
5 Penn Plaza,
19th Floor,
New York, NY 10001, USA

Cataloging-in-Publication Data

Mobile communications and networks / edited by Kameron Smith.
    p. cm.
Includes bibliographical references and index.
ISBN 978-1-63549-188-3
1. Mobile communication systems. 2. Cell phone systems.
3. Wireless communication systems. 4. Telecommunication.
I. Smith, Kameron.
TK5103.2 .M63 2017
621.384--dc23

The publisher's policy is to use permanent paper from mills that operate a sustainable forestry policy. Furthermore, the publisher ensures that the text paper and cover boards used have met acceptable environmental accreditation standards.

Printed and bound in the United States of America.

For more information regarding Larsen and Keller Education and its products, please visit the publisher's website www.larsen-keller.com

Table of Contents

Preface

This book provides comprehensive insights into the field of mobile communications and networks. It elucidates the different theories and practices related to this field. Mobile communications refer to the science and technology of wireless telephone services, which is used with the help of satellites. This text is a valuable compilation of topics, ranging from the basic to the most complex theories and principles in the field of mobile communications. Those in search of information to further their knowledge will be greatly assisted by this textbook. It explores all the important aspects of this field in the present day scenario. It aims to serve as a resource guide for students and facilitate the growth of the discipline.

A short introduction to every chapter is written below to provide an overview of the content of the book:

Chapter 1 - Mobile telephony is the provision of telephone services which are not fixed in one place rather one can move around freely with them. Mobile phones are connected to a base station unlike satellite phones which are connected to satellites. This chapter will provide an integrated understanding of mobile communications; **Chapter 2 -** The essential concepts of mobile communications explained in the text are duplex, channel capacity, tethering, near field communication, voice over IP and etiquette in technology. A duplex communication system is a system that has two connected devices that communicate which each other in both directions. This text is a compilation of the various concepts of mobile communications; **Chapter 3 -** Modulation is the process of varying the properties of periodic waveform. The topics explained in this text are amplitude modulation, frequency modulation, phase modulation, frequency-shift keying and amplitude-shift keying. The technology that is used in the recent times for transmitting information is known as amplitude modulation. This chapter serves as a source to understand the concept of modulation and the techniques related to modulation; **Chapter 4 -** A cellular network is where the last connection is wireless. Cellular frequencies are a set of frequencies that are used for cellular phones. These frequencies are ultra high frequency bands. The types of frequency bands explained in this section are GSM frequency bands, UMTS frequency bands and E-UTRA. This text helps the reader in developing an integrated understanding of the topic; **Chapter 5 -** Channel access method, which spreads the area covered, permits several terminals to be connected to the same multi-point transmission medium. Some of the examples of channel access methods are wireless networks, bus networks, ring networks and point-to-point links. This chapter also focuses on topics like time division multiple access, code division multiple access, space-division multiple access and ALOHAnet; **Chapter 6 -** Mobile security is very important in contemporary times. The concern of security arises with the concern for the security of personal and business information that is stored in the smartphones. The following section helps the reader in understanding the importance of security in mobile phones; **Chapter 7 -** The effect that technology has on human health is one of the biggest concerns of our time. Wireless electronic devices cause radiation that are known to cause issues in humans. One of the concerns related to mobile phones is also the number of accidents that are caused by using mobile phones while driving. The topics discussed in the text are of great importance to broaden the existing knowledge on the subject matter;

Chapter 8 - Mobile communications has evolved over a period of decades. This section focuses on the history of telephones and the history of mobile phones. In order to have a profound understanding of mobile communications it is very essential to develop and to understand the history of the subject concerned.

Finally, I would like to thank my fellow scholars who gave constructive feedback and my family members who supported me at every step.

Editor

Introduction to Mobile Communications

Mobile telephony is the provision of telephone services which are not fixed in one place rather one can move around freely with them. Mobile phones are connected to a base station unlike satellite phones which are connected to satellites. This chapter will provide an integrated understanding of mobile communications.

Mobile telephony is the provision of telephone services to phones which may move around freely rather than stay fixed in one location. Mobile phones connect to a terrestrial cellular network of base stations (cell sites), whereas satellite phones connect to orbiting satellites. Both networks are interconnected to the public switched telephone network (PSTN) to allow any phone in the world to be dialed.

Mobile phone tower

In 2010 there were estimated to be five billion mobile cellular subscriptions in the world.

History

According to internal memos, American Telephone & Telegraph discussed developing a wireless phone in 1915, but were afraid that deployment of the technology could undermine its monopoly on wired service in the U.S.

Public mobile phone systems were first introduced in the years after the Second World War and made use of technology developed before and during the conflict. The first system opened in St Louis, Missouri, USA in 1946 whilst other countries followed in the succeeding decades. The UK introduced its 'System 1' manual radiotelephone service as the South Lancashire Radiophone Service in 1958. Calls were made via an operator using handsets identical to ordinary phone handsets. The phone itself was a large box located in the boot (trunk) of the vehicle containing valves and other early electronic components. Although an uprated manual service ('System 3') was extended to cover most of the UK, automation did not arrive until 1981 with 'System 4'. Although this non-cellular service, based on German B-Netz technology, was expanded rapidly throughout the UK between 1982 and 1985 and continued in operation for several years before finally closing in Scotland, it was overtaken by the introduction in January 1985 of two cellular systems - the British Telecom/Securicor 'Cellnet' service and the Racal/Millicom/Barclays 'Vodafone' (from voice + data + phone) service. These cellular systems were based on US Advanced Mobile Phone Service (AMPS) technology, the modified technology being named Total Access Communication System (TACS).

In 1947 Bell Labs was the first to propose a cellular radio telephone network. The primary innovation was the development of a network of small overlapping cell sites supported by a call switching infrastructure that tracks users as they move through a network and passes their calls from one site to another without dropping the connection. In 1956 the MTA system was launched in Sweden. The early efforts to develop mobile telephony faced two significant challenges: allowing a great number of callers to use the comparatively few available frequencies simultaneously and allowing users to seamlessly move from one area to another without having their calls dropped. Both problems were solved by Bell Labs employee Amos Joel who, in 1970 applied for a patent for a mobile communications system. However, a business consulting firm calculated the entire U.S. market for mobile telephones at 100,000 units and the entire worldwide market at no more than 200,000 units based on the ready availability of pay telephones and the high cost of constructing cell towers. As a consequence, Bell Labs concluded that the invention was "of little or no consequence," leading it not to attempt to commercialize the invention. The invention earned Joel induction into the National Inventors Hall of Fame in 2008. The first call on a handheld mobile phone was made on April 3, 1973 by Martin Cooper, then of Motorola to his opposite number in Bell Labs who were also racing to be first. Bell Labs went on to install the first trial cellular network in Chicago in 1978. This trial system was licensed by the FCC to ATT for commercial use in 1982 and, as part of the divestiture arrangements for the breakup of ATT, the AMPS technology was distributed to local telcos. The first commercial system opened in Chicago in October 1983. A system designed by Motorola also operated in the Washington D.C./Baltimore area from summer 1982 and became a full public service later the following year. Japan's first commercial radiotelephony service was launched by NTT in 1978.

The first fully automatic first generation cellular system was the Nordic Mobile Telephone (NMT) system, simultaneously launched in 1981 in Denmark, Finland, Norway and Sweden. NMT was the first mobile phone network featuring international roaming. The Swedish electrical engineer Östen Mäkitalo started to work on this vision in 1966, and is considered as the father of the NMT system and some consider him also the father of the cellular phone.

The advent of cellular technology encouraged European countries to co-operate in the development of a pan-European cellular technology to rival those of the US and Japan. This resulted in the GSM system, the initials originally from the *Groupe Spécial Mobile* that was charged with the specification and development tasks but latterly as the 'Global System for Mobile Communications'. The GSM standard eventually spread outside Europe and is now the most widely used cellular technology in the world and the de facto standard. The industry association, the GSMA, now represents 219 countries and nearly 800 mobile network operators. There are now estimated to be over 5 billion phone subscriptions according to the "List of countries by number of mobile phones in use" (although some users have multiple subscriptions, or inactive subscriptions), which also makes the mobile phone the most widely spread technology and the most common electronic device in the world.

The first mobile phone to enable internet connectivity and wireless email, the Nokia Communicator, was released in 1996, creating a new category of multi-use devices called smartphones. In 1999 the first mobile internet service was launched by NTT DoCoMo in Japan under the i-Mode service. By 2007 over 798 million people around the world accessed the internet or equivalent mobile internet services such as WAP and i-Mode at least occasionally using a mobile phone rather than a personal computer.

Cellular Systems

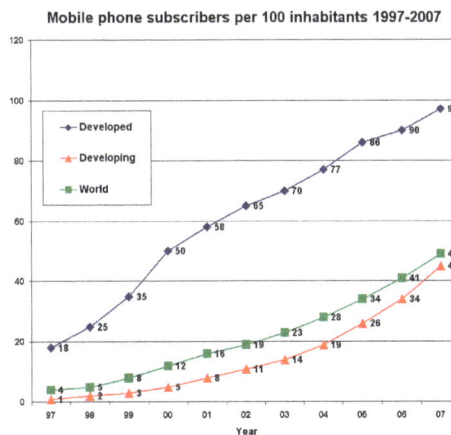

Mobile phone subscriptions, not subscribers, per 100 inhabitants 1997-2007

Mobile phones send and receive radio signals with any number of cell site base stations fitted with microwave antennas. These sites are usually mounted on a tower, pole or building, located throughout populated areas, then connected to a cabled communication network and switching system. The phones have a low-power transceiver that transmits voice and data to the nearest cell sites, normally not more than 8 to 13 km (approximately 5 to 8 miles) away. In areas of low coverage, a cellular repeater may be used, which uses a long distance high-gain dish antenna or yagi antenna to communicate with a cell tower far outside of normal range, and a repeater to re-broadcast on a small short-range local antenna that allows any cellphone within a few meters to function properly.

When the mobile phone or data device is turned on, it registers with the mobile telephone exchange, or switch, with its unique identifiers, and can then be alerted by the mobile switch when there is an incoming telephone call. The handset constantly listens for the strongest signal being received from the surrounding base stations, and is able to switch seamlessly between sites. As the user moves around the network, the "handoffs" are performed to allow the device to switch sites without interrupting the call.

Cell sites have relatively low-power (often only one or two watts) radio transmitters which broadcast their presence and relay communications between the mobile handsets and the switch. The switch in turn connects the call to another subscriber of the same wireless service provider or to the public telephone network, which includes the networks of other wireless carriers. Many of these sites are camouflaged to blend with existing environments, particularly in scenic areas.

The dialogue between the handset and the cell site is a stream of digital data that includes digitised audio (except for the first generation analog networks). The technology that achieves this depends on the system which the mobile phone operator has adopted. The technologies are grouped by generation. The first-generation systems started in 1979 with Japan, are all analog and include AMPS and NMT. Second-generation systems, started in 1991 in Finland, are all digital and include GSM, CDMA and TDMA.

The nature of cellular technology renders many phones vulnerable to 'cloning': anytime a cell phone moves out of coverage (for example, in a road tunnel), when the signal is re-established, the phone sends out a 're-connect' signal to the nearest cell-tower, identifying itself and signalling that it is again ready to transmit. With the proper equipment, it's possible to intercept the re-connect signal and encode the data it contains into a 'blank' phone—in all respects, the 'blank' is then an exact duplicate of the real phone and any calls made on the 'clone' will be charged to the original account. This problem was widespread with the first generation analogue technology, however the modern digital standards such as GSM greatly improve security and make cloning hard to achieve.

In an effort to limit the potential harm from having a transmitter close to the user's body, the first fixed/mobile cellular phones that had a separate transmitter, vehicle-mounted antenna, and handset (known as *car phones* and *bag phones*) were limited to a maximum 3 watts Effective Radiated Power. Modern *handheld* cellphones which must have the transmission antenna held inches from the user's skull are limited to a maximum transmission power of 0.6 watts ERP. Regardless of the potential biological effects, the reduced transmission range of modern handheld phones limits their usefulness in rural locations as compared to car/bag phones, and handhelds require that cell towers are spaced much closer together to compensate for their lack of transmission power.

Usage

By Civilians

This Railfone found on some Amtrak trains in North America uses cellular technology.

An increasing number of countries, particularly in Europe, now have more mobile phones than people. According to the figures from Eurostat, the European Union's in-house statistical office, Luxembourg had the highest mobile phone penetration rate at 158 mobile subscriptions per 100 people, closely followed by Lithuania and Italy. In Hong Kong the penetration rate reached 139.8% of the population in July 2007. Over 50 countries have mobile phone subscription penetration rates higher than that of the population and the Western European average penetration rate was 110% in 2007 (source Informa 2007). Canada currently has the lowest rates of mobile phone penetrations in the industrialised world at 58%.

There are over five hundred million active mobile phone accounts in China, as of 2007, but the total penetration rate there still stands below 50%. The total number of mobile phone subscribers in the world was estimated at 2.14 billion in 2005. The subscriber count reached 2.7 billion by end of 2006 according to Informa, and 3.3 billion by November, 2007, thus reaching an equivalent of over half the planet's population. Around 80% of the world's population has access to mobile phone coverage, as of 2006. This figure is expected to increase to 90% by the year 2010.

In some developing countries with little "landline" telephone infrastructure, mobile phone use has quadrupled in the last decade. The rise of mobile phone technology in developing countries is often cited as an example of the leapfrog effect. Many remote regions in the third world went from having no telecommunications infrastructure to having satellite based communications systems. At present, Africa has the largest growth rate of cellular subscribers in the world, its markets expanding nearly twice as fast as Asian markets. The availability of prepaid or 'pay-as-you-go' services, where the subscriber is not committed to a long term contract, has helped fuel this growth in Africa as well as in other continents.

On a numerical basis, India is the largest growth market, adding about 6 million mobile phones every month. It currently has a mobile subscriber base of 937.06 million mobile phones.

Traffic

Since the world is operating quickly to 3G and 4G networks, mobile traffic through video is heading high. It is expected that by end of 2018, the global traffic will reach an annual rate of 190 exabytes/year. This is the result of people shifting to smart phones now-a-days. It is predicted by 2018, mobile traffic will reach by 10 billion connections with 94% traffic comes from Smartphones, laptops and tablets. Also 69% of mobile traffic from Videos since we have high definition screens available in smart phones and 176.9 wearable devices to be at use. Apparently, 4G will be dominating the traffic by 51% of total mobile data by 2018.

By Government Agencies

Law Enforcement

Law enforcement have used mobile phone evidence in a number of different ways. Evidence about the physical location of an individual at a given time can be obtained by triangulating the individual's cellphone between several cellphone towers. This triangulation technique can be used to show that an individual's cellphone was at a certain location at a certain time. The concerns over terrorism and terrorist use of technology prompted an inquiry by the British House of Commons

Home Affairs Select Committee into the use of evidence from mobile phone devices, prompting leading mobile telephone forensic specialists to identify forensic techniques available in this area. NIST have published guidelines and procedures for the preservation, acquisition, examination, analysis, and reporting of digital information present on mobile phones can be found under the NIST Publication SP800-101.

In the UK in 2000 it was claimed that recordings of mobile phone conversations made on the day of the Omagh bombing were crucial to the police investigation. In particular, calls made on two mobile phones which were tracked from south of the Irish border to Omagh and back on the day of the bombing, were considered of vital importance.

Further example of criminal investigations using mobile phones is the initial location and ultimate identification of the terrorists of the 2004 Madrid train bombings. In the attacks, mobile phones had been used to detonate the bombs. However, one of the bombs failed to detonate, and the SIM card in the corresponding mobile phone gave the first serious lead about the terrorists to investigators. By tracking the whereabouts of the SIM card and correlating other mobile phones that had been registered in those areas, police were able to locate the terrorists.

Disaster Response

The Finnish government decided in 2005 that the fastest way to warn citizens of disasters was the mobile phone network. In Japan, mobile phone companies provide immediate notification of earthquakes and other natural disasters to their customers free of charge. In the event of an emergency, disaster response crews can locate trapped or injured people using the signals from their mobile phones. An interactive menu accessible through the phone's Internet browser notifies the company if the user is safe or in distress. In Finland rescue services suggest hikers carry mobile phones in case of emergency even when deep in the forests beyond cellular coverage, as the radio signal of a cellphone attempting to connect to a base station can be detected by overflying rescue aircraft with special detection gear. Also, users in the United States can sign up through their provider for free text messages when an AMBER Alert goes out for a missing person in their area.

However, most mobile phone networks operate close to capacity during normal times, and spikes in call volumes caused by widespread emergencies often overload the system just when it is needed the most. Examples reported in the media where this has occurred include the September 11, 2001 attacks, the 2003 Northeast blackouts, the 2005 London Tube bombings, Hurricane Katrina, the 2006 Hawaii earthquake, and the 2007 Minnesota bridge collapse.

Under FCC regulations, all mobile telephones must be capable of dialing emergency telephone numbers, regardless of the presence of a SIM card or the payment status of the account.

Impact on Society

Human Health

Since the introduction of mobile phones, concerns (both scientific and public) have been raised about the potential health impacts from regular use. But by 2008, American mobile phones transmitted and received more text messages than phone calls. Numerous studies have reported no significant relationship between mobile phone use and health, but the effect of mobile phone usage on health continues to be an area of public concern.

For example, at the request of some of their customers, Verizon created usage controls that meter service and can switch phones off, so that children could get some sleep. There have also been attempts to limit use by persons operating moving trains or automobiles, coaches when writing to potential players on their teams, and movie theater audiences. By one measure, nearly 40% of automobile drivers aged 16 to 30 years old text while driving, and by another, 40% of teenagers said they could text blindfolded.

18 studies have been conducted on the link between cell phones and brain cancer; A review of these studies found that cell phone use of 10 years or more "give a consistent pattern of an increased risk for acoustic neuroma and glioma". The tumors are found mostly on the side of the head that the mobile phone is in contact with. In July 2008, Dr. Ronald Herberman, director of the University of Pittsburgh Cancer Institute, warned about the radiation from mobile phones. He stated that there was no definitive proof of the link between mobile phones and brain tumors but there was enough studies that mobile phone usage should be reduced as a precaution. To reduce the amount of radiation being absorbed hands free devices can be used or texting could supplement calls. Calls could also be shortened or limit mobile phone usage in rural areas. Radiation is found to be higher in areas that are located away from mobile phone towers.

According to Reuters, The British Association of Dermatologists is warning of a rash occurring on people's ears or cheeks caused by an allergic reaction from the nickel surface commonly found on mobile devices' exteriors. There is also a theory it could even occur on the fingers if someone spends a lot of time text messaging on metal menu buttons. In 2008, Lionel Bercovitch of Brown University in Providence, Rhode Island, and his colleagues tested 22 popular handsets from eight different manufacturers and found nickel on 10 of the devices.

Human Behaviour

Culture and Customs

Between the 1980s and the 2000s, the mobile phone has gone from being an expensive item used by the business elite to a pervasive, personal communications tool for the general population. In most countries, mobile phones outnumber land-line phones, with fixed landlines numbering 1.3 billion but mobile subscriptions 3.3 billion at the end of 2007.

Cellular phones allow people to communicate from almost anywhere at their leisure.

In many markets from Japan and South Korea, to Europe, to Malaysia, Singapore, Taiwan and Hong Kong, most children age 8-9 have mobile phones and the new accounts are now opened for customers aged 6 and 7. Where mostly parents tend to give hand-me-down used phones to their youngest children, in Japan already new cameraphones are on the market whose target age group is under 10 years of age, introduced by KDDI in February 2007. The USA also lags on this measure, as in the US so far, about half of all children have mobile phones. In many young adults' households it has supplanted the land-line phone. Mobile phone usage is banned in some countries, such as North Korea and restricted in some other countries such as Burma.

Given the high levels of societal mobile phone service penetration, it is a key means for people to communicate with each other. The SMS feature spawned the "texting" sub-culture amongst younger users. In December 1993, the first person-to-person SMS text message was transmitted in Finland. Currently, texting is the most widely used data service; 1.8 billion users generated $80 billion of revenue in 2006 (source ITU). Many phones offer Instant Messenger services for simple, easy texting. Mobile phones have Internet service (e.g. NTT DoCoMo's i-mode), offering text messaging via e-mail in Japan, South Korea, China, and India. Most mobile internet access is much different from computer access, featuring alerts, weather data, e-mail, search engines, instant messages, and game and music downloading; most mobile internet access is hurried and short.

Because mobile phones are often used publicly, social norms have been shown to play a major role in the usage of mobile phones. Furthermore, the mobile phone can be a fashion totem custom-decorated to reflect the owner's personality and may be a part of their self-identity. This aspect of the mobile telephony business is, in itself, an industry, e.g. ringtone sales amounted to $3.5 billion in 2005. Mobile phone use on aircraft is starting to be allowed with several airlines already offering the ability to use phones during flights. Mobile phone use during flights used to be prohibited and many airlines still claim in their in-plane announcements that this prohibition is due to possible interference with aircraft radio communications. Shut-off mobile phones do not interfere with aircraft avionics. The recommendation why phones should not be used during take-off and landing, even on planes that allow calls or messaging, is so that passengers pay attention to the crew for any possible accident situations, as most aircraft accidents happen on take-off and landing.

Etiquette

Mobile phone use can be an important matter of social discourtesy: phones ringing during funerals or weddings; in toilets, cinemas and theatres. Some book shops, libraries, bathrooms, cinemas, doctors' offices and places of worship prohibit their use, so that other patrons will not be disturbed by conversations. Some facilities install signal-jamming equipment to prevent their use, although in many countries, including the US, such equipment is illegal.

Many US cities with subway transit systems underground are studying or have implemented mobile phone reception in their underground tunnels for their riders, and trains, particularly those involving long-distance services, often offer a "quiet carriage" where phone use is prohibited, much like the designated non-smoking carriage of the past. Most schools in the United States and Europe and Canada have prohibited mobile phones in the classroom, or in school in an effort to limit class disruptions.

A working group made up of Finnish telephone companies, public transport operators and communications authorities has launched a campaign to remind mobile phone users of courtesy, especially when using mass transit—what to talk about on the phone, and how to. In particular, the campaign wants to impact loud mobile phone usage as well as calls regarding sensitive matters.

Use by Drivers

The use of mobile phones by people who are driving has become increasingly common, for example as part of their job, as in the case of delivery drivers who are calling a client, or socially as for commuters who are chatting with a friend. While many drivers have embraced the convenience of using their cellphone while driving, some jurisdictions have made the practice against the law, such as Australia, the Canadian provinces of British Columbia, Quebec, Ontario, Nova Scotia, and Newfoundland and Labrador as well as the United Kingdom, consisting of a zero-tolerance system operated in Scotland and a warning system operated in England, Wales, and Northern Ireland. Officials from these jurisdictions argue that using a mobile phone while driving is an impediment to vehicle operation that can increase the risk of road traffic accidents.

Studies have found vastly different relative risks (RR). Two separate studies using case-crossover analysis each calculated RR at 4, while an epidemiological cohort study found RR, when adjusted for crash-risk exposure, of 1.11 for men and 1.21 for women.

A simulation study from the University of Utah Professor David Strayer compared drivers with a blood alcohol content of 0.08% to those conversing on a cell phone, and after controlling for driving difficulty and time on task, the study concluded that cell phone drivers exhibited greater impairment than intoxicated drivers. Meta-analysis by The Canadian Automobile Association and The University of Illinois found that response time while using both hands-free and hand-held phones was approximately 0.5 standard deviations higher than normal driving (i.e., an average driver, while talking on a cell phone, has response times of a driver in roughly the 40th percentile).

Driving while using a hands-free device is not safer than driving while using a hand-held phone, as concluded by case-crossover studies. epidemiological studies, simulation studies, and meta-analysis. Even with this information, California initiated new Wireless Communications Device Law (effective January 1, 2009) makes it an infraction to write, send, or read text-based communication on an electronic wireless communications device, such as a cell phone, while driving a motor vehicle. Two additional laws dealing with the use of wireless telephones while driving went into effect July 1, 2008. The first law prohibits all drivers from using a handheld wireless telephone while operating a motor vehicle. The law allows a driver to use a wireless telephone to make emergency calls to a law enforcement agency, a medical provider, the fire department, or other emergency services agency. The base fine for the FIRST offense is $20 and $50 for subsequent convictions. With penalty assessments, the fine can be more than triple the base fine amount. videos about California cellular phone laws; with captions (California Vehicle Code [VC] §23123). Motorists 18 and over may use a "hands-free device. The second law effective July 1, 2008, prohibits drivers under the age of 18 from using a wireless telephone or hands-free device while operating a motor vehicle (VC §23124)The consistency of increased crash risk between hands-free and hand-held phone use is at odds with legislation in over 30 countries that prohibit hand-held phone use but allow hands-free. Scientific literature is mixed on the dangers of talking on a phone versus those of talking with a passenger, with the Accident Research Unit at the University of Nottingham finding that the number of utterances was usually higher for mobile calls when compared to blindfolded and non-blindfolded passengers, but the University of Illinois meta-analysis concluding that passenger conversations were just as costly to driving performance as cell phone ones.

Use on Aircraft

As of 2007, several airlines are experimenting with base station and antenna systems installed on the airplane, allowing low power, short-range connection of any phones aboard to remain connected to the aircraft's base station. Thus, they would not attempt connection to the ground base stations as during take off and landing. Simultaneously, airlines may offer phone services to their travelling passengers either as full voice and data services, or initially only as SMS text messaging and similar services. The Australian airline Qantas is the first airline to run a test aeroplane in this configuration in the autumn of 2007. Emirates has announced plans to allow limited mobile phone usage on some flights. However, in the past, commercial airlines have prevented the use of cell phones and laptops, due to the assertion that the frequencies emitted from these devices may disturb the radio waves contact of the airplane.

On March 20, 2008, an Emirates flight was the first time voice calls have been allowed in-flight on commercial airline flights. The breakthrough came after the European Aviation Safety Agency (EASA) and the United Arab Emirates-based General Civil Aviation Authority (GCAA) granted full approval for the AeroMobile system to be used on Emirates. Passengers were able to make and receive voice calls as well as use text messaging. The system automatically came into operation as the Airbus A340-300 reached cruise altitude. Passengers wanting to use the service received a text message welcoming them to the AeroMobile system when they first switched their phones on. The approval by EASA has established that GSM phones are safe to use on airplanes, as the AeroMobile system does not require the modification of aircraft components deemed "sensitive," nor does it require the use of modified phones.

In any case, there are inconsistencies between practices allowed by different airlines and even on the same airline in different countries. For example, Delta Air Lines may allow the use of mobile phones immediately after landing on a domestic flight within the US, whereas they may state "not until the doors are open" on an international flight arriving in the Netherlands. In April 2007 the US Federal Communications Commission officially prohibited passengers' use of cell phones during a flight.

In a similar vein, signs are put up in many countries, such as Canada, the UK and the U.S., at petrol stations prohibiting the use of mobile phones, due to possible safety issues. However, it is unlikely that mobile phone use can cause any problems, and in fact "petrol station employees have themselves spread the rumour about alleged incidents."

Environmental Impacts

Like all high structures, cellular antenna masts pose a hazard to low flying aircraft. Towers over a certain height or towers that are close to airports or heliports are normally required to have warning lights. There have been reports that warning lights on cellular masts, TV-towers and other high structures can attract and confuse birds. US authorities estimate that millions of birds are killed near communication towers in the country each year.

Cellular antenna disguised to look like a tree

Some cellular antenna towers have been camouflaged to make them less obvious on the horizon, and make them look more like a tree.

An example of the way mobile phones and mobile networks have sometimes been perceived as a threat is the widely reported and later discredited claim that mobile phone masts are associated with the Colony Collapse Disorder (CCD) which has reduced bee hive numbers by up to 75% in many areas, especially near cities in the US. The Independent newspaper cited a scientific study claiming it provided evidence for the theory that mobile phone masts *are* a major cause in the collapse of bee populations, with controlled experiments demonstrating a rapid and catastrophic effect on individual hives near masts. Mobile phones were in fact not covered in the study, and the original researchers have since emphatically disavowed any connection between their research, mobile phones, and CCD, specifically indicating that the Independent article had misinterpreted their results and created "a horror story". While the initial claim of damage to bees was widely reported, the corrections to the story were almost non-existent in the media.

There are more than 500 million used mobile phones in the US sitting on shelves or in landfills, and it is estimated that over 125 million will be discarded this year alone. The problem is growing at a rate of more than two million phones per week, putting tons of toxic waste into landfills daily. Several companies offer to buy back and recycle mobile phones from users. In the United States many unwanted but working mobile phones are donated to women's shelters to allow emergency communication.

Tariff Models

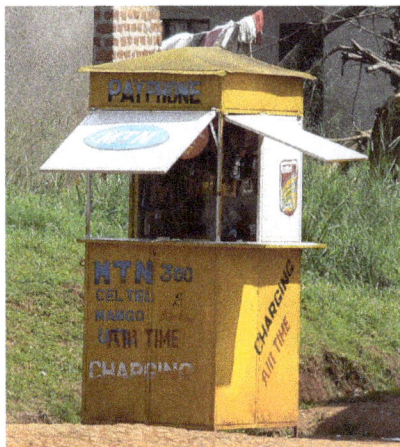

Mobile phone shop in Uganda

Payment Methods

There are two principal ways to pay for mobile telephony: the 'pay-as-you-go' model where conversation time is purchased and added to a phone unit via an Internet account or in shops or ATMs, or the contract model where bills are paid by regular intervals after the service has been consumed. It is increasingly common for a consumer to purchase a basic package and then bolt-on services and functionality to create a subscription customised to the users needs.

Pay as you go (also known as "pre-pay" or "prepaid") accounts were invented simultaneously in Portugal and Italy and today form more than half of all mobile phone subscriptions. USA, Canada, Costa Rica, Japan, Israel and Finland are among the rare countries left where most phones are still contract-based.

Incoming Call Charges

In the early days of mobile telephony, the operators (carriers) charged for all air time consumed by the mobile phone user, which included both outbound and inbound telephone calls. As mobile phone adoption rates increased, competition between operators meant that some decided not to charge for incoming calls in some markets (also called "calling party pays").

The European market adopted a calling party pays model throughout the GSM environment and soon various other GSM markets also started to emulate this model.

In Hong Kong, Singapore, Canada, and the United States, it is common for the party receiving the call to be charged per minute, although a few carriers are beginning to offer unlimited received phone calls. This is called the "Receiving Party Pays" model. In China, it was reported that both of its two operators will adopt the caller-pays approach as early as January 2007.

One disadvantage of the receiving party pays systems is that phone owners keep their phones turned off to avoid receiving unwanted calls, which results in the total voice usage rates (and profits) in Calling Party Pays countries outperform those in Receiving Party Pays countries. To avoid the problem of users keeping their phone turned off, most Receiving Party Pays countries have either switched to Calling Party Pays, or their carriers offer additional incentives such as a large number of monthly minutes at a sufficiently discounted rate to compensate for the inconvenience.

Note that when a user roaming in another country, international roaming tariffs apply to all calls received, regardless of the model adopted in the home country.

References

- Domain, Public. "Text Messaging Law Effective January 1, 2009 Cellular Phone Laws Effective July 1, 2008". California Department of Motor Vehicles. California, USA: State of California. Retrieved 19 August 2013.

- Rachel Lieberman, Brandel France de Bravo, MPH, and Diana Zuckerman, Ph.D. "Can Cell Phones Harm Our Health?" National Research Center for Women and Families. August 2008. Retrieved August 13, 2013.

- "The BBC. "US Cancer Boss in Mobiles Warning." BBC News 24 July 2008. Retrieved November 20, 2008". BBC News. 2008-07-24. Retrieved 2011-07-11.

- The Committee Office, House of Commons. "Supplementary memorandum submitted by Gregory Smith". Publications.parliament.uk. Retrieved 2011-07-11.

- Lean, Geoffrey; Shawcross, Harriet (2007-04-15). "Are mobile phones wiping out our bees?". The Independent. UK. Retrieved 2010-05-12.

2

Essential Concepts of Mobile Communications

The essential concepts of mobile communications explained in the text are duplex, channel capacity, tethering, near field communication, voice over IP and etiquette in technology. A duplex communication system is a system that has two connected devices that communicate which each other in both directions. This text is a compilation of the various concepts of mobile communications.

Duplex (Telecommunications)

A duplex communication system is a point-to-point system composed of two connected parties or devices that can communicate with one another in both directions. "Du" comes from "duo" that means "double", and "plex" that means "structure" or "parts of"; thus, a duplex system has two clearly defined data transmissions, with each path carrying information in only one direction: A to B over one path, and B to A over the other. There are two types of duplex communication systems: full-Duplex and half-Duplex.

In a full duplex system, both parties can communicate with each other simultaneously. An example of a full-duplex device is a telephone; the parties at both ends of a call can speak and be heard by the other party simultaneously. The earphone reproduces the speech of the remote party as the microphone transmits the speech of the local party, because there is a two-way communication channel between them, or more strictly speaking, because there are two communication paths/channels between them.

In a half-duplex system, there are still two clearly defined paths/channels, and each party can communicate with the other but not simultaneously; the communication is one direction at a time. An example of a half-duplex device is a walkie-talkie two-way radio that has a "push-to-talk" button; when the local user wants to speak to the remote person they push this button, which turns on the transmitter but turns off the receiver, so they cannot hear the remote person. To listen to the other person they release the button, which turns on the receiver but turns off the transmitter.

Duplex systems are employed in many communications networks, either to allow for a communication "two-way street" between two connected parties or to provide a "reverse path" for the monitoring and remote adjustment of equipment in the field.

Systems that do not need the duplex capability may instead use simplex communication, in which one device transmits and the others can only "listen". Examples are broadcast radio and television, garage door openers, baby monitors, wireless microphones, and surveillance cameras. In these devices the communication is only in one direction.

Half-duplex

A *half-duplex* (HDX) system provides communication in both directions, but only one direction at a time (not simultaneously). Typically, once a party begins receiving a signal, it must wait for the transmitter to stop transmitting, before replying.

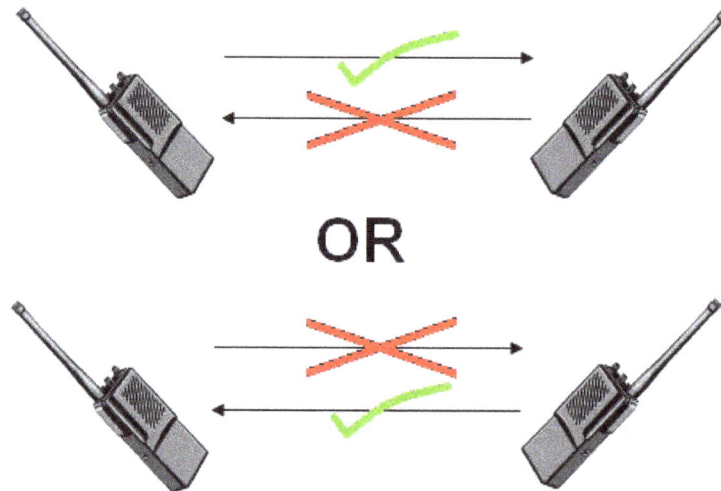

A simple illustration of a half-duplex communication system

An example of a half-duplex system is a two-party system such as a walkie-talkie, wherein one must use "over" or another previously designated keyword to indicate the end of transmission, and ensure that only one party transmits at a time, because both parties transmit and receive on the same frequency. A good analogy for a half-duplex system would be a one-lane road with traffic controllers at each end, such as a two-lane bridge under re-construction. Traffic can flow in both directions, but only one direction at a time, regulated by the traffic controllers.

Half-duplex systems are usually used to conserve bandwidth, since only a single communication channel is needed, which is shared alternately between the two directions. For example, a walkie-talkie requires only a single frequency for bidirectional communication, while a cell phone, which is a full-duplex device, requires two frequencies to carry the two simultaneous voice channels, one in each direction.

In automatically run communications systems, such as two-way data-links, the time allocations for communications in a half-duplex system can be firmly controlled by the hardware. Thus, there is no waste of the channel for switching. For example, station A on one end of the data link could be allowed to transmit for exactly one second, then station B on the other end could be allowed to transmit for exactly one second, and then the cycle repeats.

In half-duplex systems, if more than one party transmits at the same time, a collision occurs, resulting in lost messages.

Full-duplex

A *full-duplex* (FDX) system, or sometimes called *double-duplex*, allows communication in both directions, and, unlike half-duplex, allows this to happen simultaneously. Land-line telephone networks are full-duplex, since they allow both callers to speak and be heard at the same time, with the transition from four to two wires being achieved by a hybrid coil in a telephone hybrid. Modern cell phones are also full-duplex.

A simple illustration of a full-duplex communication system. Full-duplex is not common in handheld radios as shown here due to the cost and complexity of common duplexing methods, but is used in telephones, cellphones and cordless phones.

A good analogy for a full-duplex system is a two-lane road with one lane for each direction. Moreover, in most full-duplex mode systems carrying computer data, transmitted data does not appear to be sent until it has been received and an acknowledgment is sent back by the other party; that way, such systems implement reliable transmission methods.

Two-way radios can be designed as full-duplex systems, transmitting on one frequency and receiving on another; this is also called frequency-division duplex. Frequency-division duplex systems can extend their range by using sets of simple repeater stations because the communications transmitted on any single frequency always travel in the same direction.

Full-duplex Ethernet connections work by making simultaneous use of two physical twisted pairs inside the same jacket, which are directly connected to each networked device: one pair is for receiving packets, while the other pair is for sending packets. This effectively makes the cable itself a collision-free environment and doubles the maximum total transmission capacity supported by each Ethernet connection.

Full-duplex has also several benefits over the use of half-duplex. First, there are no collisions so time is not wasted by having to retransmit frames. Second, full transmission capacity is available in both directions because the send and receive functions are separate. Third, since there is only one transmitter on each twisted pair, stations (nodes) do not need to wait for others to complete their transmissions.

Some computer-based systems of the 1960s and 1970s required full-duplex facilities, even for half-duplex operation, since their poll-and-response schemes could not tolerate the slight delays in reversing the direction of transmission in a half-duplex line.

Full-duplex Emulation

Where channel access methods are used in point-to-multipoint networks (such as cellular networks) for dividing forward and reverse communication channels on the same physical communications medium, they are known as duplexing methods, such as *time-division duplexing* and *frequency-division duplexing*.

Time-division Duplexing

Time-division duplexing (TDD) is the application of time-division multiplexing to separate outward and return signals. It emulates full duplex communication over a half duplex communication link.

Time-division duplexing has a strong advantage in the case where there is asymmetry of the uplink and downlink data rates. As the amount of uplink data increases, more communication capacity can be dynamically allocated, and as the traffic load becomes lighter, capacity can be taken away. The same applies in the downlink direction.

For radio systems that aren't moving quickly, another advantage is that the uplink and downlink radio paths are likely to be very similar. This means that techniques such as beamforming work well with TDD systems.

Examples of time-division duplexing systems are:

- UMTS 3G supplementary air interfaces TD-CDMA for indoor mobile telecommunications.

- The Chinese TD-LTE 4-G, TD-SCDMA 3-G mobile communications air interface.

- DECT wireless telephony

- Half-duplex packet switched networks based on carrier sense multiple access, for example 2-wire or hubbed Ethernet, Wireless local area networks and Bluetooth, can be considered as time-division duplexing systems, albeit not TDMA with fixed frame-lengths.

- IEEE 802.16 WiMAX

- PACTOR

- ISDN BRI U interface, variants using the time-compression multiplex (TCM) line system

- G.fast, a digital subscriber line (DSL) standard under development by the ITU-T

Frequency-division Duplexing

Frequency-division duplexing (FDD) means that the transmitter and receiver operate at different carrier frequencies. The term is frequently used in ham radio operation, where an operator is attempting to contact a repeater station. The station must be able to send and receive a transmission at the same time, and does so by slightly altering the frequency at which it sends and receives. This mode of operation is referred to as *duplex mode* or *offset mode*.

Uplink and downlink sub-bands are said to be separated by the *frequency offset*. Frequency-division duplexing can be efficient in the case of symmetric traffic. In this case time-division duplexing tends to waste bandwidth during the switch-over from transmitting to receiving, has greater inherent latency, and may require more complex circuitry.

Another advantage of frequency-division duplexing is that it makes radio planning easier and more efficient, since base stations do not "hear" each other (as they transmit and receive in different sub-bands) and therefore will normally not interfere with each other. On the converse, with time-division duplexing systems, care must be taken to keep guard times between neighboring base stations (which decreases spectral efficiency) or to synchronize base stations, so that they will transmit and receive at the same time (which increases network complexity and therefore cost, and reduces bandwidth allocation flexibility as all base stations and sectors will be forced to use the same uplink/downlink ratio)

Examples of frequency-division duplexing systems are:

- ADSL and VDSL

- Most cellular systems, including the UMTS/WCDMA use frequency-division duplexing mode and the cdma2000 system.

- IEEE 802.16 WiMax also uses frequency-division duplexing mode.

Echo Cancellation

Full-duplex audio systems like telephones can create echo, which needs to be removed. Echo occurs when the sound coming out of the speaker, originating from the far end, re-enters the microphone and is sent back to the far end. The sound then reappears at the original source end, but delayed. This feedback path may be acoustic, through the air, or it may be mechanically coupled, for example in a telephone handset. Echo cancellation is a signal-processing operation that subtracts the far-end signal from the microphone signal before it is sent back over the network.

Echo cancellation is important to the V.32, V.34, V.56, and V.90 modem standards.

Echo cancelers are available as both software and hardware implementations. They can be independent components in a communications system or integrated into the communication system's central processing unit. Devices that do not eliminate echo sometimes will not produce good full-duplex performance.

Channel Capacity

In electrical engineering, computer science and information theory, channel capacity is the tight upper bound on the rate at which information can be reliably transmitted over a communications channel.

By the noisy-channel coding theorem, the channel capacity of a given channel is the limiting information rate (in units of information per unit time) that can be achieved with arbitrarily small error probability.

Information theory, developed by Claude E. Shannon during World War II, defines the notion of channel capacity and provides a mathematical model by which one can compute it. The key result states that the capacity of the channel, as defined above, is given by the maximum of the mutual information between the input and output of the channel, where the maximization is with respect to the input distribution.

The notion of channel capacity has been central to the development of modern wireline and wireless communication systems, with the advent of novel error correction coding mechanisms that have resulted in achieving performance very close to the limits promised by channel capacity.

Formal Definition

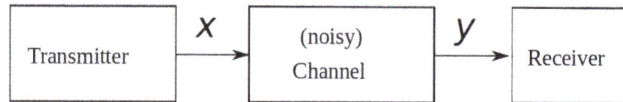

Let X and Y be the random variables representing the input and output of the channel, respectively. Let $p_{Y|X}(y\,|\,x)$ be the conditional distribution function of Y given X, which is an inherent fixed property of the communications channel. Then the choice of the marginal distribution $p_X(x)$ completely determines the joint distribution $p_{X,Y}(x,y)$ due to the identity

$$p_{X,Y}(x,y) = p_{Y|X}(y\,|\,x)p_X(x)$$

which, in turn, induces a mutual information $I(X;Y)$. The channel capacity is defined as

$$C = \sup_{p_X(x)} I(X;Y)$$

where the supremum is taken over all possible choices of $p_X(x)$.

Shannon Capacity of a Graph

If G is an undirected graph, it can be used to define a communications channel in which the symbols are the graph vertices, and two codewords may be confused with each other if their symbols in each position are equal or adjacent. The computational complexity of finding the Shannon capacity of such a channel remains open, but it can be upper bounded by another important graph invariant, the Lovász number.

Noisy-channel Coding Theorem

The noisy-channel coding theorem states that for any $\varepsilon > 0$ and for any transmission rate R less than the channel capacity C, there is an encoding and decoding scheme transmitting data at rate R whose error probability is less than ε, for a sufficiently large block length. Also, for any rate greater than the channel capacity, the probability of error at the receiver goes to one as the block length goes to infinity.

Example Application

An application of the channel capacity concept to an additive white Gaussian noise (AWGN) channel with B Hz bandwidth and signal-to-noise ratio S/N is the Shannon–Hartley theorem:

$$C = B\log_2\left(1 + \frac{S}{N}\right)$$

C is measured in bits per second if the logarithm is taken in base 2, or nats per second if the natural logarithm is used, assuming B is in hertz; the signal and noise powers S and N are measured in watts or volts², so the signal-to-noise ratio here is expressed as a power ratio, *not* in decibels (dB); since figures are often cited in dB, a conversion may be needed. For example, 30 dB is a power ratio of $10^{30/10} = 10^3 = 1000$.

Channel Capacity in Wireless Communications

This section focuses on the single-antenna, point-to-point scenario. For channel capacity in systems with multiple antennas.

AWGN Channel

If the average received power is $\bar{P}$ [W] and the noise power spectral density is N_0 [W/Hz], the AWGN channel capacity is

$$C_{\text{AWGN}} = W \log_2 \left(1 + \frac{\bar{P}}{N_0 W} \right) \text{ [bits/s]},$$

where $\dfrac{\bar{P}}{N_0 W}$ is the received signal-to-noise ratio (SNR). This result is known as the Shannon–Hartley theorem.

When the SNR is large (SNR >> 0 dB), the capacity $C \approx W \log_2 \dfrac{\bar{P}}{N_0 W}$ is logarithmic in power and approximately linear in bandwidth. This is called the *bandwidth-limited regime*.

When the SNR is small (SNR << 0 dB), the capacity $C \approx \dfrac{\bar{P}}{N_0} \log_2$ is linear in power but insensitive to bandwidth. This is called the *power-limited regime*.

The bandwidth-limited regime and power-limited regime are illustrated in the figure.

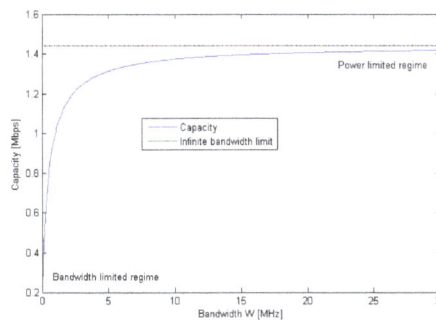

AWGN channel capacity with the power-limited regime and bandwidth-limited regime indicated.

$$\text{Here,} \quad \frac{\bar{P}}{N_0} = 10^6.$$

Frequency-selective Channel

The capacity of the frequency-selective channel is given by so-called water filling power allocation,

$$C_{N_c} = \sum_{n=0}^{N_c-1} \log_2 \left(1 + \frac{P_n^* \, |\bar{h}_n|^2}{N_0} \right),$$

where $P_n^* = \max\left(\left(\dfrac{1}{\lambda} - \dfrac{N_0}{|\bar{h}_n|^2} \right), 0 \right)$ and $|\bar{h}_n|^2$ is the gain of subchannel n, with $\quad$ chosen to meet the power constraint.

Slow-fading Channel

In a slow-fading channel, where the coherence time is greater than the latency requirement, there is no definite capacity as the maximum rate of reliable communications supported by the channel, $\log_2(1+|h|^2\ SNR)$, depends on the random channel gain $|h|^2$, which is unknown to the transmitter. If the transmitter encodes data at rate R [bits/s/Hz], there is a non-zero probability that the decoding error probability cannot be made arbitrarily small,

$$p_{out} = \mathbb{P}(\log(1+|h|^2\ SNR) < R),$$

in which case the system is said to be in outage. With a non-zero probability that the channel is in deep fade, the capacity of the slow-fading channel in strict sense is zero. However, it is possible to determine the largest value of R such that the outage probability p_{out} is less than ϵ. This value is known as the ϵ-outage capacity.

Fast-fading Channel

In a fast-fading channel, where the latency requirement is greater than the coherence time and the codeword length spans many coherence periods, one can average over many independent channel fades by coding over a large number of coherence time intervals. Thus, it is possible to achieve a reliable rate of communication of $(\log(1\ |h|\ SNR))$ [bits/s/Hz] and it is meaningful to speak of this value as the capacity of the fast-fading channel.

Tethering

Tethering is connecting one device to another. In the context of mobile phones and tablet computers, tethering allows sharing the Internet connection of the phone or tablet with other devices such as laptops. Connection of the phone or tablet with other devices can be done over wireless LAN (Wi-Fi), over Bluetooth or by physical connection using a cable, for example through USB.

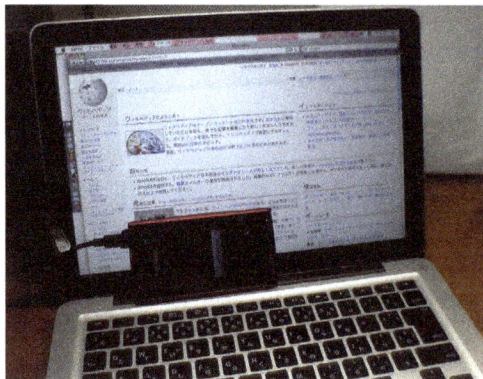

A phone tethered to a laptop

If tethering is done over WLAN, the feature may be branded as a mobile hotspot, which allows the smartphone to serve as a portable router; mobile hotspots may be protected by a PIN or password. The Internet-connected mobile device can act as a portable wireless access point and router for devices connected to it.

Mobile Device's OS Support

Many mobile phones are equipped with software to offer tethered Internet access. Windows Mobile 6.5, Windows Phone 7, Android (starting from version 2.2), and iOS 3.0 (or later) offer tethering over a Bluetooth PAN or a USB connection. Tethering over Wi-Fi, also known as Personal Hotspot, is available on iOS starting with iOS 4.2.5 (or later) on iPhone 4, 4S (2010), 5, iPad (3rd generation), certain Windows Mobile 6.5 devices like the HTC HD2, Windows Phone 7, 8 and 8.1 devices (varies by manufacturer and model), and certain Android phones (varies widely depending on carrier, manufacturer, and software version).

For IPv4 networks, the tethering normally works via NAT on the handset's existing data connection, so from the network point of view, there is just one device with a single IPv4 network address, though it is technically possible to attempt to identify multiple machines.

On some networks, this feature is only contractually available by paying to add a tethering package to a data plan or choosing a data plan that includes tethering, such as Lycamobile MVNO. This is done primarily because with a computer sharing the network connection, there may well be a substantial increase in the customer's mobile data use, for which the network may not have budgeted in their network design and pricing structures.

Some network-provided handsets have carrier-specific software that may deny the inbuilt tethering ability normally available on the handset, or only enable it if the subscriber pays an additional fee. Some operators have asked Google or any mobile producer using Android to completely remove tethering support from the operating system on certain handsets. Handsets purchased SIM-free, without a network provider subsidy, are often unhindered with regards to tethering.

There are, however, several ways to enable tethering on restricted devices without paying the carrier for it, including 3rd party USB Tethering apps such as PDAnet, rooting Android devices or jailbreaking iOS devices and installing a tethering application on the device. Tethering is also available as a downloadable third-party application on most Symbian mobile phones as well as on the MeeGo platform and on WebOS mobiles phones.

In Carriers' Contracts

Depending on the wireless carrier, a user's phone may have restricted functionality. While tethering may be allowed at no extra cost, some carriers impose a one-time charge to enable tethering and others forbid tethering or impose added data charges. Contracts that advertise "unlimited" data usage often have limits detailed in a Fair usage policy.

New Zealand

In New Zealand, tethering is permitted on all carriers since 2011 after the disaster in Christchurch.

United Kingdom

In the UK, two tethering-permitted mobile plans offered unlimited data: *The Full Monty* on T-Mobile, and *The One Plan* on Three. Three offered tethering as a standard feature until early 2012, retaining it on selected plans. T-Mobile dropped tethering on its unlimited data plans in late 2012.

United States

As cited in Sprint Nextel's "Terms of Service":

Except with Phone-as-Modem plans, you may not use a phone (including a Bluetooth phone) as a modem in connection with a computer, PDA, or similar device. We reserve the right to deny or terminate service without notice for any misuse or any use that adversely affects network performance.

T-Mobile USA has a similar clause in its "Terms & Conditions":

Unless explicitly permitted by your Data Plan, other uses, including for example, using your Device as a modem or tethering your Device to a personal computer or other hardware, are not permitted.

T-Mobile's Simple Family or Simple Business plans offer "Hotspot" from devices that offer that function (such as Apple iPhone) to up to 5 devices. Since 2014-03-27, 1000 MB/month is free in the USA with cellular service. The host device has unlimited slow internet for the rest of the month, and all month while roaming in 100 countries, but with no tethering. For $10 or $20/month more per host device, the amount of data available for tethering can be increased markedly. The host device cellular services can be canceled, added, or changed at any time, pro-rated, data tethering levels can be changed month-to-month, and T-Mobile no longer requires any long-term service contracts, allowing users to bring their own devices or buy devices from them, independent of whether they continue service with them.

As of 2013 Verizon Wireless and AT&T Mobility offer wired tethering to their plans for a fee, while Sprint Nextel offers a Wi-Fi connected "mobile hotspot" tethering feature at an added charge. However, actions by the FCC and a small claims court in California may make it easier for consumers to tether. On July 31, 2012, the FCC released an unofficial announcement of Commission action, decreeing Verizon Wireless must pay US$1.25 million to resolve the investigation regarding compliance of the C Block Spectrum. The announcement also stated that "(Verizon) recently revised its service offerings such that consumers on usage-based pricing plans may tether, using any application, without paying an additional fee." After that judgement Verizon release "Share Everything" plans that enable tethering, however users must drop old plans they were grandfathered under (such as the Unlimited Data plans) and switch, or pay a tethering fee.

In another instance, Judge Russell Nadel awarded Matt Spaccarelli US$850 via the Ventura Superior Court, despite the fact that Spaccarelli had violated his terms of service by jailbreaking his iPhone in order to fully utilize his iPhone's hardware. Spaccarelli demonstrated that AT&T had unfairly throttled his data connection. His data shows that AT&T had been throttling his connection after approximately 2GB of data was used. Spaccarelli responded by creating a personal web page in order to provide information that allows others to file a similar lawsuit.

Hopefully with all this concrete data and the courts on our side, AT&T will be forced to change something. Let's just hope it chooses to go the way of Sprint, not T-Mobile.

While T-Mobile did eventually allow tethering, on August 31, 2015 the company announced it will punish users who abuse its unlimited data by violating T-Mobile's rules on tethering (which unlike standard data does carry a 7 GB cap before throttling takes effect) by permanently kicking them off the unlimited plans and making users sign up for tiered data plans. T-Mobile mentioned that it was only a small handful of users who abused the tethering rules by using an Android app that masks T-Mobile's tethering monitoring and uses as much as 2 TB's per month, causing speed issues for most customers who don't abuse the rules.

Near Field Communication

Near-field communication (NFC) is a set of communication protocols that enable two electronic devices, one of which is usually a portable device such as a smartphone, to establish communication by bringing them within 4 cm (2 in) of each other.

NFC devices are used in contactless payment systems, similar to those used in credit cards and electronic ticket smartcards and allow mobile payment to replace/supplement these systems. NFC is used for social networking, for sharing contacts, photos, videos or files. NFC-enabled devices can act as electronic identity documents and keycards. NFC offers a low-speed connection with simple setup that can be used to bootstrap more capable wireless connections.

Overview

Similar ideas in advertising and industrial applications were not generally successful commercially, outpaced by technologies such as barcodes and UHF RFID tags. NFC protocols established a generally-supported standard. When one of the connected devices has Internet connectivity, the other can exchange data with online services.

NFC-enabled portable devices can be provided with apps, for example to read electronic tags or make payments when connected to an NFC-compliant apparatus. Earlier close-range communication used technology that was proprietary to the manufacturer, for applications such as stock ticket, access control and payment readers.

Like other "proximity card" technologies, NFC employs electromagnetic induction between two loop antennae when NFC devices—for example a smartphone and a "smartposter"—exchange information, operating within the globally available unlicensed radio frequency ISM band of 13.56 MHz on ISO/IEC 18000-3 air interface at rates ranging from 106 to 424 kbit/s.

Each full NFC device can work in three modes:

- NFC card emulation—enables NFC-enabled devices such as smartphones to act like smart cards, allowing users to perform transactions such as payment or ticketing.
- NFC reader/writer—enables NFC-enabled devices to read information stored on inexpensive NFC tags embedded in labels or smart posters.
- NFC peer-to-peer—enables two NFC-enabled devices to communicate with each other to exchange information in an adhoc fashion.

NFC tags are passive data stores which can be read, and under some circumstances written to, by an NFC device. They typically contain data (as of 2015 between 96 and 8,192 bytes) and are read-only in normal use, but may be rewritable. Applications include secure personal data storage (e.g. debit or credit card information, loyalty program data, personal identification numbers (PINs), contacts). NFC tags can be custom-encoded by their manufacturers or use the industry specifications.

The standards were provided by the NFC Forum. The forum was responsible for promoting the technology and setting standards and certifies device compliance. Secure communications are available by applying encryption algorithms as is done for credit cards and if they fit the criteria for being considered a personal area network.

NFC standards cover communications protocols and data exchange formats and are based on existing radio-frequency identification (RFID) standards including ISO/IEC 14443 and FeliCa. The standards include ISO/IEC 18092 and those defined by the NFC Forum. In addition to the NFC Forum, the GSMA group defined a platform for the deployment of GSMA NFC Standards within mobile handsets. GSMA's efforts include Trusted Services Manager, Single Wire Protocol, testing/certification and secure element.

A patent licensing program for NFC is under deployment by France Brevets, a patent fund created in 2011. This program was under development by Via Licensing Corporation, an independent subsidiary of Dolby Laboratories, and was terminated in May 2012. A platform-independent free and open source NFC library, libnfc, is available under the GNU Lesser General Public License.

Present and anticipated applications include contactless transactions, data exchange and simplified setup of more complex communications such as Wi-Fi.

History

NFC is rooted in radio-frequency identification technology (known as RFID) which allows compatible hardware to both supply power to and communicate with an otherwise unpowered and passive electronic tag using radio waves. This is used for identification, authentication and tracking.

- 1983 The first patent to be associated with the abbreviation "RFID" was granted to Charles Walton.

- 1997 Early form patented and first used in *Star Wars* character toys for Hasbro. The patent was originally held by Andrew White and Marc Borrett at Innovision Research and Technology (Patent WO9723060). The device allowed data communication between two units in close proximity.

- 2002 Sony and Philips agreed to establish a technology specification and created a technical outline on March 25, 2002.

- 2003 NFC was approved as an ISO/IEC standard on December 8, and later as an ECMA standard.

- 2004 Nokia, Philips and Sony established the NFC Forum

- 2006 Initial specifications for NFC Tags

- 2006 Specification for "SmartPoster" records

- 2007 Innovision's NFC tags used in the first consumer trial in the UK, in the Nokia 6131 handset.

- 2009 In January, NFC Forum released Peer-to-Peer standards to transfer contacts, URLs, initiate Bluetooth, etc.

- 2010 Innovision released a suite of designs and patents for low cost, mass-market mobile phones and other devices.

- 2010 Samsung Nexus S: First Android NFC phone shown

- 2010 Nice, France launches the "Nice City of contactless mobile" project, providing inhabitants with NFC mobile phones and bank cards, and a "bouquet of services" covering transportation, tourism and student's services

- 2011 Tapit Media launches in Sydney, Australia as the first specialized NFC marketing company

- 2011 Google I/O "How to NFC" demonstrates NFC to initiate a game and to share a contact, URL, app or video.

- 2011 NFC support becomes part of the Symbian mobile operating system with the release of Symbian Anna version.

- 2011 Research In Motion devices are the first ones certified by MasterCard Worldwide for their PayPass service

- 2012 UK restaurant chain EAT. and Everything Everywhere (Orange Mobile Network Operator), partner on the UK's first nationwide NFC-enabled smartposter campaign. A specially created mobile phone app is triggered when the NFC-enabled mobile phone comes into contact with the smartposter.

- 2012 Sony introduced NFC "Smart Tags" to change modes and profiles on a Sony smartphone at close range, included with the Sony Xperia P Smartphone released the same year.

- 2013 Samsung and VISA announce their partnership to develop mobile payments.

- 2013 IBM scientists, in an effort to curb fraud and security breaches, develop an NFC-based mobile authentication security technology. This technology works on similar principles to dual-factor authentication security.

- 2014 AT&T, Verizon and T-Mobile released Softcard (formally ISIS mobile wallet). It runs on NFC-enabled Android phones and iPhone 4 and iPhone 5 when an external NFC case is attached. The technology was purchased by Google and the service ended on March 31, 2015.

- 2014 Apple introduced Apple Pay for NFC-enabled mobile payment on iPhone 6 and 6 Plus, and the Apple Watch, which was released on April 24, 2015.

- In November 2015, Swatch and Visa Inc. announced a partnership to enable NFC financial transactions using the "Swatch Bellamy" wristwatch. The system is currently online in Asia thanks to a partnership with China UnionPay and Bank of Communications. The partnership will bring the technology to the US, Brazil, and Switzerland.

- November 2015, Google's Android Pay function was launched, a direct rival to Apple Pay, and it started rolling out across the US.

Design

NFC is a set of short-range wireless technologies, typically requiring a separation of 10 cm or less. NFC operates at 13.56 MHz on ISO/IEC 18000-3 air interface and at rates ranging from 106 kbit/s to 424 kbit/s. NFC always involves an initiator and a target; the initiator actively generates an RF field that can power a passive target. This enables NFC targets to take very simple form factors such as unpowered tags, stickers, key fobs, or cards. NFC peer-to-peer communication is possible, provided both devices are powered.

NFC tags contain data and are typically read-only, but may be writeable. They can be custom-encoded by their manufacturers or use NFC Forum specifications. The tags can securely store personal data such as debit and credit card information, loyalty program data, PINs and networking contacts, among other information. The NFC Forum defines four types of tags that provide different communication speeds and capabilities in terms of configurability, memory, security, data retention and write endurance. Tags currently offer between 96 and 4,096 bytes of memory.

As with proximity card technology, near-field communication uses magnetic induction between two loop antennas located within each other's near field, effectively forming an air-core transformer. It operates within the globally available and unlicensed radio frequency ISM band of 13.56 MHz. Most of the RF energy is concentrated in the allowed ±7 kHz bandwidth range, but the full spectral envelope may be as wide as 1.8 MHz when using ASK modulation.

Theoretical working distance with compact standard antennas: up to 20 cm (practical working distance of about 10 cm).

Supported data rates: 106, 212 or 424 kbit/s (the bit rate 848 kbit/s is not compliant with the standard ISO/IEC 18092)

The two modes are:

- Passive—The initiator device provides a carrier field and the target device answers by modulating the existing field. In this mode, the target device may draw its operating power from the initiator-provided electromagnetic field, thus making the target device a transponder.

- Active—Both initiator and target device communicate by alternately generating their own fields. A device deactivates its RF field while it is waiting for data. In this mode, both devices typically have power supplies.

Speed	Active device	Passive device
424 kbit/s	Man, 10% ASK	Man, 10% ASK
212 kbit/s	Man, 10% ASK	Man, 10% ASK
106 kbit/s	Modified Miller, 100% ASK	Man, 10% ASK

NFC employs two different codings to transfer data. If an active device transfers data at 106 kbit/s, a modified Miller coding with 100% modulation is used. In all other cases Manchester coding is used with a modulation ratio of 10%.

NFC devices are full-duplex—they are able to receive and transmit data at the same time. Thus, they can check for potential collisions if the received signal frequency does not match the transmitted signal's frequency.

Although the range of NFC is limited to a few centimeters, plain NFC does not ensure secure communications. In 2006, Ernst Haselsteiner and Klemens Breitfuß described possible attacks and detailed how to leverage NFC's resistance to man-in-the-middle attacks to establish a specific key. As this technique is not part of the ISO standard, NFC offers no protection against eavesdropping and can be vulnerable to data modifications. Applications may use higher-layer cryptographic protocols (e.g. SSL) to establish a secure channel.

The RF signal for the wireless data transfer can be picked up with antennas. The distance from which an attacker is able to eavesdrop the RF signal depends on multiple parameters, but is typically less than 10 meters. Also, eavesdropping is highly affected by the communication mode. A passive device that doesn't generate its own RF field is much harder to eavesdrop on than an active device. An attacker can typically eavesdrop within 10 m and 1 m for active devices and passive devices, respectively.

Because NFC devices usually include ISO/IEC 14443 protocols, relay attacks are feasible.[page needed] For this attack the adversary forwards the request of the reader to the victim and relays its answer to the reader in real time, pretending to be the owner of the victim's smart card. This is similar to a man-in-the-middle attack. One libnfc code example demonstrates a relay attack using two stock commercial NFC devices. This attack can be implemented using only two NFC-enabled mobile phones.

Standards

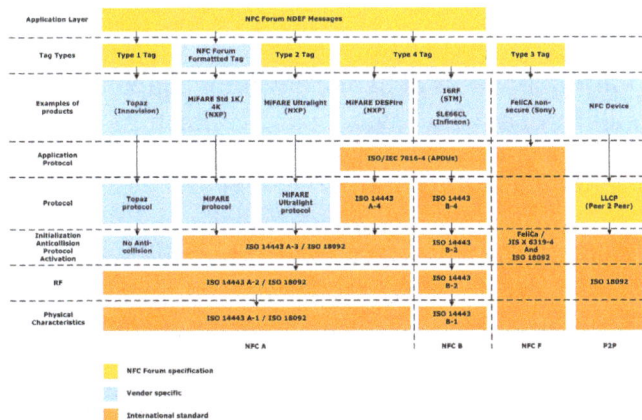

NFC Protocol stack overview

NFC standards cover communications protocols and data exchange formats, and are based on existing RFID standards including ISO/IEC 14443 and FeliCa. The standards include ISO/IEC 18092 and those defined by the NFC Forum.

ISO / IEC

NFC is standardized in ECMA-340 and ISO/IEC 18092. These standards specify the modulation schemes, coding, transfer speeds and frame format of the RF interface of NFC devices, as well as initialization schemes and conditions required for data collision-control during initialization for both passive and active NFC modes. They also define the transport protocol, including protocol activation and data-exchange methods. The air interface for NFC is standardized in:

- ISO/IEC 18092 / ECMA-340—*Near Field Communication Interface and Protocol-1* (NF-CIP-1)

- ISO/IEC 21481 / ECMA-352—*Near Field Communication Interface and Protocol-2* (NF-CIP-2)

NFC incorporates a variety of existing standards including ISO/IEC 14443 Type A and Type B, and FeliCa. NFC-enabled phones work at a basic level with existing readers. In "card emulation mode" an NFC device should transmit, at a minimum, a unique ID number to a reader. In addition, NFC Forum defined a common data format called *NFC Data Exchange Format* (NDEF) that can store and transport items ranging from any MIME-typed object to ultra-short RTD-documents, such as URLs. The NFC Forum added the *Simple NDEF Exchange Protocol* (SNEP) to the spec that allows sending and receiving messages between two NFC devices.

GSMA

The GSM Association (GSMA) is a trade association representing nearly 800 mobile telephony operators and more than 200 product and service companies across 219 countries. Many of its members have led NFC trials and are preparing services for commercial launch.

GSM is involved with several initiatives:

- Standards: GSMA is developing certification and testing standards to ensure global interoperability of NFC services.

- *Pay-Buy-Mobile initiative*: Seeks to define a common global approach to using NFC technology to link mobile devices with payment and contactless systems.

- On November 17, 2010, after two years of discussions, AT&T, Verizon and T-Mobile launched a joint venture to develop a platform through which point of sale payments could be made using NFC in cell phones. Initially known as Isis Mobile Wallet and later as Softcard, the venture was designed to usher in broad deployment of NFC technology, allowing their customers' NFC-enabled cell phones to function similarly to credit cards throughout the US. Following an agreement with—and IP purchase by—Google, the Softcard payment system was shuttered in March, 2015, with an endorsement for its earlier rival, Google Wallet.

StoLPaN

StoLPaN ('Store Logistics and Payment with NFC') is a pan-European consortium supported by the European Commission's Information Society Technologies program. StoLPaN will examine the potential for NFC local wireless mobile communication.

NFC Forum

NFC Forum is a non-profit industry association formed on March 18, 2004, by NXP Semiconductors, Sony and Nokia to advance the use of NFC wireless interaction in consumer electronics, mobile devices and PCs. Standards include the four distinct tag types that provide different communication speeds and capabilities covering flexibility, memory, security, data retention and write endurance. NFC Forum promotes implementation and standardization of NFC technology to ensure interoperability between devices and services. As of June 2013, the NFC Forum had over 190 member companies.

NFC Forum promotes NFC and certifies device compliance and whether it fits in a personal area network.

Other Standardization Bodies

GSMA defined a platform for the deployment of GSMA NFC Standards within mobile handsets. GSMA's efforts include, Single Wire Protocol, testing and certification and secure element. The GSMA standards surrounding the deployment of NFC protocols (governed by NFC Forum) on mobile handsets are neither exclusive nor universally accepted. For example, Google's deployment of Host Card Emulation on Android KitKat provides for software control of a universal radio. In this HCE Deployment the NFC protocol is leveraged without the GSMA standards.

Other standardization bodies involved in NFC include:

- ETSI / SCP (Smart Card Platform) to specify the interface between the SIM card and the NFC chipset.

- GlobalPlatform to specify a multi-application architecture of the secure element.

- EMVCo for the impacts on the EMV payment applications

Applications

N-Mark logo for NFC-enabled devices

NFC allows one- and two-way communication between endpoints, suitable for many applications.

Commerce

NFC devices can be used in contactless payment systems, similar to those used in credit cards and electronic ticket smartcards and allow mobile payment to replace/supplement these systems.

In Android 4.4, Google introduced platform support for secure NFC-based transactions through Host Card Emulation (HCE), for payments, loyalty programs, card access, transit passes and other custom services. HCE allows any Android 4.4 app to emulate an NFC smart card, letting users initiate transactions with their device. Apps can use a new Reader Mode to act as readers for HCE cards and other NFC-based transactions.

On September 9, 2014, Apple announced support for NFC-powered transactions as part of Apple Pay. Apple stated that their approach to NFC payment is more secure because Apple Pay tokenizes its data to encrypt and protect it from unauthorized use.

Bootstrapping other Connections

NFC offers a low-speed connection with simple setup that can be used to bootstrap more capable wireless connections. For example, Android Beam software uses NFC to enable pairing and establish a Bluetooth connection when doing a file transfer and then disabling Bluetooth on both devices upon completion. Nokia, Samsung, BlackBerry and Sony have used NFC technology to pair Bluetooth headsets, media players and speakers with one tap. The same principle can be applied to the configuration of Wi-Fi networks. Samsung Galaxy devices have a feature named S-Beam—an extension of Android Beam that uses NFC (to share MAC Address and IP addresses) and then uses Wi-Fi Direct to share files and documents. The advantage of using Wi-Fi Direct over Bluetooth is that it permits much faster data transfers, running up to 300Mbit/s.

Social Networking

NFC can be used for social networking, for sharing contacts, photos, videos or files and entering multiplayer mobile games.

Identity and Access Tokens

NFC-enabled devices can act as electronic identity documents and keycards. NFC's short range and encryption support make it more suitable than less private RFID systems.

Smartphone Automation and NFC Tags

NFC-equipped smartphones can be paired with NFC Tags or stickers that can be programmed by NFC apps. These programs can allow a change of phone settings, texting, app launching, or command execution.

Such apps do not rely on a company or manufacturer, but can be utilized immediately with an NFC-equipped smartphone and an NFC tag.

The NFC Forum published the Signature Record Type Definition (RTD) 2.0 in 2015 to add integrity and authenticity for NFC Tags. This specification allows an NFC device to verify tag data and identify the tag author.

Gaming

NFC was used in video games starting with Skylanders: Spyro's Adventure. With it you buy figurines that are customizable and contain personal data with each figure, so no two figures are exactly alike. The Wii U was the first system to include NFC technology out of the box via the GamePad. It was later included in the New Nintendo 3DS range. The Amiibo range of accessories utilises NFC technology to unlock features.

Bluetooth Comparison

Aspect	NFC	Bluetooth	Bluetooth Low Energy
Tag requires power	No	Yes	Yes
Cost of Tag	$0.10 USD	$5.00 USD	$5.00 USD
RFID compatible	ISO 18000-3	Active	Active
Standardisation body	ISO/IEC	Bluetooth SIG	Bluetooth SIG
Network standard	ISO 13157 etc.	IEEE 802.15.1 (no longer maintained)	IEEE 802.15.1 (no longer maintained)
Network type	Point-to-point	WPAN	WPAN
Cryptography	Not with RFID	Available	Available
Range	< 20 cm	~100 m (class 1)	~50 m
Frequency	13.56 MHz	2.4–2.5 GHz	2.4–2.5 GHz
Bit rate	424 kbit/s	2.1 Mbit/s	1 Mbit/s
Set-up time	< 0.1 s	< 6 s	< 0.006 s
Current consumption	< 15mA (read)	Varies with class	< 15 mA (read and transmit)

NFC and Bluetooth are both short-range communication technologies available on mobile phones. NFC operates at slower speeds than Bluetooth, but consumes far less power and doesn't require pairing.

NFC sets up more quickly than standard Bluetooth, but has a lower transfer rate than Bluetooth low energy. With NFC, instead of performing manual configurations to identify devices, the connection between two NFC devices is automatically established in less than .1 second. The maximum data transfer rate of NFC (424 kbit/s) is slower than that of Bluetooth V2.1 (2.1 Mbit/s).

With a maximum working distance of less than 20 cm, NFC has a shorter range, which reduces the likelihood of unwanted interception. That makes NFC particularly suitable for crowded areas that complicate correlating a signal with its transmitting physical device (and by extension, its user).

NFC is compatible with existing passive RFID (13.56 MHz ISO/IEC 18000-3) infrastructures. NFC requires comparatively low power, similar to the Bluetooth V4.0 low energy protocol. When NFC works with an unpowered device (e.g. on a phone that may be turned off, a contactless smart credit card, a smart poster), however, the NFC power consumption is greater than that of Bluetooth V4.0 Low Energy, since illuminating the passive tag needs extra power.

Devices

In 2011, handset vendors released more than 40 NFC-enabled handsets with the Android mobile operating system.The iPhone 6 line is the first set of handsets from Apple to support NFC. BlackBerry devices support NFC using BlackBerry Tag on devices running BlackBerry OS 7.0 and greater.

Mastercard added further NFC support for PayPass for the Android and BlackBerry platforms, enabling PayPass users to make payments using their Android or BlackBerry smartphones. A partnership between Samsung and Visa added a 'payWave' application on the Galaxy S4 smartphone.

Microsoft added native NFC functionality in their mobile OS with Windows Phone 8, as well as the Windows 8 operating system. Microsoft provides the "Wallet hub" in Windows Phone 8 for NFC payment, and can integrate multiple NFC payment services within a single application.

Deployments

As of April 2011, hundreds of NFC trials had been conducted. Some firms moved to full-scale service deployments, spanning one or more countries. Multi-country deployments include Orange's rollout of NFC technology to banks, retailers, transport, and service providers in multiple European countries, and Airtel Africa and Oberthur Technologies deploying to 15 countries throughout Africa.

- China telecom (China's 3rd largest mobile operator) made its NFC rollout in November 2013. The company signed up multiple banks to make their payment apps available on its SIM Cards. China telecom stated that the wallet would support coupons, membership cards, fuel cards and boarding passes. The company planned to achieve targets of rolling out 40 NFC phone models and 30 Mn NFC SIMs by 2014.

- Softcard (formerly Isis Mobile Wallet), a joint venture from Verizon Wireless, AT&T and T-Mobile, focuses on in-store payments making use of NFC technology. After doing pilots in some regions, they launched across the US.

- Vodafone launched the NFC-based Vodafone SmartPass mobile payment service in Spain in partnership with Visa. It enables consumers with an NFC-enabled mobile device to make contactless payments via their SmartPass credit balance at any POS.

- OTI, an Israeli company that designs and develops contactless microprocessor based smart card technology, contracted to supply NFC-readers to one of its channel partners in the US. The partner was required to buy $10MM worth of OTI NFC readers over 3 years.

- Rogers Communications launched virtual wallet Suretap to enable users to make payments with their phone in Canada in April 2014. Suretap users can load up gift cards and pre-paid MasterCards from national retailers.

- Sri Lanka's first workforce smartcard, uses NFC.

- As of December 13, 2013 Tim Hortons TimmyME BlackBerry 10 Application allowed users to link their prepaid Tim Card to the app, allowing payment by tapping the NFC-enabled device to a standard contactless terminal.

- Google Wallet allows consumers to store credit card and store loyalty card information in a virtual wallet and then use an NFC-enabled device at terminals that also accept MasterCard PayPass transactions.

- Germany, Austria, Finland, New Zealand, Italy, Iran, and Turkey trialed NFC ticketing systems for public transport. The Lithuanian capital of Vilnius fully replaced paper tickets for public transportation with ISO/IEC 14443 Type A cards on July 1, 2013.

- NFC sticker-based payments in Australia's Bankmecu and card issuer Cuscal completed an NFC payment sticker trial, enabling consumers to make contactless payments at Visa payWave terminals using a smart sticker stuck to their phone.

- India was implementing NFC based transactions in box offices for ticketing purposes.

- A partnership of Google and Equity Bank in Kenya introduced NFC payment systems for public transport in the Capital city Nairobi under the branding "Beba Pay".

Subscriber Identity Module

A subscriber identity module or subscriber identification module (SIM) is an integrated circuit chip that is intended to securely store the international mobile subscriber identity (IMSI) number and its related key, which are used to identify and authenticate subscribers on mobile telephony devices (such as mobile phones and computers). It is also possible to store contact information on many SIM cards. SIM cards are always used on GSM phones; for CDMA phones, they are only needed for newer LTE-capable handsets. SIM cards can also be used in satellite phones.

A mini-SIM card next to its electrical contacts in a Nokia 6233

A TracFone Wireless SIM card has no distinctive carrier markings and is only marked as a "SIM CARD"

The SIM circuit is part of the function of a Universal Integrated Circuit Card (UICC) physical smart card, which is usually made of PVC with embedded contacts and semiconductors. "SIM cards" are transferable between different mobile devices. The first UICC smart cards were the size of credit and bank cards; sizes were reduced several times over the years, usually keeping electrical contacts the same, so that a larger card could be cut down to a smaller size.

A SIM card contains its unique serial number (ICCID), international mobile subscriber identity (IMSI) number, security authentication and ciphering information, temporary information related to the local network, a list of the services the user has access to, and two passwords: a personal identification number (PIN) for ordinary use, and a personal unblocking code (PUK) for PIN unlocking.

History and Procurement

The SIM was initially specified by the European Telecommunications Standards Institute in the specification with the number TS 11.11. This specification describes the physical and logical behaviour of the SIM. With the development of UMTS the specification work was partially transferred to 3GPP. 3GPP is now responsible for the further development of applications like SIM (TS 51.011) and USIM (TS 31.102) and ETSI for the further development of the physical card UICC.

The first SIM card was developed in 1991 by Munich smart-card maker Giesecke & Devrient, who sold the first 300 SIM cards to the Finnish wireless network operator Radiolinja.

Inactivation

Many non-contractual "pay-as-you-go" arrangements require you to actively use credit at least once every three months lest your account expires. This is sometimes associated with the SIM card being made "inactive" by the network.

Design

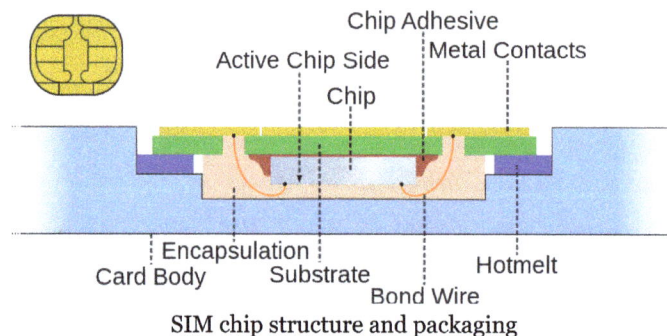

SIM chip structure and packaging

There are three operating voltages for SIM cards: 5 V, 3 V and 1.8 V (ISO/IEC 7816-3 classes A, B and C, respectively). The operating voltage of the majority of SIM cards launched before 1998 was 5 V. SIM cards produced subsequently are compatible with 3 V and 5 V. Modern cards support 5 V, 3 V and 1.8 V.

Modern SIM cards allow applications to be loaded when the SIM is in use by the subscriber. These applications communicate with the handset or a server using SIM application toolkit, which was initially specified by 3GPP in TS 11.14 (there is an identical ETSI specification with different numbering). ETSI and 3GPP maintain the SIM specifications; the main specifications are: ETSI TS 102 223, ETSI TS 102 241, ETSI TS 102 588, and ETSI TS 131 111. SIM toolkit applications were initially written in native code using proprietary APIs. In order to allow interoperability of the applications, Java Card was taken as the solution of choice by ETSI. Additional standards and specifications of interest are maintained by GlobalPlatform.

Data

SIM cards store network-specific information used to authenticate and identify subscribers on the network. The most important of these are the ICCID, IMSI, Authentication Key (Ki), Local Area Identity (LAI) and Operator-Specific Emergency Number. The SIM also stores other carrier-specific data such as the SMSC (Short Message Service Center) number, Service Provider Name (SPN), Service Dialing Numbers (SDN), Advice-Of-Charge parameters and Value Added Service (VAS) applications. (Refer to GSM 11.11)

SIM cards can come in various data capacities, from 8 KB to at least 256 KB. All allow a maximum of 250 contacts to be stored on the SIM, but while the 32 KB has room for 33 Mobile Network Codes (MNCs) or "network identifiers", the 64 KB version has room for 80 MNCs. This is used by network operators to store information on preferred networks, mostly used when the SIM is not in its home network but is roaming. The network operator that issued the SIM card can use this to have a phone connect to a preferred network, in order to make use of the best commercial agreement for the original network company instead of having to pay the network operator that the phone 'saw' first. This does not mean that a phone containing this SIM card can connect to a maximum of only 33 or 80 networks, but it means that the SIM card issuer can specify only up to that number of preferred networks; if a SIM is outside these preferred networks it will use the first or best available network.

ICCID

Each SIM is internationally identified by its integrated circuit card identifier (ICCID). ICCIDs are stored in the SIM cards and are also engraved or printed on the SIM card body during a process called personalisation. The ICCID is defined by the ITU-T recommendation E.118 as the *Primary Account Number*. Its layout is based on ISO/IEC 7812. According to E.118, the number is up to 22 digits long, including a single check digit calculated using the Luhn algorithm. However, the GSM Phase 1 defined the ICCID length as 10 octets (20 digits) with operator-specific structure.

The number is composed of the following subparts:

Issuer Identification Number (IIN)

Maximum of seven digits:

- Major industry identifier (MII), 2 fixed digits, *89* for telecommunication purposes.
- Country code, 1–3 digits, as defined by ITU-T recommendation E.164.
- Issuer identifier, 1–4 digits.

Individual Account Identification

- Individual account identification number. Its length is variable, but every number under one IIN will have the same length.

Check Digit

- Single digit calculated from the other digits using the Luhn algorithm.

With the GSM Phase 1 specification using 10 octets into which ICCID is stored as packed BCD, the data field has room for 20 digits with hexadecimal digit "F" being used as filler when necessary.

In practice, this means that on GSM SIM cards there are 20-digit (19+1) and 19-digit (18+1) ICCIDs in use, depending upon the issuer. However, a single issuer always uses the same size for its ICCIDs.

To confuse matters more, SIM factories seem to have varying ways of delivering electronic copies of SIM personalization datasets. Some datasets are without the ICCID checksum digit, others are with the digit.

As required by E.118, The ITU publishes a list of all current internationally assigned IIN codes in its Operational Bulletins. As of October 2016 the list issued on 15 November 2013 was current.

International Mobile Subscriber Identity (IMSI)

SIM cards are identified on their individual operator networks by a unique International Mobile Subscriber Identity (IMSI). Mobile network operators connect mobile phone calls and communicate with their market SIM cards using their IMSIs. The format is:

- The first three digits represent the Mobile Country Code (MCC).

- The next two or three digits represent the Mobile Network Code (MNC). Three-digit MNC codes are allowed by E.212 but are mainly used in the United States and Canada.

- The next digits represent the Mobile Subscriber Identification Number (MSIN). Normally there will be 10 digits but would be fewer in the case of a 3-digit MNC or if national regulations indicate that the total length of the IMSI should be less than 15 digits.

- Digits are different from country to country.

Authentication Key (K_i)

The Kn_i is a 128-bit value used in authenticating the SIMs on the mobile network. Each SIM holds a unique K_i assigned to it by the operator during the personalization process. The K_i is also stored in a database (termed authentication center or AuC) on the carrier's network.

The SIM card is designed not to allow the K_i to be obtained using the smart-card interface. Instead, the SIM card provides a function, *Run GSM Algorithm*, that allows the phone to pass data to the SIM card to be signed with the K_i. This, by design, makes usage of the SIM card mandatory unless the K_i can be extracted from the SIM card, or the carrier is willing to reveal the K_i. In practice, the GSM cryptographic algorithm for computing SRES_2 from the K_i has certain vulnerabilities that can allow the extraction of the K_i from a SIM card and the making of a duplicate SIM card.

Authentication process:

1. When the Mobile Equipment starts up, it obtains the International Mobile Subscriber Identity (IMSI) from the SIM card, and passes this to the mobile operator requesting access and authentication. The Mobile Equipment may have to pass a PIN to the SIM card before the SIM card will reveal this information.

2. The operator network searches its database for the incoming IMSI and its associated K_i.

3. The operator network then generates a Random Number (RAND, which is a nonce) and signs it with the K_i associated with the IMSI (and stored on the SIM card), computing another number, that is split into the Signed Response 1 (SRES_1, 32 bits) and the encryption key K_c (64 bits).

4. The operator network then sends the RAND to the Mobile Equipment, which passes it to the SIM card. The SIM card signs it with its K_i, producing SRES_2 and K_c, which it gives to the Mobile Equipment. The Mobile Equipment passes SRES_2 on to the operator network.

5. The operator network then compares its computed SRES_1 with the computed SRES_2 that the Mobile Equipment returned. If the two numbers match, the SIM is authenticated and the Mobile Equipment is granted access to the operator's network. K_c is used to encrypt all further communications between the Mobile Equipment and the network.

Location Area Identity

The SIM stores network state information, which is received from the Location Area Identity (LAI). Operator networks are divided into Location Areas, each having a unique LAI number. When the device changes locations, it stores the new LAI to the SIM and sends it back to the operator network with its new location. If the device is power cycled, it will take data off the SIM, and search for the prior LAI.

SMS messages and Contacts

Most SIM cards will orthogonally store a number of SMS messages and phone book contacts. The contacts are stored in simple "name and number" pairs: entries containing multiple phone numbers and additional phone numbers will usually not be stored on the SIM card. When a user tries to copy such entries to a SIM the handset's software will break them up into multiple entries, discarding any information that is not a phone number. The number of contacts and messages stored depends on the SIM; early models would store as few as five messages and 20 contacts while modern SIM cards can usually store over 250 contacts.

Formats

SIM cards have been made smaller over the years; functionality is independent of format. Full-size SIM were followed by mini-SIM, micro-SIM, and nano-SIM. SIM cards are also made to be embedded in devices.

The memory film from a micro SIM card without the plastic backing plate, next to a US dime, which is approx. 18 mm in diameter.

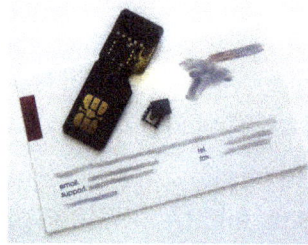

Embedded SIM from M2M supplier Eseye with an adapter board for evaluation in a Mini-SIM socket

SIM card sizes						
SIM card	**Introduced**	**Standard reference**	**Length (mm)**	**Width (mm)**	**Thickness (mm)**	**Volume (mm³)**
Full-size (1FF)	1991	ISO/IEC 7810:2003, ID-1	85.60	53.98	0.76	3511.72
Mini-SIM (2FF)	1996	ISO/IEC 7810:2003, ID-000	25.00	15.00	0.76	285.00
Micro-SIM (3FF)	2003	ETSI TS 102 221 V9.0.0, Mini-UICC	15.00	12.00	0.76	136.80
Nano-SIM (4FF)	early 2012	ETSI TS 102 221 V11.0.0	12.30	8.80	0.67	72.52
Embedded-SIM (eSIM)		JEDEC Design Guide 4.8, SON-8 ETSI TS 103 383 V12.0.0 GSMA SGP.22 V1.0	6.00	5.00	<1.00	<30.00

Full-size SIM

The *full-size SIM* (or 1FF, 1st form factor) was the first form factor to appear. It has the size of a credit card (85.60 mm × 53.98 mm × 0.76 mm). Later smaller SIMs are often supplied embedded in a full-size card that they can be pushed out of.

Full-size SIM (1FF), mini-SIM (2FF), micro-SIM (3FF) and nano-SIM (4FF)

Mini-SIM

The *mini-SIM* (or 2FF) card has the same contact arrangement as the full-size SIM card and is normally supplied within a full-size card carrier, attached by a number of linking pieces. This arrangement (defined in ISO/IEC 7810 as ID-1/000) allows such a card to be used in a device requiring a full-size card, or in a device requiring a mini-SIM card after breaking the linking pieces. As the full-size SIM is no longer used, some suppliers refer to this form factor as a standard or regular SIM.

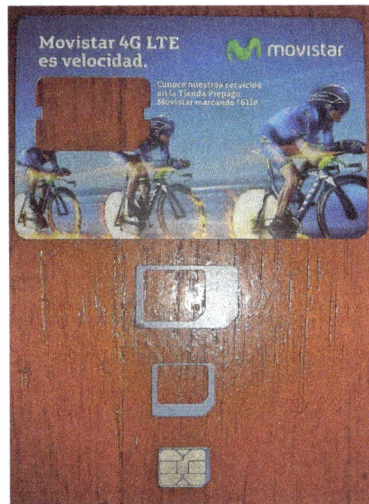

Nano SIM (in bottom), Micro SIM and Mini SIM, SIM brackets from Movistar 4G LTE in Colombia

Micro-SIM

The *micro-SIM* (or 3FF) card has the same thickness and contact arrangements, but reduced length and width as shown in the table above.

The micro-SIM was introduced by the European Telecommunications Standards Institute (ETSI) along with SCP, 3GPP (UTRAN/GERAN), 3GPP2 (CDMA2000), ARIB, GSM Association (GSMA SCaG and GSMNA), GlobalPlatform, Liberty Alliance, and the Open Mobile Alliance (OMA) for the purpose of fitting into devices too small for a mini-SIM card.

The form factor was mentioned in the December 1998 3GPP SMG9 UMTS Working Party, which is the standards-setting body for GSM SIM cards, and the form factor was agreed upon in late 2003.

The micro-SIM was designed for backward compatibility. The major issue for backward compatibility was the contact area of the chip. Retaining the same contact area allows the micro-SIM to be compatible with the prior, larger SIM readers through the use of plastic cutout surrounds. The SIM was also designed to run at the same speed (5 MHz) as the prior version. The same size and positions of pins resulted in numerous "How-to" tutorials and YouTube video with detailed instructions how to cut a mini-SIM card to micro-SIM size.

The chairman of EP SCP, Dr. Klaus Vedder, said

> "ETSI has responded to a market need from ETSI customers, but additionally there is a strong desire not to invalidate, overnight, the existing interface, nor reduce the performance of the cards."

Micro-SIM cards were introduced by various mobile service providers for the launch of the original iPad, and later for smartphones, from April 2010. The iPhone 4 was the first smartphone to use a micro-SIM card in June 2010, followed by many others.

Nano-SIM

The *nano-SIM* (or 4FF) card was introduced on 11 October 2012, when mobile service providers in various countries started to supply it for phones that supported the format. The nano-SIM measures 12.3 × 8.8 × 0.67 mm and reduces the previous format to the contact area while maintaining the existing contact arrangements. A small rim of isolating material is left around the contact area to avoid short circuits with the socket. The nano-SIM is 0.67 mm thick, compared to the 0.76 mm of its predecessor. 4FF can be put into adapters for use with devices designed for 2FF or 3FF SIMs, and is made thinner for that purpose, but many phone companies do not recommend this.

The iPhone 5, released in September 2012, was the first device to use a nano-SIM card, followed by other handsets.

Embedded-SIM / Embedded Universal Integrated Circuit Card (eUICC)

SIMs for M2M applications are available in a surface mount SON-8 package which may be soldered directly onto a circuit board.

The surface mount format provides the same electrical interface as the full size, 2FF and 3FF SIM cards, but is soldered to the circuit board as part of the manufacturing process. In M2M applications where there is no requirement to change the SIM card, this avoids the requirement for a connector, improving reliability and security. GSMA has been discussing the possibilities of a software based SIM card since 2010. While Motorola noted that eUICC is geared at industrial devices, Apple "disagreed that there is any statement forbidding the use of an embedded UICC in a consumer product." in 2012, The European Commission has selected the Embedded UICC format for its in-vehicle emergency call service known as eCall. All new car models in the EU will need to have one by 2015 to instantly connect the car to the emergency services in case of an accident. In Russia there is a similar plan with the ERA-GLONASS regional satellite positioning system and in Brazil with the SIMRAV anti-theft system.

Security

In July 2013, it was revealed that Karsten Nohl, a cryptographer and security researcher from SR-Labs, had discovered vulnerabilities in some SIM cards that enabled them to be hacked to provide root access. The cards affected use the Data Encryption Standard (DES) which, despite its age, is still used by some operators. Cards using the more recent Advanced Encryption Standard (AES) or Triple DES standards are not affected. Among other risks, the hack could lead to the phone being remotely cloned or allow payment credentials from the SIM to be stolen. Further details of the research were to be given at BlackHat on July 31, 2013.

In response, the International Telecommunication Union said that the development was "hugely significant" and that it would be contacting its members.

In February, 2015 it was reported by The Intercept that the NSA and GCHQ had stolen the encryption keys (Ki's) used by Gemalto (the manufacturer of 2 billion SIM cards annually), enabling these intelligence agencies to monitor voice and data communications without the knowledge or approval of cellular network providers or judicial oversight. Having finished its investigation, Gemalto claimed that it has "reasonable grounds" to believe that the NSA and GCHQ carried out an operation to hack its network in 2010 and 2011, but says the number of possibly stolen keys would not have been massive.

Developments

When GSM was already in use, the specifications were further developed and enhanced with functionality like SMS, GPRS, etc. These development steps are referred as releases by ETSI. Within these development cycles, the SIM specification was enhanced as well: new voltage classes, formats and files were introduced.

USIM

In GSM-only times, the SIM consisted of the hardware and the software. With the advent of UMTS this naming was split: the SIM was now an application and hence only software. The hardware part was called UICC. This split was necessary because UMTS introduced a new application, the Universal Subscriber Identity Module (USIM). The USIM brought, among other things, security improvements like the mutual authentication and longer encryption keys and an improved address book.

UICC

"SIM cards" in developed countries are today usually UICCs containing at least a SIM and a USIM application. This configuration is necessary because older GSM only handsets are solely compatible with the SIM [application] and some UMTS security enhancements do rely on the USIM [application].

Other Variants

The equivalent of SIM on CDMA networks is the R-UIM (and the equivalent of USIM is CSIM).

A *virtual SIM* is a mobile phone number provided by a mobile network operator that does not require a SIM card to connect phone calls to a user's mobile phone.

At the 2015 Mobile World Congress in Barcelona, Simless, Inc., a US-based startup unveiled world's first GSM phone without a SIM card slot. The reference phone was capable of downloading multiple virtual SIM cards over-the-air.

Virtual SIM are called e-SIM or eSIM.

Usage in mobile phone standards

The use of SIM cards is mandatory in GSM devices.

The satellite phone networks Iridium, Thuraya and Inmarsat's BGAN also use SIM cards. Sometimes, these SIM cards work in regular GSM phones and also allow GSM customers to roam in satellite networks by using their own SIM card in a satellite phone.

Japan's 2G PDC system (which was shut down in 2012; SoftBank Mobile has already shut down PDC from March 31, 2010) also specifies a SIM, but this has never been implemented commercially. The specification of the interface between the Mobile Equipment and the SIM is given in the RCR STD-27 annex 4. The Subscriber Identity Module Expert Group was a committee of specialists assembled by the European Telecommunications Standards Institute (ETSI) to draw up the specifications (GSM 11.11) for interfacing between smart cards and mobile telephones. In 1994, the name SIMEG was changed to SMG9.

Japan's current and next generation cellular systems are based on W-CDMA (UMTS) and CDMA2000 and all use SIM cards. However, Japanese CDMA2000-based phones are locked to the R-UIM they are associated with and thus, the cards are not interchangeable with other Japanese CDMA2000 handsets (though they may be inserted into GSM/WCDMA handsets for roaming purposes outside Japan).

CDMA-based devices originally did not use a removable card, and the service for these phones bound to a unique identifier contained in the handset itself. This is most prevalent in operators in the Americas. The first publication of the TIA-820 standard (also known as 3GPP2 C.S0023) in 2000 defined the Removable User Identity Module (R-UIM). Card-based CDMA devices are most prevalent in Asia.

The equivalent of a SIM in UMTS is called the Universal Integrated Circuit Card (UICC), which runs a USIM application. The UICC is still colloquially called a *SIM card*.

Chunghwa Telecom's LTE Card	KDDI's au IC-Card	NTT DoCoMo's FOMA Card	*Three* UK SIM Card with Packaging
China Mobile's LTE SIM card	H2O Wireless prepaid SIM card	Hotlink (Maxis) Nano SIM card	SIM card for Thuraya satellite phone

SIM and Carriers

The SIM card introduced a new and significant business opportunity for MVNOs — mobile virtual network operators — who lease capacity from one of the network operators rather than owning or operating a cellular telecoms network, and only provide a SIM card to their customers. MVNOs first appeared in Denmark, Hong Kong, Finland and the UK. Today they exist in over 50 countries, including most of Europe, United States, Canada, Mexico, Australia and parts of Asia, and account for approximately 10% of all mobile phone subscribers around the world.

On some networks, the mobile phone is locked to its carrier SIM card, meaning that the phone only works with SIM cards from the specific carrier. This is more common in markets where mobile phones are heavily subsidised by the carriers, and the business model depends on the customer staying with the service provider for a minimum term (typically 12, 18 or 24 months). SIM cards that are issued by providers with an associated contract are called SIM only deals. Common examples are the GSM networks in the United States, Canada, Australia, the UK and Poland. Many businesses offer the ability to remove the SIM lock from a phone, effectively making it possible to then use the phone on any network by inserting a different SIM card. Mostly, GSM and 3G mobile handsets can easily be unlocked and used on any suitable network with any SIM card.

In countries where the phones are not subsidised, *e.g.*, India, Israel and Belgium, all phones are unlocked. Where the phone is not locked to its SIM card, the users can easily switch networks by simply replacing the SIM card of one network with that of another while using only one phone. This is typical, for example, among users who may want to optimise their carrier's traffic by different tariffs to different friends on different networks, or when traveling internationally.

SIM Only

Commonly sold as a product by mobile telecommunications companies, the term SIM only refers to a type of legally binding contract between a mobile network provider and a customer. The contract itself takes the form of a credit agreement and is subject to a credit check.

Within a SIM only contract the mobile network provider supplies their customer with just one piece of hardware, a SIM card, which includes an agreed amount network usage in exchange of set monthly payment to be made by the customer. Network usage within a SIM only contract can be measured in minutes, text, data or any combination of these. The duration of a SIM only contract can vary depending on the deal selected by the customer but in the UK they are available over 1, 3, 6 and 12 month periods.

SIM only contracts differ from mobile phone contracts in that they do not include any hardware other than a SIM card. In terms of network usage SIM only is typically more cost effective than other mobile device contracts because unlike these other contracts a SIM only deal does not offset the cost of a mobile device over the contract period. Short contract length is one of the key features of SIM only which are made possible due to the absence of a mobile device.

SIM only is increasing in popularity very quickly. In 2010 pay monthly based mobile phone subscriptions grew from 41 percent to 49 percent of all UK mobile phone subscriptions. According to German research company Gfk, 250,000 SIM-only mobile contracts were taken up in the UK during July 2012 alone, the highest figure since GfK began keeping records.

Increasing smartphone penetration combined with financial concerns are leading customers to save money by moving onto a SIM-only when their initial contract term is over.

Multiple-SIM Devices

Devices with two SIM slots are known as dual SIMs. Dual-SIM mobile phones usually come with two slots for SIMs, one behind the battery and another on the side of the phone, though in some devices both slots can be found on the battery tray, or on the side of the phone if the device does not have a removable battery. Multiple-SIM devices are commonplace in developing markets such as in Africa, East Asia, the Indian subcontinent and South East Asia, where billing rates and variable network coverage make it desirable for consumers to use multiple SIMs from competing networks. They are not common in the Western world.

Voice Over IP

Voice over Internet Protocol (Voice over IP, VoIP and IP telephony) is a methodology and group of technologies for the delivery of voice communications and multimedia sessions over Internet Protocol (IP) networks, such as the Internet. The terms Internet telephony, broadband telephony, and broadband phone service specifically refer to the provisioning of communications services (voice, fax, SMS, voice-messaging) over the public Internet, rather than via the public switched telephone network (PSTN).

The steps and principles involved in originating VoIP telephone calls are similar to traditional digital telephony and involve signaling, channel setup, digitization of the analog voice signals, and encoding. Instead of being transmitted over a circuit-switched network; however, the digital information is packetized, and transmission occurs as IP packets over a packet-switched network. They transport audio streams using special media delivery protocols that encode audio and video with audio codecs, and video codecs. Various codecs exist that optimize the media stream based on application requirements and network bandwidth; some implementations rely on narrowband and compressed speech, while others support high fidelity stereo codecs. Some popular codecs include μ-law and a-law versions of G.711, G.722, a popular open source voice codec known as iLBC, a codec that only uses 8 kbit/s each way called G.729, and many others.

Early providers of voice-over-IP services offered business models and technical solutions that mirrored the architecture of the legacy telephone network. Second-generation providers, such as Skype, have built closed networks for private user bases, offering the benefit of free calls and convenience while potentially charging for access to other communication networks, such as the PSTN. This has limited the freedom of users to mix-and-match third-party hardware and software. Third-generation providers, such as Google Talk, have adopted the concept of federated VoIP—which is a departure from the architecture of the legacy networks. These solutions typically allow dynamic interconnection between users on any two domains on the Internet when a user wishes to place a call.

In addition to VoIP phones, VoIP is available on many smartphones, personal computers, and on Internet access devices. Calls and SMS text messages may be sent over 3G/4G or Wi-Fi.

Protocols

Voice over IP has been implemented in various ways using both proprietary protocols and protocols based on open standards. VoIP protocols include:

- Session Initiation Protocol (SIP)

- H.323

- Media Gateway Control Protocol (MGCP)

- Gateway Control Protocol (Megaco, H.248)

- Real-time Transport Protocol (RTP)

- Real-time Transport Control Protocol (RTCP)

- Secure Real-time Transport Protocol (SRTP)

- Session Description Protocol (SDP)

- Inter-Asterisk eXchange (IAX)

- Jingle XMPP VoIP extensions

- Skype protocol

The H.323 protocol was one of the first VoIP protocols that found widespread implementation for long-distance traffic, as well as local area network services. However, since the development of newer, less complex protocols such as MGCP and SIP, H.323 deployments are increasingly limited to carrying existing long-haul network traffic.

These protocols can be used by special-purpose software, such as Jitsi, or integrated into a web page (web-based VoIP), like Google Talk.

Adoption

Consumer Market

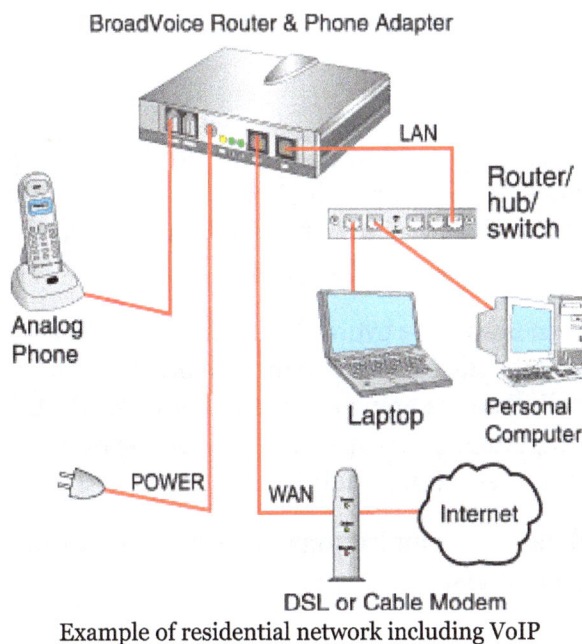

Example of residential network including VoIP

A major development that started in 2004 was the introduction of mass-market VoIP services that utilize existing broadband Internet access, by which subscribers place and receive telephone calls in much the same manner as they would via the public switched telephone network (PSTN). Full-service VoIP phone companies provide inbound and outbound service with direct inbound dialing. Many offer unlimited domestic calling for a flat monthly subscription fee. This sometimes includes international calls to certain countries. Phone calls between subscribers of the same provider are usually free when flat-fee service is not available. A VoIP phone is necessary to connect to a VoIP service provider. This can be implemented in several ways:

- Dedicated VoIP phones connect directly to the IP network using technologies such as wired Ethernet or Wi-Fi. They are typically designed in the style of traditional digital business telephones.

- An analog telephone adapter is a device that connects to the network and implements the electronics and firmware to operate a conventional analog telephone attached through a modular phone jack. Some residential Internet gateways and cablemodems have this function built in.

- A softphone is application software installed on a networked computer that is equipped with a microphone and speaker, or headset. The application typically presents a dial pad and display field to the user to operate the application by mouse clicks or keyboard input.

PSTN and Mobile Network Providers

It is becoming increasingly common for telecommunications providers to use VoIP telephony over dedicated and public IP networks to connect switching centers and to interconnect with other telephony network providers; this is often referred to as "IP backhaul".

Smartphones and Wi-Fi-enabled mobile phones may have SIP clients built into the firmware or available as an application download.

Corporate Use

Because of the bandwidth efficiency and low costs that VoIP technology can provide, businesses are migrating from traditional copper-wire telephone systems to VoIP systems to reduce their monthly phone costs. In 2008, 80% of all new Private branch exchange (PBX) lines installed internationally were VoIP.

VoIP solutions aimed at businesses have evolved into unified communications services that treat all communications—phone calls, faxes, voice mail, e-mail, Web conferences, and more—as discrete units that can all be delivered via any means and to any handset, including cellphones. Two kinds of competitors are competing in this space: one set is focused on VoIP for medium to large enterprises, while another is targeting the small-to-medium business (SMB) market.

VoIP allows both voice and data communications to be run over a single network, which can significantly reduce infrastructure costs.

The prices of extensions on VoIP are lower than for PBX and key systems. VoIP switches may run on commodity hardware, such as personal computers. Rather than closed architectures, these devices rely on standard interfaces.

VoIP devices have simple, intuitive user interfaces, so users can often make simple system configuration changes. Dual-mode phones enable users to continue their conversations as they move between an outside cellular service and an internal Wi-Fi network, so that it is no longer necessary to carry both a desktop phone and a cellphone. Maintenance becomes simpler as there are fewer devices to oversee.

Skype, which originally marketed itself as a service among friends, has begun to cater to businesses, providing free-of-charge connections between any users on the Skype network and connecting to and from ordinary PSTN telephones for a charge.

In the United States the Social Security Administration (SSA) is converting its field offices of 63,000 workers from traditional phone installations to a VoIP infrastructure carried over its existing data network.

Quality of Service

Communication on the IP network is perceived as less reliable in contrast to the circuit-switched public telephone network because it does not provide a network-based mechanism to ensure that data packets are not lost, and are delivered in sequential order. It is a best-effort network without fundamental Quality of Service (QoS) guarantees. Voice, and all other data, travels in packets over IP networks with fixed maximum capacity. This system may be more prone to congestion and DoS attacks than traditional circuit switched systems; a circuit switched system of insufficient capacity will refuse new connections while carrying the remainder without impairment, while the quality of real-time data such as telephone conversations on packet-switched networks degrades dramatically. Therefore, VoIP implementations may face problems with latency, packet loss, and jitter.

By default, network routers handle traffic on a first-come, first-served basis. Fixed delays cannot be controlled as they are caused by the physical distance the packets travel. They are especially problematic when satellite circuits are involved because of the long distance to a geostationary satellite and back; delays of 400–600 ms are typical. Latency can be minimized by marking voice packets as being delay-sensitive with QoS methods such as DiffServ.

Network routers on high volume traffic links may introduce latency that exceeds permissible thresholds for VoIP. When the load on a link grows so quickly that its switches experience queue overflows, congestion results and data packets are lost. This signals a transport protocol like TCP to reduce its transmission rate to alleviate the congestion. But VoIP usually uses UDP not TCP because recovering from congestion through retransmission usually entails too much latency. So QoS mechanisms can avoid the undesirable loss of VoIP packets by immediately transmitting them ahead of any queued bulk traffic on the same link, even when that bulk traffic queue is overflowing.

VoIP endpoints usually have to wait for completion of transmission of previous packets before new data may be sent. Although it is possible to preempt (abort) a less important packet in mid-transmission, this is not commonly done, especially on high-speed links where transmission times are short even for maximum-sized packets. An alternative to preemption on slower links, such as dialup and digital subscriber line (DSL), is to reduce the maximum transmission time by reducing the maximum transmission unit. But every packet must contain protocol headers, so this increases relative header overhead on every link traversed, not just the bottleneck (usually Internet access) link.

The receiver must resequence IP packets that arrive out of order and recover gracefully when packets arrive too late or not at all. Jitter results from the rapid and random (i.e. unpredictable) changes in queue lengths along a given Internet path due to competition from other users for the same transmission links. VoIP receivers counter jitter by storing incoming packets briefly in a "de-jitter" or "playout" buffer, deliberately increasing latency to improve the chance that each packet will be on hand when it is time for the voice engine to play it. The added delay is thus a compromise between excessive latency and excessive dropout, i.e. momentary audio interruptions.

Although jitter is a random variable, it is the sum of several other random variables that are at least somewhat independent: the individual queuing delays of the routers along the Internet path in question. Thus according to the central limit theorem, we can model jitter as a gaussian random variable. This suggests continually estimating the mean delay and its standard deviation and setting the playout delay so that only packets delayed more than several standard deviations above the mean will arrive too late to be useful. In practice, however, the variance in latency of many Internet paths is dominated by a small number (often one) of relatively slow and congested "bottleneck" links. Most Internet backbone links are now so fast (e.g. 10 Gbit/s) that their delays are dominated by the transmission medium (e.g. optical fiber) and the routers driving them do not have enough buffering for queuing delays to be significant.

It has been suggested to rely on the packetized nature of media in VoIP communications and transmit the stream of packets from the source phone to the destination phone simultaneously across different routes (multi-path routing). In such a way, temporary failures have less impact on the communication quality. In capillary routing it has been suggested to use at the packet level Fountain codes or particularly raptor codes for transmitting extra redundant packets making the communication more reliable.

A number of protocols have been defined to support the reporting of quality of service (QoS) and quality of experience (QoE) for VoIP calls. These include RTCP Extended Report (RFC 3611), SIP RTCP Summary Reports, H.460.9 Annex B (for H.323), H.248.30 and MGCP extensions. The RFC 3611 VoIP Metrics block is generated by an IP phone or gateway during a live call and contains information on packet loss rate, packet discard rate (because of jitter), packet loss/discard burst metrics (burst length/ density, gap length/density), network delay, end system delay, signal / noise / echo level, Mean Opinion Scores (MOS) and R factors and configuration information related to the jitter buffer.

RFC 3611 VoIP metrics reports are exchanged between IP endpoints on an occasional basis during a call, and an end of call message sent via SIP RTCP Summary Report or one of the other signaling protocol extensions. RFC 3611 VoIP metrics reports are intended to support real time feedback related to QoS problems, the exchange of information between the endpoints for improved call quality calculation and a variety of other applications.

Rural areas in particular are greatly hindered in their ability to choose a VoIP system over PBX. This is generally down to the poor access to superfast broadband in rural country areas. With the release of 4G data, there is a potential for corporate users based outside of populated areas to switch their internet connection to 4G data, which is comparatively as fast as a regular superfast broadband connection. This greatly enhances the overall quality and user experience of a VoIP system in these areas. This method was already trialled in rural Germany, surpassing all expectations.

DSL and ATM

DSL modems provide Ethernet (or Ethernet over USB) connections to local equipment, but inside they are actually Asynchronous Transfer Mode (ATM) modems. They use ATM Adaptation Layer 5 (AAL5) to segment each Ethernet packet into a series of 53-byte ATM cells for transmission, reassembling them back into Ethernet frames at the receiving end. A virtual circuit identifier (VCI) is part of the 5-byte header on every ATM cell, so the transmitter can multiplex the active virtual circuits (VCs) in any arbitrary order. Cells from the *same* VC are always sent sequentially.

However, a majority of DSL providers use only one VC for each customer, even those with bundled VoIP service. Every Ethernet frame must be completely transmitted before another can begin. If a second VC were established, given high priority and reserved for VoIP, then a low priority data packet could be suspended in mid-transmission and a VoIP packet sent right away on the high priority VC. Then the link would pick up the low priority VC where it left off. Because ATM links are multiplexed on a cell-by-cell basis, a high priority packet would have to wait at most 53 byte times to begin transmission. There would be no need to reduce the interface MTU and accept the resulting increase in higher layer protocol overhead, and no need to abort a low priority packet and resend it later.

ATM has substantial header overhead: $5/53 = 9.4\%$, roughly twice the total header overhead of a 1500 byte Ethernet frame. This "ATM tax" is incurred by every DSL user whether or not they take advantage of multiple virtual circuits - and few can.

ATM's potential for latency reduction is greatest on slow links, because worst-case latency decreases with increasing link speed. A full-size (1500 byte) Ethernet frame takes 94 ms to transmit at 128 kbit/s but only 8 ms at 1.5 Mbit/s. If this is the bottleneck link, this latency is probably small enough to ensure good VoIP performance without MTU reductions or multiple ATM VCs. The latest generations of DSL, VDSL and VDSL2, carry Ethernet without intermediate ATM/AAL5 layers, and they generally support IEEE 802.1p priority tagging so that VoIP can be queued ahead of less time-critical traffic.

Layer 2

A number of protocols that deal with the data link layer and physical layer include quality-of-service mechanisms that can be used to ensure that applications like VoIP work well even in congested scenarios. Some examples include:

- IEEE 802.11e is an approved amendment to the IEEE 802.11 standard that defines a set of quality-of-service enhancements for wireless LAN applications through modifications to the Media Access Control (MAC) layer. The standard is considered of critical importance for delay-sensitive applications, such as voice over wireless IP.

- IEEE 802.1p defines 8 different classes of service (including one dedicated to voice) for traffic on layer-2 wired Ethernet.

- The ITU-T G.hn standard, which provides a way to create a high-speed (up to 1 gigabit per second) Local area network (LAN) using existing home wiring (power lines, phone lines and coaxial cables). G.hn provides QoS by means of "Contention-Free Transmission Opportunities" (CFTXOPs) which are allocated to flows (such as a VoIP call) which require QoS and which have negotiated a "contract" with the network controllers.

VoIP Performance Metrics

The quality of voice transmission is characterized by several metrics that may be monitored by network elements, by the user agent hardware or software. Such metrics include network packet loss, packet jitter, packet latency (delay), post-dial delay, and echo. The metrics are determined by VoIP performance testing and monitoring.

PSTN Integration

The Media VoIP Gateway connects the digital media stream, so as to complete creating the path for voice as well as data media. It includes the interface for connecting the standard PSTN networks with the ATM and Inter Protocol networks. The Ethernet interfaces are also included in the modern systems, which are specially designed to link calls that are passed via the VoIP.

E.164 is a global FGFnumbering standard for both the PSTN and PLMN. Most VoIP implementations support E.164 to allow calls to be routed to and from VoIP subscribers and the PSTN/PLMN. VoIP implementations can also allow other identification techniques to be used. For example, Skype allows subscribers to choose "Skype names" (usernames) whereas SIP implementations can use URIs similar to email addresses. Often VoIP implementations employ methods of translating non-E.164 identifiers to E.164 numbers and vice versa, such as the Skype-In service provided by Skype and the ENUM service in IMS and SIP.

Echo can also be an issue for PSTN integration. Common causes of echo include impedance mismatches in analog circuitry and acoustic coupling of the transmit and receive signal at the receiving end.

Number Portability

Local number portability (LNP) and Mobile number portability (MNP) also impact VoIP business. In November 2007, the Federal Communications Commission in the United States released an order extending number portability obligations to interconnected VoIP providers and carriers that support VoIP providers. Number portability is a service that allows a subscriber to select a new telephone carrier without requiring a new number to be issued. Typically, it is the responsibility of the former carrier to "map" the old number to the undisclosed number assigned by the new carrier. This is achieved by maintaining a database of numbers. A dialed number is initially received by the original carrier and quickly rerouted to the new carrier. Multiple porting references must be maintained even if the subscriber returns to the original carrier. The FCC mandates carrier compliance with these consumer-protection stipulations.

A voice call originating in the VoIP environment also faces challenges to reach its destination if the number is routed to a mobile phone number on a traditional mobile carrier. VoIP has been identified in the past as a Least Cost Routing (LCR) system, which is based on checking the destination of each telephone call as it is made, and then sending the call via the network that will cost the customer the least. This rating is subject to some debate given the complexity of call routing created by number portability. With GSM number portability now in place, LCR providers can no longer rely on using the network root prefix to determine how to route a call. Instead, they must now determine the actual network of every number before routing the call.

Therefore, VoIP solutions also need to handle MNP when routing a voice call. In countries without a central database, like the UK, it might be necessary to query the GSM network about which home network a mobile phone number belongs to. As the popularity of VoIP increases in the enterprise markets because of least cost routing options, it needs to provide a certain level of reliability when handling calls.

MNP checks are important to assure that this quality of service is met. Handling MNP lookups before routing a call provides some assurance that the voice call will actually work.

Emergency Calls

A telephone connected to a land line has a direct relationship between a telephone number and a physical location, which is maintained by the telephone company and available to emergency responders via the national emergency response service centers in form of emergency subscriber lists. When an emergency call is received by a center the location is automatically determined from its databases and displayed on the operator console.

In IP telephony, no such direct link between location and communications end point exists. Even a provider having hardware infrastructure, such as a DSL provider, may only know the approximate location of the device, based on the IP address allocated to the network router and the known service address. However, some ISPs do not track the automatic assignment of IP addresses to customer equipment.

IP communication provides for device mobility. For example, a residential broadband connection may be used as a link to a virtual private network of a corporate entity, in which case the IP address being used for customer communications may belong to the enterprise, not being the network address of the residential ISP. Such off-premises extensions may appear as part of an upstream IP PBX. On mobile devices, e.g., a 3G handset or USB wireless broadband adapter, the IP address has no relationship with any physical location known to the telephony service provider, since a mobile user could be anywhere in a region with network coverage, even roaming via another cellular company.

At the VoIP level, a phone or gateway may identify itself with a Session Initiation Protocol (SIP) registrar by its account credentials. In such cases, the Internet telephony service provider (ITSP) only knows that a particular user's equipment is active. Service providers often provide emergency response services by agreement with the user who registers a physical location and agrees that emergency services are only provided to that address if an emergency number is called from the IP device.

Such emergency services are provided by VoIP vendors in the United States by a system called Enhanced 911 (E911), based on the Wireless Communications and Public Safety Act of 1999. The VoIP E911 emergency-calling system associates a physical address with the calling party's telephone number. All VoIP providers that provide access to the public switched telephone network are required to implement E911, a service for which the subscriber may be charged. However, end-customer participation in E911 is not mandatory and customers may opt out of the service.

The VoIP E911 system is based on a static table lookup. Unlike in cellular phones, where the location of an E911 call can be traced using assisted GPS or other methods, the VoIP E911 information is only accurate so long as subscribers, who have the legal responsibility, are diligent in keeping their emergency address information current.

Fax Support

Support for fax has been problematic in many VoIP implementations, as most voice digitization and compression codecs are optimized for the representation of the human voice and the proper timing of the modem signals cannot be guaranteed in a packet-based, connection-less network. An alternative IP-based solution for delivering fax-over-IP called T.38 is available. Sending faxes using VoIP is sometimes referred to as FoIP, or Fax over IP.

The T.38 protocol is designed to compensate for the differences between traditional packet-less communications over analog lines and packet-based transmissions which are the basis for IP communications. The fax machine could be a traditional fax machine connected to the PSTN, or an ATA box (or similar). It could be a fax machine with an RJ-45 connector plugged straight into an IP network, or it could be a computer pretending to be a fax machine. Originally, T.38 was designed to use UDP and TCP transmission methods across an IP network. TCP is better suited for use between two IP devices. However, older fax machines, connected to an analog system, benefit from UDP near real-time characteristics due to the "no recovery rule" when a UDP packet is lost or an error occurs during transmission. UDP transmissions are preferred as they do not require testing for dropped packets and as such since each T.38 packet transmission includes a majority of the data sent in the prior packet, a T.38 termination point has a higher degree of success in re-assembling the fax transmission back into its original form for interpretation by the end device. This in an attempt to overcome the obstacles of simulating real time transmissions using packet based protocol.

There have been updated versions of T.30 to resolve the fax over IP issues, which is the core fax protocol. Some newer high end fax machines have T.38 built-in capabilities which allow the user to plug right into the network and transmit/receive faxes in native T.38 like the Ricoh 4410NF Fax Machine. A unique feature of T.38 is that each packet contains a portion of the main data sent in the previous packet. With T.38, two successive lost packets are needed to actually lose any data. The data one will lose will only be a small piece, but with the right settings and error correction mode, there is an increased likelihood that they will receive enough of the transmission to satisfy the requirements of the fax machine for output of the sent document.

While many late-model analog telephone adapters (ATAs) support T.38, uptake has been limited as many voice-over-IP providers perform least-cost routing which selects the least expensive PSTN gateway in the called city for an outbound message. There is typically no means to ensure that that gateway is T.38 capable. Providers often place their own equipment (such as an Asterisk PBX installation) in the signal path, which creates additional issues as every link in the chain must be T.38 aware for the protocol to work. Similar issues arise if a provider is purchasing local direct inward dial numbers from the lowest bidder in each city, as many of these may not be T.38 enabled.

Power Requirements

Telephones for traditional residential analog service are usually connected directly to telephone company phone lines which provide direct current to power most basic analog handsets independently of locally available electrical power.

IP Phones and VoIP telephone adapters connect to routers or cable modems which typically depend on the availability of mains electricity or locally generated power. Some VoIP service providers use customer premises equipment (e.g., cablemodems) with battery-backed power supplies to assure uninterrupted service for up to several hours in case of local power failures. Such battery-backed devices typically are designed for use with analog handsets.

Some VoIP service providers implement services to route calls to other telephone services of the subscriber, such a cellular phone, in the event that the customer's network device is inaccessible to terminate the call.

The susceptibility of phone service to power failures is a common problem even with traditional analog service in areas where many customers purchase modern telephone units that operate with wireless handsets to a base station, or that have other modern phone features, such as built-in voicemail or phone book features.

Security

The security concerns of VoIP telephone systems are similar to those of any Internet-connected device. This means that hackers who know about these vulnerabilities can institute denial-of-service attacks, harvest customer data, record conversations and compromise voicemail messages. The quality of internet connection determines the quality of the calls. VoIP phone service also will not work if there is power outage and when the internet connection is down. The 9-1-1 or 112 service provided by VoIP phone service is also different from analog phone which is associated with a fixed address. The emergency center may not be able to determine your location based on your virtual phone number. Compromised VoIP user account or session credentials may enable an attacker to incur substantial charges from third-party services, such as long-distance or international telephone calling.

The technical details of many VoIP protocols create challenges in routing VoIP traffic through firewalls and network address translators, used to interconnect to transit networks or the Internet. Private session border controllers are often employed to enable VoIP calls to and from protected networks. Other methods to traverse NAT devices involve assistive protocols such as STUN and Interactive Connectivity Establishment (ICE).

Many consumer VoIP solutions do not support encryption of the signaling path or the media, however securing a VoIP phone is conceptually easier to implement than on traditional telephone circuits. A result of the lack of encryption is a relative easy to eavesdrop on VoIP calls when access to the data network is possible. Free open-source solutions, such as Wireshark, facilitate capturing VoIP conversations.

Standards for securing VoIP are available in the Secure Real-time Transport Protocol (SRTP) and the ZRTP protocol for analog telephony adapters as well as for some softphones. IPsec is available to secure point-to-point VoIP at the transport level by using opportunistic encryption.

Government and military organizations use various security measures to protect VoIP traffic, such as voice over secure IP (VoSIP), secure voice over IP (SVoIP), and secure voice over secure IP (SVoSIP). The distinction lies in whether encryption is applied in the telephone or in the network or both. Secure voice over secure IP is accomplished by encrypting VoIP with protocols such as SRTP or ZRTP. Secure voice over IP is accomplished by using Type 1 encryption on a classified network, like SIPRNet. Public Secure VoIP is also available with free GNU programs and in many popular commercial VoIP programs via libraries such as ZRTP.

Caller ID

Voice over IP protocols and equipment provide caller ID support that is compatible with the facility provided in the public switched telephone network (PSTN). Many VoIP service providers also allow callers to configure arbitrary caller ID information.

Compatibility with Traditional Analog Telephone Sets

Most analog telephone adapters do not decode dial pulses generated by rotary dial telephones, supporting only touch-tone signaling, but pulse-to-tone converters are commercially available.

Support for other Telephony Devices

Some special telephony services, such as those that operate in conjunction with digital video recorders, satellite television receivers, alarm systems, conventional modems over PSTN lines, may be impaired when operated over VoIP services, because of incompatibilities in design.

Operational Cost

VoIP has drastically reduced the cost of communication by sharing network infrastructure between data and voice. A single broad-band connection has the ability to transmit more than one telephone call. Secure calls using standardized protocols, such as Secure Real-time Transport Protocol, as most of the facilities of creating a secure telephone connection over traditional phone lines, such as digitizing and digital transmission, are already in place with VoIP. It is only necessary to encrypt and authenticate the existing data stream. Automated software, such as a virtual PBX, may eliminate the need of personnel to greet and switch incoming calls.

Regulatory and Legal Issues

As the popularity of VoIP grows, governments are becoming more interested in regulating VoIP in a manner similar to PSTN services.

Throughout the developing world, countries where regulation is weak or captured by the dominant operator, restrictions on the use of VoIP are imposed, including in Panama where VoIP is taxed, Guyana where VoIP is prohibited and India where its retail commercial sales is allowed but only for long distance service. In Ethiopia, where the government is nationalising telecommunication service, it is a criminal offence to offer services using VoIP. The country has installed firewalls to prevent international calls being made using VoIP. These measures were taken after the popularity of VoIP reduced the income generated by the state owned telecommunication company.

European Union

In the European Union, the treatment of VoIP service providers is a decision for each national telecommunications regulator, which must use competition law to define relevant national markets and then determine whether any service provider on those national markets has "significant market power" (and so should be subject to certain obligations). A general distinction is usually made between VoIP services that function over managed networks (via broadband connections) and VoIP services that function over unmanaged networks (essentially, the Internet).

The relevant EU Directive is not clearly drafted concerning obligations which can exist independently of market power (e.g., the obligation to offer access to emergency calls), and it is impossible to say definitively whether VoIP service providers of either type are bound by them. A review of the EU Directive is under way and should be complete by 2007.

Middle East

In the UAE and Oman it is illegal to use any form of VoIP, to the extent that Web sites of Gizmo5 are blocked. Providing or using VoIP services is illegal in Oman. Those who violate the law stand to be fined 50,000 Omani Rial (about 130,317 US dollars) or spend two years in jail or both. In 2009, police in Oman have raided 121 Internet cafes throughout the country and arrested 212 people for using/providing VoIP services.

India

In India, it is legal to use VoIP, but it is illegal to have VoIP gateways inside India. This effectively means that people who have PCs can use them to make a VoIP call to any number, but if the remote side is a normal phone, the gateway that converts the VoIP call to a POTS call is not permitted by law to be inside India. Foreign based Voip server services are illegal to use in India.

In the interest of the Access Service Providers and International Long Distance Operators the Internet telephony was permitted to the ISP with restrictions. Internet Telephony is considered to be different service in its scope, nature and kind from real time voice as offered by other Access Service Providers and Long Distance Carriers. Hence the following type of Internet Telephony are permitted in India:

(a) PC to PC; within or outside India

(b) PC / a device / Adapter conforming to standard of any international agencies like-ITU or IETF etc. in India to PSTN/PLMN abroad.

(c) Any device / Adapter conforming to standards of International agencies like ITU, IETF etc. connected to ISP node with static IP address to similar device / Adapter; within or outside India.

(d) Except whatever is described in condition (ii) above, no other form of Internet Telephony is permitted.

(e) In India no Separate Numbering Scheme is provided to the Internet Telephony. Presently the 10 digit Numbering allocation based on E.164 is permitted to the Fixed Telephony, GSM, CDMA wireless service. For Internet Telephony the numbering scheme shall only conform to IP addressing Scheme of Internet Assigned Numbers Authority (IANA). Translation of E.164 number / private number to IP address allotted to any device and vice versa, by ISP to show compliance with IANA numbering scheme is not permitted.

(f) The Internet Service Licensee is not permitted to have PSTN/PLMN connectivity. Voice communication to and from a telephone connected to PSTN/PLMN and following E.164 numbering is prohibited in India.

South Korea

In South Korea, only providers registered with the government are authorized to offer VoIP services. Unlike many VoIP providers, most of whom offer flat rates, Korean VoIP services are generally metered and charged at rates similar to terrestrial calling. Foreign VoIP providers encounter high barriers to government registration. This issue came to a head in 2006 when Internet service providers providing personal Internet services by contract to United States Forces Korea members residing on USFK bases threatened to block off access to VoIP services used by USFK members as an economical way to keep in contact with their families in the United States, on the grounds that the service members' VoIP providers were not registered. A compromise was reached between USFK and Korean telecommunications officials in January 2007, wherein USFK service members arriving in Korea before June 1, 2007, and subscribing to the ISP services provided on base may continue to use their US-based VoIP subscription, but later arrivals must use a Korean-based VoIP provider, which by contract will offer pricing similar to the flat rates offered by US VoIP providers.

United States

In the United States, the Federal Communications Commission requires all interconnected VoIP service providers to comply with requirements comparable to those for traditional telecommunications service providers. VoIP operators in the US are required to support local number portability; make service accessible to people with disabilities; pay regulatory fees, universal service contributions, and other mandated payments; and enable law enforcement authorities to conduct surveillance pursuant to the Communications Assistance for Law Enforcement Act (CALEA).

Operators of "Interconnected" VoIP (fully connected to the PSTN) are mandated to provide Enhanced 911 service without special request, provide for customer location updates, clearly disclose any limitations on their E-911 functionality to their consumers, obtain affirmative acknowledgements of these disclosures from all consumers, and 'may not allow their customers to "opt-out" of 911 service.' VoIP operators also receive the benefit of certain US telecommunications regulations, including an entitlement to interconnection and exchange of traffic with incumbent local exchange carriers via wholesale carriers. Providers of "nomadic" VoIP service—those who are unable to determine the location of their users—are exempt from state telecommunications regulation.

Another legal issue that the US Congress is debating concerns changes to the Foreign Intelligence Surveillance Act. The issue in question is calls between Americans and foreigners. The National Security Agency (NSA) is not authorized to tap Americans' conversations without a warrant—but the Internet, and specifically VoIP does not draw as clear a line to the location of a caller or a call's recipient as the traditional phone system does. As VoIP's low cost and flexibility convinces more and more organizations to adopt the technology, the surveillance for law enforcement agencies becomes more difficult. VoIP technology has also increased security concerns because VoIP and similar technologies have made it more difficult for the government to determine where a target is physically located when communications are being intercepted, and that creates a whole set of new legal challenges.

Historical Milestones

- 1973: Network Voice Protocol (NVP) developed by Danny Cohen and others to carry real time voice over Arpanet.

- 1974: The Institute of Electrical and Electronic Engineers (IEEE) published a paper titled "A Protocol for Packet Network Interconnection".

- 1974: Network Voice Protocol (NVP) first tested over Arpanet in August 1974, carrying 16k CVSD encoded voice – first implementation of Voice over IP

- 1977: Danny Cohen, Vint Cerf, Jon Postel agree to separate IP from TCP, and create UDP for carrying real time traffic

- 1981: IPv4 is described in RFC 791.

- 1985: The National Science Foundation commissions the creation of NSFNET.

- 1986: Proposals from various standards organizations[specify] for Voice over ATM, in addition to commercial packet voice products from companies such as StrataCom

- 1991: First Voice Over IP application, Speak Freely, released as public domain. Originally written by John Walker and further developed by Brian C. Wiles.

- 1992: Voice over Frame Relay standards development within Frame Relay Forum

- 1994: MTALK, a freeware VoIP application for Linux

- 1995: VocalTec releases the first commercial Internet phone software.

  o Beginning in 1995, Intel, Microsoft and Radvision initiated standardization activities for VoIP communications system.

- 1996:

  o ITU-T begins development of standards for the transmission and signaling of voice communications over Internet Protocol networks with the H.323 standard.

  o US telecommunication companies petition the US Congress to ban Internet phone technology.

- 1997: Level 3 began development of its first softswitch, a term they coined in 1998.

- 1999:

  o The Session Initiation Protocol (SIP) specification RFC 2543 is released.

- o Mark Spencer of Digium develops the first open source private branch exchange (PBX) software (Asterisk).

- 2004: Commercial VoIP service providers proliferate.

- 2007: VOIP device manufacturers and sellers boom in Asia, specifically in the Philippines where many families of overseas workers reside.

- 2011: Raise of WebRTC technology which allows VoIP directly in browsers

- 2015: Trend of using VoIP services in cloud: PBXes and contact centers, it means higher requirements to IP network to achieve good quality of service and reliability

Etiquette in Technology

Etiquette in technology governs what conduct is socially acceptable in an online or digital situation. While etiquette is ingrained into culture, etiquette in technology is a fairly recent concept. The rules of etiquette that apply when communicating over the Internet or social networks or devices are different from those applying when communicating in person or by audio (such as telephone) or videophone (such as Skype video). It is a social code of network communication.

Communicating with others via the Internet without misunderstandings in the heat of the moment can be challenging, mainly because facial expressions and body language cannot be interpreted in cyberspace. Therefore, several recommendations to attempt to safeguard against these misunderstandings have been proposed.

Netiquette, a colloquial portmanteau of *network etiquette* or 'etiquette, *is a set of social conventions that facilitate interaction over networks, ranging from and to and*

Like the network itself, these developing norms remain in a state of flux and vary from community to community. The points most strongly emphasized about Usenet netiquette often include using simple electronic signatures, and avoiding multiposting, cross-posting, off-topic posting, hijacking a discussion thread, and other techniques used to minimize the effort required to read a post or a thread. Similarly, some Usenet guidelines call for use of unabbreviated English while users of instant messaging protocols like SMS occasionally encourage just the opposite, bolstering use of SMS language. However, many online communities frown upon this practice.

Common rules for e-mail and Usenet such as avoiding flamewars and spam are constant across most mediums and communities. Another rule is to avoid typing in all caps or grossly enlarging script for emphasis, which is considered to be the equivalent of shouting or yelling. Other commonly shared points, such as remembering that one's posts are (or can easily be made) public, are generally intuitively understood by publishers of Web pages and posters to Usenet, although this rule is somewhat flexible depending on the environment. On more private protocols, however, such as e-mail and SMS, some users take the privacy of their posts for granted. One-on-one communications, such as private messages on chat forums and direct SMSs, may be considered more private than other such protocols, but infamous breaches surround even these relatively private media. For example, Paris Hilton's Sidekick PDA was cracked in 2005, resulting in the publication of her private photos, SMS history, address book, etc.

A group e-mail sent by Cerner CEO Neal Patterson to managers of a facility in Kansas City concerning "Cerner's declining work ethic" read, in part, "The parking lot is sparsely used at 8 A.M.; likewise at 5 P.M. As managers—you either do not know what your EMPLOYEES are doing; or YOU do not CARE ... In either case, you have a problem and you will fix it or I will replace you." After the e-mail was forwarded to hundreds of other employees, it quickly leaked to the public. On the day that the e-mail was posted to Yahoo!, Cerner's stock price fell by over 22% from a high market capitalization of US$1.5 billion.

Beyond matters of basic courtesy and privacy, e-mail syntax (defined by RFC 2822) allows for different types of recipients. The primary recipient, defined by the To: line, can reasonably be expected to respond, but recipients of carbon copies cannot be, although they still might. Likewise, misuse of the CC: functions in lieu of traditional mailing lists can result in serious technical issues. In late 2007, employees of the United States Department of Homeland Security used large CC: lists in place of a mailing list to broadcast messages to several hundred users. Misuse of the "reply to all" caused the number of responses to that message to quickly expand to some two million messages, bringing down their mail server. In cases like this, rules of netiquette have more to do with efficient sharing of resources—ensuring that the associated technology continues to function—rather than more basic etiquette. On Usenet, cross-posting, in which a single copy of a message is posted to multiple groups is intended to prevent this from happening, but many newsgroups frown on the practice, as it means users must sometimes read many copies of a message in multiple groups.

"When someone makes a mistake – whether it's a spelling error or a spelling flame, a stupid question or an unnecessarily long answer – be kind about it. If it's a minor error, you may not need to say anything. Even if you feel strongly about it, think twice before reacting. Having good manners yourself doesn't give you license to correct everyone else. If you do decide to inform someone of a mistake, point it out politely, and preferably by private email rather than in public. Give people the benefit of the doubt; assume they just don't know any better. And never be arrogant or self-righteous about it. Just as it's a law of nature that spelling flames always contain spelling errors, notes pointing out Netiquette violations are often examples of poor Netiquette."

Due to the large variation between what is considered acceptable behavior in various professional environments and between professional and social networks, codified internal manuals of style can help clarify acceptable limits and boundaries for user behavior. For instance, failure to publish such a guide for e-mail style was cited among the reasons for a NZ$17,000 wrongful dismissal finding against a firm that fired a woman for misuse of all caps in company-wide e-mail traffic.

Online Etiquette

Digital citizenship is a term that describes how a person should act while using digital technology online and has also been defined as "the ability to participate in society online". The term is often mentioned in relation to Internet safety and netiquette.

The term has been used as early as 1998 and has gone through several changes in description as newer technological advances have changed the method and frequency of how people interact with one another online. Classes on digital citizenship have been taught in some public education systems and some argue that the term can be "measured in terms of economic and political activities online".

Cell Phone Etiquette

A headrest cover in the "quiet carriage" of a British intercity train, reminding passengers that mobile phones must not be used in this carriage

The issue of mobile communication and etiquette has also become an issue of academic interest. The rapid adoption of the device has resulted in the intrusion of telephony into situations where it was previously not used. This has exposed the implicit rules of courtesy and opened them to reevaluation.

Cell Phone Etiquette in the Education System

Most schools in the United States and Europe and Canada have prohibited mobile phones in the classroom, citing class disruptions and the potential for cheating via text messaging. In the UK, possession of a mobile phone in an examination can result in immediate disqualification from that subject or from all that student's subjects. This still applies even if the mobile phone was not turned on at the time. In New York City, students are banned from taking cell phones to school. This has been a debate for several years, but finally passed legislature in 2008.

"Most schools allow students to have cell phones for safety purposes"—a reaction to the Littleton, Colorado, high school shooting incident of 1999 (Lipscomb 2007: 50). Apart from emergency situations, most schools don't officially allow students to use cell phones during class time.

Cell Phone Etiquette in the Public Sphere

Talking or texting on a cell phone in public may seem a distraction for many individuals. When in public there are two times when one uses a phone. The first is when someone is alone and the other is when he/she is in a group. The main issue for most people is when they are in a group, and the cell phone becomes a distraction or a barrier for successful socialization among family and friends. In the past few years, society has become less tolerant of cell phone use in public areas for example, public transportation, restaurants and much more. This is exemplified by the widespread recognition of campaigns such as Stop Phubbing, which prompted global discussion as to how mobile phones should be used in the presence of others. "Some have suggested that mobile phones 'affect every aspect of our personal and professional lives either directly or indirectly'" (Humphrey). Every culture's tolerance of cell phone usage varies, for instance in Western society cell phones are permissible during free time at schools, whereas in the eastern countries, cell phones are strictly prohibited on school property.

Mobile phone use can be an important matter of social discourtesy: phones ringing during funerals or weddings; in toilets, cinemas and theatres. Some book shops, libraries, bathrooms, cinemas, doctors' offices and places of worship prohibit their use, so that other patrons will not be disturbed by conversations. Some facilities install signal-jamming equipment to prevent their use, although in many countries, including the US, such equipment is illegal. Some new auditoriums have installed wire mesh in the walls to make a Faraday cage, which prevents signal penetration without violating signal jamming laws.

A working group made up of Finnish telephone companies, public transport operators and communications authorities has launched a campaign to remind mobile phone users of courtesy, especially when using mass transit—what to talk about on the phone, and how to. In particular, the campaign wants to impact loud mobile phone usage as well as calls regarding sensitive matters.

Trains, particularly those involving long-distance services, often offer a "quiet carriage" where phone use is prohibited, much like the designated non-smoking carriage of the past. In the UK however many users tend to ignore this as it is rarely enforced, especially if the other carriages are crowded and they have no choice but to go in the "quiet carriage". In Japan, it is generally considered impolite to talk using a phone on any train—e-mailing is generally the mode of mobile communication. Mobile phone usage on local public transport is also increasingly seen as a nuisance; the city of Graz, for instance, has mandated a total ban of mobile phones on its tram and bus network in 2008 (though texting and emailing is still allowed).

Nancy J. Friedman has spoken widely about landline and cell phone etiquette. Emily Post has also written on her essential rules for using a cell phone.

Cell Phone Etiquette within Social Relationships

When critically assessing the family structure, it is important to examine the parent/child negotiations which occur in the household, in relation to the increased use of cell phones. Teenagers use their cell phones as a way to negotiate spatial boundaries with their parents (Williams 2005:316). This includes extending curfews in the public space and allowing more freedom for the teenagers when they are outside of the home (Williams 2005:318). More importantly, cell phone etiquette relates to kinship groups and the family as an institution. This is because cell phones act as a threat due to the rapid disconnect within families. Children are often so closely affiliated with their technological gadgets, and they tend to interact with their friends constantly and this has a negative impact on their relationship with their parents (Williams 2005:326). Teenagers see themselves as gaining a sense of empowerment from the mobile phone. Cell phone etiquette in the household from an anthropological perspective has shown an evolution in the institution of family. The mobile phone has now been integrated into family practices and perpetuated a wider concern which is the fracture between parent and child relationships. We are able to see the traditional values disappearing; however, reflexive monitoring is occurring (Williams 2005:320). Through this, parents are becoming friendlier with their children and critics emphasize that this change is problematic because children should be subjected to social control. One way of social control is limiting the time spent interacting with friends, which is difficult to do in today's society because of the rapid use of cell phones.

Netiquette vs. Cell Phone Etiquette

Cell phone etiquette is largely dependent on the cultural context and what is deemed to be socially acceptable. For instance, in certain cultures using your hand held devices while interacting in a group environment is considered bad manners, whereas, in other cultures around the world it may be viewed differently. In addition, cell phone etiquette also encompasses the various types of activities which are occurring and the nature of the messages which are being sent. More importantly, messages of an inappropriate nature can be sent to an individual and this could potentially orchestrate problems such as verbal/ cyber abuse.

New Technology and Behaviors

Perhaps the biggest obstacle to communication in online settings is the lack of emotional cues. Facial cues dictate the mood and corresponding diction of two people in a conversation. During phone conversations, tone of voice communicates the emotions of the person on the other line. But with chat rooms, instant messaging apps and texting, any signals that would indicate the tone of a person's words or their state of emotion are absent. Because of this, there have been some interesting accommodations. Perhaps the two most prevalent compensating behaviors are the use of emoticons and abbreviations. Emoticons use punctuation marks to illustrate common symbols that pertain to facial cues. For example, one would combine a colon and parenthesis to recreate the symbol of the smiley face indicating the happiness or satisfaction of the other person. To symbolize laughter, the abbreviation "LOL" standing for "laughing out loud" developed. Along with these, countless other symbols and abbreviations have developed including, "BRB" ("be right back"), "TTYL" (talk to you later) and specific designs incorporated by apps of a laughing face, sad face, crying face, angry face etc.

Now, as newer modes of communication are becoming more common, the rules of communication must adapt as fast as the technology. For example, one of the most popular new apps, Snapchat, is growing to have its own rules and etiquette. This app lets users send pictures or videos to friends that disappear after a couple seconds of viewing it. Initially, the thought that occurs to people when confronted by this app is its implications for sexting. Although it's entirely possible to make use of Snapchat for that purpose, what the app has developed into is a form of communication that shares funny or interesting moments. Originally compared to Instagram by way of the app's ability to broadcast pictures to many people, it has now become standard to communicate through Snapchat by sending pictures back and forth and using the caption bar for messages. The reply option on Snapchat specifically promotes this behavior, but Snapchat etiquette is not set in stone. It is becoming clear that Snaps personalized for the receiver expect a reply, but where ends this obligation? Some people use Snapchat specifically for the purpose of communication, while some use it to simply provide a visual update of their day. The newest update of Snapchat, an instant messaging add-on, seems to be catered to those who use the app to send messages back and forth. This new messaging add-on, along with the video chat feature will warrant new forms of social construct and expectations of behavior in accordance with this application.

References

- Greenstein, Shane; Stango, Victor (2006). Standards and Public Policy. Cambridge University Press. pp. 129–132. ISBN 978-1-139-46075-0

- The Handbook of Electrical Engineering. Research & Education Association. 1996. p. D-149. ISBN 9780878919819.

- Asif, Saad Z. (2011). Next Generation Mobile Communications Ecosystem. John Wiley & Sons. p. 306. ISBN 1119995817.

- Ribble, Mike (2011). Digital Citizenship in Schools. International Society for Technology in Education. ISBN 1564843017.

- Bebo White, Irwin King, Philip Tsang (2011). Social Media Tools and Platforms in Learning Environments. Springer. pp. 406–407. ISBN 3642203914.

- Karen Mossberger, Caroline J. Tolbert, William Franko (2012). Digital Cities: The Internet and the Geography of Opportunity. Oxford University Press. pp. 64–65. ISBN 0199812950.

- Doane, Darryl S.; Sloat, Rose D (2003-09-01). 50 Activities for Achieving Excellent Customer Service. pp. 6, 24, 85. ISBN 9780874257373. Retrieved 8 May 2012.

- "WIRELESS: Carriers look to IP for back haul". www.eetimes.com. EE Times. Archived from the original on August 9, 2011. Retrieved 8 April 2015. CS1 maint: Unfit url (link)

- "Mobile's IP challenge". www.totaltele.com. Total Telecom Online. Archived from the original on February 17, 2006. Retrieved 8 April 2015. CS1 maint: Unfit url (link)

- "The Great SIM Heist - How Spies Stole the Keys to the Encryption Castle". The Intercept. The Intercept (First Look Media). February 19, 2015. Retrieved February 19, 2015.

- "Gemalto: NSA/GCHQ Hack 'Probably Happened' But Didn't Include Mass SIM Key Theft". techcrunch.com. February 25, 2015. Retrieved April 2, 2015.

3

Modulation and its Techniques

Modulation is the process of varying the properties of periodic waveform. The topics explained in this text are amplitude modulation, frequency modulation, phase modulation, frequency-shift keying and amplitude-shift keying. The technology that is used in the recent times for transmitting information is known as amplitude modulation. This chapter serves as a source to understand the concept of modulation and the techniques related to modulation.

Modulation

In electronics and telecommunications, modulation is the process of varying one or more properties of a periodic waveform, called the *carrier signal*, with a modulating signal that typically contains information to be transmitted.

In telecommunications, modulation is the process of conveying a message signal, for example a digital bit stream or an analog audio signal, inside another signal that can be physically transmitted. Modulation of a sine waveform transforms a baseband message signal into a passband signal.

A modulator is a device that performs modulation. A demodulator (sometimes *detector* or *demod*) is a device that performs demodulation, the inverse of modulation. A modem (from modulator–demodulator) can perform both operations.

The aim of analog modulation is to transfer an analog baseband (or lowpass) signal, for example an audio signal or TV signal, over an analog bandpass channel at a different frequency, for example over a limited radio frequency band or a cable TV network channel.

The aim of digital modulation is to transfer a digital bit stream over an analog bandpass channel, for example over the public switched telephone network (where a bandpass filter limits the frequency range to 300–3400 Hz) or over a limited radio frequency band.

Analog and digital modulation facilitate frequency division multiplexing (FDM), where several low pass information signals are transferred simultaneously over the same shared physical medium, using separate passband channels (several different carrier frequencies).

The aim of digital baseband modulation methods, also known as line coding, is to transfer a digital bit stream over a baseband channel, typically a non-filtered copper wire such as a serial bus or a wired local area network.

The aim of pulse modulation methods is to transfer a narrowband analog signal, for example a phone call over a wideband baseband channel or, in some of the schemes, as a bit stream over another digital transmission system.

In music synthesizers, modulation may be used to synthesise waveforms with an extensive overtone spectrum using a small number of oscillators. In this case the carrier frequency is typically in the same order or much lower than the modulating waveform.

Analog Modulation Methods

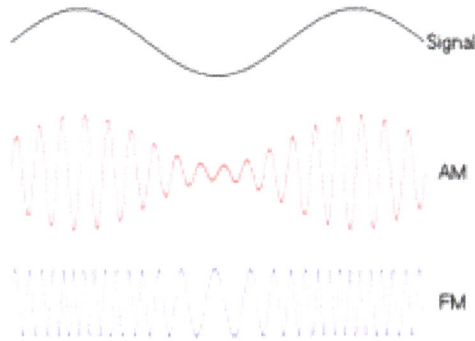

A low-frequency message signal (top) may be carried by an AM or FM radio wave.

In analog modulation, the modulation is applied continuously in response to the analog information signal.

List of Analog Modulation Techniques

Common analog modulation techniques are:

- Amplitude modulation (AM) (here the amplitude of the carrier signal is varied in accordance to the instantaneous amplitude of the modulating signal)
    - Double-sideband modulation (DSB)
        - Double-sideband modulation with carrier (DSB-WC) (used on the AM radio broadcasting band)
        - Double-sideband suppressed-carrier transmission (DSB-SC)
        - Double-sideband reduced carrier transmission (DSB-RC)
    - Single-sideband modulation (SSB, or SSB-AM)
        - Single-sideband modulation with carrier (SSB-WC)
        - Single-sideband modulation suppressed carrier modulation (SSB-SC)
    - Vestigial sideband modulation (VSB, or VSB-AM)
    - Quadrature amplitude modulation (QAM)
- Angle modulation, which is approximately constant envelope
    - Frequency modulation (FM) (here the frequency of the carrier signal is varied in accordance to the instantaneous amplitude of the modulating signal)

- o Phase modulation (PM) (here the phase shift of the carrier signal is varied in accordance with the instantaneous amplitude of the modulating signal)

- o Transpositional Modulation (TM), in which the waveform inflection is modified resulting in a signal where each quarter cycle is transposed in the modulation process. TM is a pesudo-analog modulation (AM). Where an AM carrier also carries a phase variable phase f(ǿ). TM is f(AM,ǿ)

Digital Modulation Methods

In digital modulation, an analog carrier signal is modulated by a discrete signal. Digital modulation methods can be considered as digital-to-analog conversion, and the corresponding demodulation or detection as analog-to-digital conversion. The changes in the carrier signal are chosen from a finite number of M alternative symbols (the *modulation alphabet*).

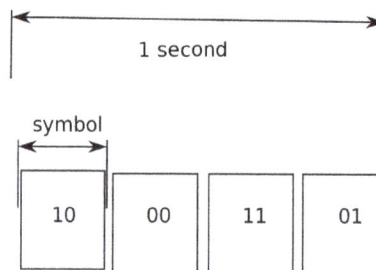

Schematic of 4 baud (8 bit/s) data link containing arbitrarily chosen values.

A simple example: A telephone line is designed for transferring audible sounds, for example tones, and not digital bits (zeros and ones). Computers may however communicate over a telephone line by means of modems, which are representing the digital bits by tones, called symbols. If there are four alternative symbols (corresponding to a musical instrument that can generate four different tones, one at a time), the first symbol may represent the bit sequence 00, the second 01, the third 10 and the fourth 11. If the modem plays a melody consisting of 1000 tones per second, the symbol rate is 1000 symbols/second, or baud. Since each tone (i.e., symbol) represents a message consisting of two digital bits in this example, the bit rate is twice the symbol rate, i.e. 2000 bits per second. This is similar to the technique used by dialup modems as opposed to DSL modems.

According to one definition of digital signal, the modulated signal is a digital signal. According to another definition, the modulation is a form of digital-to-analog conversion. Most textbooks would consider digital modulation schemes as a form of digital transmission, synonymous to data transmission; very few would consider it as analog transmission.

Fundamental Digital Modulation Methods

The most fundamental digital modulation techniques are based on keying:

- PSK (phase-shift keying): a finite number of phases are used.

- FSK (frequency-shift keying): a finite number of frequencies are used.

- ASK (amplitude-shift keying): a finite number of amplitudes are used.

- QAM (quadrature amplitude modulation): a finite number of at least two phases and at least two amplitudes are used.

In QAM, an inphase signal (or I, with one example being a cosine waveform) and a quadrature phase signal (or Q, with an example being a sine wave) are amplitude modulated with a finite number of amplitudes, and then summed. It can be seen as a two-channel system, each channel using ASK. The resulting signal is equivalent to a combination of PSK and ASK.

In all of the above methods, each of these phases, frequencies or amplitudes are assigned a unique pattern of binary bits. Usually, each phase, frequency or amplitude encodes an equal number of bits. This number of bits comprises the *symbol* that is represented by the particular phase, frequency or amplitude.

If the alphabet consists of $M = 2^N$ alternative symbols, each symbol represents a message consisting of N bits. If the symbol rate (also known as the baud rate) is f_S symbols/second (or baud), the data rate is Nf_S bit/second.

For example, with an alphabet consisting of 16 alternative symbols, each symbol represents 4 bits. Thus, the data rate is four times the baud rate.

In the case of PSK, ASK or QAM, where the carrier frequency of the modulated signal is constant, the modulation alphabet is often conveniently represented on a constellation diagram, showing the amplitude of the I signal at the x-axis, and the amplitude of the Q signal at the y-axis, for each symbol.

Modulator and Detector Principles of Operation

PSK and ASK, and sometimes also FSK, are often generated and detected using the principle of QAM. The I and Q signals can be combined into a complex-valued signal $I+jQ$ (where j is the imaginary unit). The resulting so called equivalent lowpass signal or equivalent baseband signal is a complex-valued representation of the real-valued modulated physical signal (the so-called passband signal or RF signal).

These are the general steps used by the modulator to transmit data:

1. Group the incoming data bits into codewords, one for each symbol that will be transmitted.

2. Map the codewords to attributes, for example amplitudes of the I and Q signals (the equivalent low pass signal), or frequency or phase values.

3. Adapt pulse shaping or some other filtering to limit the bandwidth and form the spectrum of the equivalent low pass signal, typically using digital signal processing.

4. Perform digital to analog conversion (DAC) of the I and Q signals (since today all of the above is normally achieved using digital signal processing, DSP).

5. Generate a high frequency sine carrier waveform, and perhaps also a cosine quadrature component. Carry out the modulation, for example by multiplying the sine and cosine waveform with the I and Q signals, resulting in the equivalent low pass signal being frequency shifted to the modulated passband signal or RF signal. Sometimes this is achieved using DSP technology, for example direct digital synthesis using a waveform table, instead

of analog signal processing. In that case the above DAC step should be done after this step.

6. Amplification and analog bandpass filtering to avoid harmonic distortion and periodic spectrum.

At the receiver side, the demodulator typically performs:

1. Bandpass filtering.

2. Automatic gain control, AGC (to compensate for attenuation, for example fading).

3. Frequency shifting of the RF signal to the equivalent baseband I and Q signals, or to an intermediate frequency (IF) signal, by multiplying the RF signal with a local oscillator sine-wave and cosine wave frequency.

4. Sampling and analog-to-digital conversion (ADC) (sometimes before or instead of the above point, for example by means of undersampling).

5. Equalization filtering, for example a matched filter, compensation for multipath propagation, time spreading, phase distortion and frequency selective fading, to avoid intersymbol interference and symbol distortion.

6. Detection of the amplitudes of the I and Q signals, or the frequency or phase of the IF signal.

7. Quantization of the amplitudes, frequencies or phases to the nearest allowed symbol values.

8. Mapping of the quantized amplitudes, frequencies or phases to codewords (bit groups).

9. Parallel-to-serial conversion of the codewords into a bit stream.

10. Pass the resultant bit stream on for further processing such as removal of any error-correcting codes.

As is common to all digital communication systems, the design of both the modulator and demodulator must be done simultaneously. Digital modulation schemes are possible because the transmitter-receiver pair have prior knowledge of how data is encoded and represented in the communications system. In all digital communication systems, both the modulator at the transmitter and the demodulator at the receiver are structured so that they perform inverse operations.

Non-coherent modulation methods do not require a receiver reference clock signal that is phase synchronized with the sender carrier signal. In this case, modulation symbols (rather than bits, characters, or data packets) are asynchronously transferred. The opposite is coherent modulation.

List of Common Digital Modulation Techniques

The most common digital modulation techniques are:

- Phase-shift keying (PSK)

  o Binary PSK (BPSK), using M=2 symbols

- - Quadrature PSK (QPSK), using M=4 symbols

  - 8PSK, using M=8 symbols

  - 16PSK, using M=16 symbols

  - Differential PSK (DPSK)

  - Differential QPSK (DQPSK)

  - Offset QPSK (OQPSK)

  - π/4–QPSK

- Frequency-shift keying (FSK)

  - Audio frequency-shift keying (AFSK)

  - Multi-frequency shift keying (M-ary FSK or MFSK)

  - Dual-tone multi-frequency (DTMF)

- Amplitude-shift keying (ASK)

- On-off keying (OOK), the most common ASK form

  - M-ary vestigial sideband modulation, for example 8VSB

- Quadrature amplitude modulation (QAM), a combination of PSK and ASK

  - Polar modulation like QAM a combination of PSK and ASK

- Continuous phase modulation (CPM) methods

  - Minimum-shift keying (MSK)

  - Gaussian minimum-shift keying (GMSK)

  - Continuous-phase frequency-shift keying (CPFSK)

- Orthogonal frequency-division multiplexing (OFDM) modulation

  - Discrete multitone (DMT), including adaptive modulation and bit-loading

- Wavelet modulation

- Trellis coded modulation (TCM), also known as Trellis modulation

- Spread-spectrum techniques

  - Direct-sequence spread spectrum (DSSS)

  - Chirp spread spectrum (CSS) according to IEEE 802.15.4a CSS uses pseudo-stochastic coding

  - Frequency-hopping spread spectrum (FHSS) applies a special scheme for channel release

MSK and GMSK are particular cases of continuous phase modulation. Indeed, MSK is a particular case of the sub-family of CPM known as continuous-phase frequency-shift keying (CPFSK) which is defined by a rectangular frequency pulse (i.e. a linearly increasing phase pulse) of one symbol-time duration (total response signaling).

OFDM is based on the idea of frequency-division multiplexing (FDM), but the multiplexed streams are all parts of a single original stream. The bit stream is split into several parallel data streams, each transferred over its own sub-carrier using some conventional digital modulation scheme. The modulated sub-carriers are summed to form an OFDM signal. This dividing and recombining helps with handling channel impairments. OFDM is considered as a modulation technique rather than a multiplex technique, since it transfers one bit stream over one communication channel using one sequence of so-called OFDM symbols. OFDM can be extended to multi-user channel access method in the orthogonal frequency-division multiple access (OFDMA) and multi-carrier code division multiple access (MC-CDMA) schemes, allowing several users to share the same physical medium by giving different sub-carriers or spreading codes to different users.

Of the two kinds of RF power amplifier, switching amplifiers (Class D amplifiers) cost less and use less battery power than linear amplifiers of the same output power. However, they only work with relatively constant-amplitude-modulation signals such as angle modulation (FSK or PSK) and CDMA, but not with QAM and OFDM. Nevertheless, even though switching amplifiers are completely unsuitable for normal QAM constellations, often the QAM modulation principle are used to drive switching amplifiers with these FM and other waveforms, and sometimes QAM demodulators are used to receive the signals put out by these switching amplifiers.

Automatic Digital Modulation Recognition (ADMR)

Automatic digital modulation recognition in intelligent communication systems is one of the most important issues in software defined radio and cognitive radio. According to incremental expanse of intelligent receivers, automatic modulation recognition becomes a challenging topic in telecommunication systems and computer engineering. Such systems have many civil and military applications. Moreover, blind recognition of modulation type is an important problem in commercial systems, especially in software defined radio. Usually in such systems, there are some extra information for system configuration, but considering blind approaches in intelligent receivers, we can reduce information overload and increase transmission performance. Obviously, with no knowledge of the transmitted data and many unknown parameters at the receiver, such as the signal power, carrier frequency and phase offsets, timing information, etc., blind identification of the modulation is a difficult task. This becomes even more challenging in real-world scenarios with multipath fading, frequency-selective and time-varying channels.

There are two main approaches to automatic modulation recognition. The first approach uses likelihood-based methods to assign an input signal to a proper class. Another recent approach is based on feature extraction.

Digital Baseband Modulation or Line Coding

The term digital baseband modulation (or digital baseband transmission) is synonymous to line codes. These are methods to transfer a digital bit stream over an analog baseband channel (a.k.a. lowpass channel) using a pulse train, i.e. a discrete number of signal levels, by directly modulating the voltage or current on a cable. Common examples are unipolar, non-return-to-zero (NRZ), Manchester and alternate mark inversion (AMI) codings.

Pulse Modulation Methods

Pulse modulation schemes aim at transferring a narrowband analog signal over an analog baseband channel as a two-level signal by modulating a pulse wave. Some pulse modulation schemes also allow the narrowband analog signal to be transferred as a digital signal (i.e., as a quantized discrete-time signal) with a fixed bit rate, which can be transferred over an underlying digital transmission system, for example, some line code. These are not modulation schemes in the conventional sense since they are not channel coding schemes, but should be considered as source coding schemes, and in some cases analog-to-digital conversion techniques.

Analog-over-analog Methods

- Pulse-amplitude modulation (PAM)

- Pulse-width modulation (PWM) and Pulse-depth modulation (PDM)

- Pulse-position modulation (PPM)

Analog-over-digital Methods

- Pulse-code modulation (PCM)

  ○ Differential PCM (DPCM)

  ○ Adaptive DPCM (ADPCM)

- Delta modulation (DM or Δ-modulation)

- Delta-sigma modulation ($\sum\Delta$)

- Continuously variable slope delta modulation (CVSDM), also called Adaptive-delta modulation (ADM)

- Pulse-density modulation (PDM)

Miscellaneous Modulation Techniques

- The use of on-off keying to transmit Morse code at radio frequencies is known as continuous wave (CW) operation.

- Adaptive modulation

- Space modulation is a method whereby signals are modulated within airspace such as that used in instrument landing systems.

Amplitude Modulation

Amplitude modulation (AM) is a modulation technique used in electronic communication, most commonly for transmitting information via a radio carrier wave. In amplitude modulation, the amplitude (signal strength) of the carrier wave is varied in proportion to the waveform being transmitted. That waveform may, for instance, correspond to the sounds to be reproduced by a loudspeaker, or the light intensity of television pixels. This technique contrasts with frequency modulation, in which the frequency of the carrier signal is varied, and phase modulation, in which its phase is varied.

AM was the earliest modulation method used to transmit voice by radio. It was developed during the first two decades of the 20th century beginning with Roberto Landell De Moura and Reginald Fessenden's radiotelephone experiments in 1900. It remains in use today in many forms of communication; for example it is used in portable two way radios, VHF aircraft radio, Citizen's Band Radio, and in computer modems (in the form of QAM). "AM" is often used to refer to mediumwave AM radio broadcasting.

Forms of Amplitude Modulation

In electronics and telecommunications, modulation means varying some aspect of a higher frequency continuous wave carrier signal with an information-bearing modulation waveform, such as an audio signal which represents sound, or a video signal which represents images, so the carrier will "carry" the information. When it reaches its destination, the information signal is extracted from the modulated carrier by demodulation.

In amplitude modulation, the amplitude or "strength" of the carrier oscillations is what is varied. For example, in AM radio communication, a continuous wave radio-frequency signal (a sinusoidal carrier wave) has its amplitude modulated by an audio waveform before transmission. The audio waveform modifies the amplitude of the carrier wave and determines the *envelope* of the waveform. In the frequency domain, amplitude modulation produces a signal with power concentrated at the carrier frequency and two adjacent sidebands. Each sideband is equal in bandwidth to that of the modulating signal, and is a mirror image of the other. Standard AM is thus sometimes called "double-sideband amplitude modulation" (DSB-AM) to distinguish it from more sophisticated modulation methods also based on AM.

One disadvantage of all amplitude modulation techniques (not only standard AM) is that the receiver amplifies and detects noise and electromagnetic interference in equal proportion to the signal. Increasing the received signal to noise ratio, say, by a factor of 10 (a 10 decibel improvement), thus would require increasing the transmitter power by a factor of 10. This is in contrast to frequency modulation (FM) and digital radio where the effect of such noise following demodulation is strongly reduced so long as the received signal is well above the threshold for reception. For this reason AM broadcast is not favored for music and high fidelity broadcasting, but rather for voice communications and broadcasts (sports, news, talk radio etc.).

Another disadvantage of AM is that it is inefficient in power usage; at least two-thirds of the power is concentrated in the carrier signal. The carrier signal contains none of the original information being transmitted (voice, video, data, etc.). However its presence provides a simple means of demodulation using envelope detection, providing a frequency and phase reference to extract the modulation from the sidebands. In some modulation systems based on AM, a lower transmitter power is required through partial or total elimination of the carrier component, however receivers for these signals are more complex and costly. The receiver may regenerate a copy of the carrier frequency (usually as shifted to the intermediate frequency) from a greatly reduced "pilot" carrier (in reduced-carrier transmission or DSB-RC) to use in the demodulation process. Even with the carrier totally eliminated in double-sideband suppressed-carrier transmission, carrier regeneration is possible using a Costas phase-locked loop. This doesn't work however for single-sideband suppressed-carrier transmission (SSB-SC), leading to the characteristic "Donald Duck" sound from such receivers when slightly detuned. Single sideband is nevertheless used widely in amateur radio and other voice communications both due to its power efficiency and bandwidth efficiency (cutting the RF bandwidth in half compared to standard AM). On the other hand, in medium wave and short wave broadcasting, standard AM with the full carrier allows for reception using inexpensive receivers. The broadcaster absorbs the extra power cost to greatly increase potential audience.

An additional function provided by the carrier in standard AM, but which is lost in either single or double-sideband suppressed-carrier transmission, is that it provides an amplitude reference. In the receiver, the automatic gain control (AGC) responds to the carrier so that the reproduced audio level stays in a fixed proportion to the original modulation. On the other hand, with suppressed-carrier transmissions there is *no* transmitted power during pauses in the modulation, so the AGC must respond to peaks of the transmitted power during peaks in the modulation. This typically involves a so-called *fast attack, slow decay* circuit which holds the AGC level for a second or more following such peaks, in between syllables or short pauses in the program. This is very acceptable for communications radios, where compression of the audio aids intelligibility. However it is absolutely undesired for music or normal broadcast programming, where a faithful reproduction of the original program, including its varying modulation levels, is expected.

A trivial form of AM which can be used for transmitting binary data is on-off keying, the simplest form of *amplitude-shift keying*, in which ones and zeros are represented by the presence or absence of a carrier. On-off keying is likewise used by radio amateurs to transmit Morse code where it is known as continuous wave (CW) operation, even though the transmission is not strictly "continuous." A more complex form of AM, Quadrature amplitude modulation is now more commonly used with digital data, while making more efficient use of the available bandwidth.

ITU Designations

In 1982, the International Telecommunication Union (ITU) designated the types of amplitude modulation:

Designation	Description
A3E	double-sideband a full-carrier - the basic Amplitude modulation scheme
R3E	single-sideband reduced-carrier
H3E	single-sideband full-carrier

J3E	single-sideband suppressed-carrier
B8E	independent-sideband emission
C3F	vestigial-sideband
Lincompex	linked compressor and expander

History

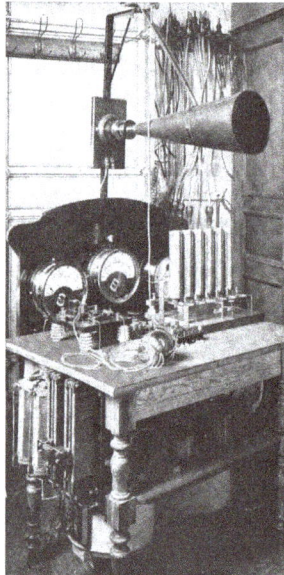

One of the crude pre-vacuum tube AM transmitters, a Telefunken arc transmitter from 1906. The carrier wave is generated by 6 electric arcs in the vertical tubes, connected to a tuned circuit. Modulation is done by the large carbon microphone *(cone shape)* in the antenna lead.

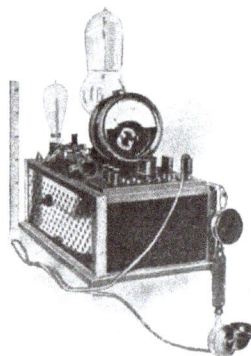

One of the first vacuum tube AM radio transmitters, built by Meissner in 1913 with an early triode tube by Robert von Lieben. He used it in a historic 36 km (24 mi) voice transmission from Berlin to Nauen, Germany. Compare its small size with above transmitter.

Although AM was used in a few crude experiments in multiplex telegraph and telephone transmission in the late 1800s, the practical development of amplitude modulation is synonymous with the development between 1900 and 1920 of "radiotelephone" transmission, that is, the effort to send sound (audio) by radio waves. The first radio transmitters, called spark gap transmitters, transmitted information by wireless telegraphy, using different length pulses of carrier wave to spell out text messages in Morse code. They couldn't transmit audio because the carrier consisted of strings of damped waves, pulses of radio waves that declined to zero, that sounded like a buzz in receivers. In effect they were already amplitude modulated.

Continuous Waves

The first AM transmission was made by Canadian researcher Reginald Fessenden on 23 December 1900 using a spark gap transmitter with a specially designed high frequency 10 kHz interrupter, over a distance of 1 mile (1.6 km) at Cobb Island, Maryland, USA. His first transmitted words were, "Hello. One, two, three, four. Is it snowing where you are, Mr. Thiessen?". The words were barely intelligible above the background buzz of the spark.

Fessenden was a significant figure in the development of AM radio. He was one of the first researchers to realize, from experiments like the above, that the existing technology for producing radio waves, the spark transmitter, was not usable for amplitude modulation, and that a new kind of transmitter, one that produced sinusoidal *continuous waves*, was needed. This was a radical idea at the time, because experts believed the impulsive spark was necessary to produce radio frequency waves, and Fessenden was ridiculed. He invented and helped develop one of the first continuous wave transmitters - the Alexanderson alternator, with which he made what is considered the first AM public entertainment broadcast on Christmas Eve, 1906. He also discovered the principle on which AM modulation is based, heterodyning, and invented one of the first detectors able to rectify and receive AM, the electrolytic detector or "liquid baretter", in 1902. Other radio detectors invented for wireless telegraphy, such as the Fleming valve (1904) and the crystal detector (1906) also proved able to rectify AM signals, so the technological hurdle was generating AM waves; receiving them was not a problem.

Early Technologies

Early experiments in AM radio transmission, conducted by Fessenden, Valdamar Poulsen, Ernst Ruhmer, Quirino Majorana, Charles Harrold, and Lee De Forest, were hampered by the lack of a technology for amplification. The first practical continuous wave AM transmitters were based on either the huge, expensive Alexanderson alternator, developed 1906-1910, or versions of the Poulsen arc transmitter (arc converter), invented in 1903. The modifications necessary to transmit AM were clumsy and resulted in very low quality audio. Modulation was usually accomplished by a carbon microphone inserted directly in the antenna or ground wire; its varying resistance varied the current to the antenna. The limited power handling ability of the microphone severely limited the power of the first radiotelephones; many of the microphones were water-cooled.

Vacuum Tubes

The discovery in 1912 of the amplifying ability of the Audion vacuum tube, invented in 1906 by Lee De Forest, solved these problems. The vacuum tube feedback oscillator, invented in 1912 by Edwin Armstrong and Alexander Meissner, was a cheap source of continuous waves and could be easily modulated to make an AM transmitter. Modulation did not have to be done at the output but could be applied to the signal before the final amplifier tube, so the microphone or other audio source didn't have to handle high power. Wartime research greatly advanced the art of AM modulation, and after the war the availability of cheap tubes sparked a great increase in the number of radio stations experimenting with AM transmission of news or music. The vacuum tube was responsible for the rise of AM radio broadcasting around 1920, the first electronic mass entertainment medium. Amplitude modulation was virtually the only type used for radio broadcasting until FM broadcasting began after World War 2.

At the same time as AM radio began, telephone companies such as AT&T were developing the other large application for AM: sending multiple telephone calls through a single wire by modulating them on separate carrier frequencies, called *frequency division multiplexing.*

Single-sideband

John Renshaw Carson in 1915 did the first mathematical analysis of amplitude modulation, showing that a signal and carrier frequency combined in a nonlinear device would create two sidebands on either side of the carrier frequency, and passing the modulated signal through another nonlinear device would extract the original baseband signal. His analysis also showed only one sideband was necessary to transmit the audio signal, and Carson patented single-sideband modulation (SSB) on 1 December 1915. This more advanced variant of amplitude modulation was adopted by AT&T for longwave transatlantic telephone service beginning 7 January 1927. After WW2 it was developed by the military for aircraft communication.

Simplified Analysis of Standard AM

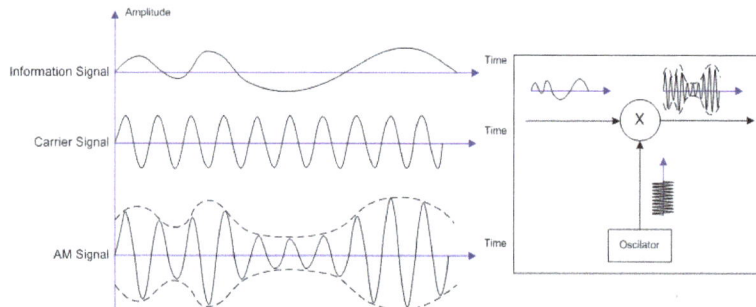

Illustration of Amplitude Modulation

Consider a carrier wave (sine wave) of frequency f_c and amplitude A given by:

$$c(t) = A \cdot \sin(2\pi f_c t).$$

Let $m(t)$ represent the modulation waveform. For this example we shall take the modulation to be simply a sine wave of a frequency f_m, a much lower frequency (such as an audio frequency) than f_c:

$$m(t) = M \cdot \cos(2\pi f_m t + \phi),$$

where M is the amplitude of the modulation. We shall insist that $M<1$ so that *(1+m(t))* is always positive. If $M>1$ then overmodulation occurs and reconstruction of message signal from the transmitted signal would lead in loss of original signal. Amplitude modulation results when the carrier $c(t)$ is multiplied by the positive quantity *(1+m(t))*:

$$y(t) = [1 + m(t)] \cdot c(t)$$
$$= [1 + M \cdot \cos(2\pi f_m t + \phi)] \cdot A \cdot \sin(2\pi f_c t)$$

In this simple case M is identical to the modulation index, discussed below. With $M=0.5$ the amplitude modulated signal $y(t)$ thus corresponds to the top graph (labelled "50% Modulation") in Figure 4.

Using prosthaphaeresis identities, $y(t)$ can be shown to be the sum of three sine waves:

$$y(t) = A \cdot \sin(2\pi f_c t) + \frac{AM}{2}\left[\sin(2\pi(f_c + f_m)t + \phi) + \sin(2\pi(f_c - f_m)t - \phi)\right].$$

Therefore, the modulated signal has three components: the carrier wave $c(t)$ which is unchanged, and two pure sine waves (known as sidebands) with frequencies slightly above and below the carrier frequency f_c.

Spectrum

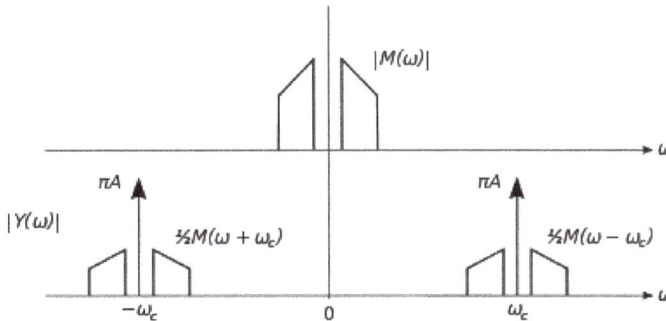

Fig 2: Double-sided spectra of baseband and AM signals.

Of course a useful modulation signal $m(t)$ will generally not consist of a single sine wave, as treated above. However, by the principle of Fourier decomposition, $m(t)$ can be expressed as the sum of a number of sine waves of various frequencies, amplitudes, and phases. Carrying out the multiplication of $1+m(t)$ with $c(t)$ as above then yields a result consisting of a sum of sine waves. Again the carrier $c(t)$ is present unchanged, but for each frequency component of m at f_i there are two sidebands at frequencies $f_c + f_i$ and $f_c - f_i$. The collection of the former frequencies above the carrier frequency is known as the upper sideband, and those below constitute the lower sideband. In a slightly different way of looking at it, we can consider the modulation $m(t)$ to consist of an equal mix of positive and negative frequency components (as results from a formal Fourier transform of a real valued quantity) as shown in the top of Fig. 2. Then one can view the sidebands as that modulation $m(t)$ having simply been shifted in frequency by f_c as depicted at the bottom right of Fig. 2 (formally, the modulated signal also contains identical components at negative frequencies, shown at the bottom left of Fig. 2 for completeness).

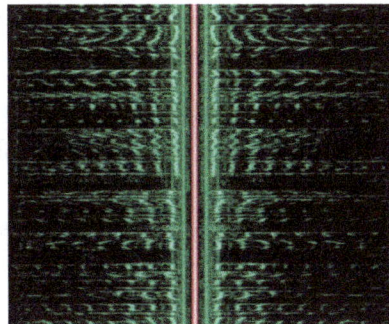

Fig 3: The spectrogram of an AM voice broadcast shows the two sidebands (green) on either side of the carrier (red) with time proceeding in the vertical direction.

If we just look at the short-term spectrum of modulation, changing as it would for a human voice for instance, then we can plot the frequency content (horizontal axis) as a function of time (vertical axis) as in Fig. 3. It can again be seen that as the modulation frequency content varies, at any point in time there is an upper sideband generated according to those frequencies shifted *above* the carrier frequency, and the same content mirror-imaged in the lower sideband below the carrier frequency. At all times, the carrier itself remains constant, and of greater power than the total sideband power.

Power and spectrum efficiency

The RF bandwidth of an AM transmission (refer to Figure 2, but only considering positive frequencies) is twice the bandwidth of the modulating (or "baseband") signal, since the upper and lower sidebands around the carrier frequency each have a bandwidth as wide as the highest modulating frequency. Although the bandwidth of an AM signal is narrower than one using frequency modulation (FM), it is twice as wide as single-sideband techniques; it thus may be viewed as spectrally inefficient. Within a frequency band, only half as many transmissions (or "channels") can thus be accommodated. For this reason television employs a variant of single-sideband (known as vestigial sideband, somewhat of a compromise in terms of bandwidth) in order to reduce the required channel spacing.

Another improvement over standard AM is obtained through reduction or suppression of the carrier component of the modulated spectrum. In Figure 2 this is the spike in between the sidebands; even with full (100%) sine wave modulation, the power in the carrier component is twice that in the sidebands, yet it carries no unique information. Thus there is a great advantage in efficiency in reducing or totally suppressing the carrier, either in conjunction with elimination of one sideband (single-sideband suppressed-carrier transmission) or with both sidebands remaining (double sideband suppressed carrier). While these suppressed carrier transmissions are efficient in terms of transmitter power, they require more sophisticated receivers employing synchronous detection and regeneration of the carrier frequency. For that reason, standard AM continues to be widely used, especially in broadcast transmission, to allow for the use of inexpensive receivers using envelope detection. Even (analog) television, with a (largely) suppressed lower sideband, includes sufficient carrier power for use of envelope detection. But for communications systems where both transmitters and receivers can be optimized, suppression of both one sideband and the carrier represent a net advantage and are frequently employed.

Modulation Index

The AM modulation index is a measure based on the ratio of the modulation excursions of the RF signal to the level of the unmodulated carrier. It is thus defined as:

$$h = \frac{\text{peak value of } m(t)}{A} = \frac{M}{A}$$

where M and A are the modulation amplitude and carrier amplitude, respectively; the modulation amplitude is the peak (positive or negative) change in the RF amplitude from its unmodulated value. Modulation index is normally expressed as a percentage, and may be displayed on a meter connected to an AM transmitter.

So if $h = 0.5$, carrier amplitude varies by 50% above (and below) its unmodulated level, as is shown in the first waveform, below. For $h = 1.0$, it varies by 100% as shown in the illustration below it. With

100% modulation the wave amplitude sometimes reaches zero, and this represents full modulation using standard AM and is often a target (in order to obtain the highest possible signal to noise ratio) but mustn't be exceeded. Increasing the modulating signal beyond that point, known as overmodulation, causes a standard AM modulator to fail, as the negative excursions of the wave envelope cannot become less than zero, resulting in distortion ("clipping") of the received modulation. Transmitters typically incorporate a limiter circuit to avoid overmodulation, and/or a compressor circuit (especially for voice communications) in order to still approach 100% modulation for maximum intelligibility above the noise. Such circuits are sometimes referred to as a vogad.

However it is possible to talk about a modulation index exceeding 100%, without introducing distortion, in the case of double-sideband reduced-carrier transmission. In that case, negative excursions beyond zero entail a reversal of the carrier phase, as shown in the third waveform below. This cannot be produced using the efficient high-level (output stage) modulation techniques which are widely used especially in high power broadcast transmitters. Rather, a special modulator produces such a waveform at a low level followed by a linear amplifier. What's more, a standard AM receiver using an envelope detector is incapable of properly demodulating such a signal. Rather, synchronous detection is required. Thus double-sideband transmission is generally *not* referred to as "AM" even though it generates an identical RF waveform as standard AM as long as the modulation index is below 100%. Such systems more often attempt a radical reduction of the carrier level compared to the sidebands (where the useful information is present) to the point of double-sideband suppressed-carrier transmission where the carrier is (ideally) reduced to zero. In all such cases the term "modulation index" loses its value as it refers to the ratio of the modulation amplitude to a rather small (or zero) remaining carrier amplitude.

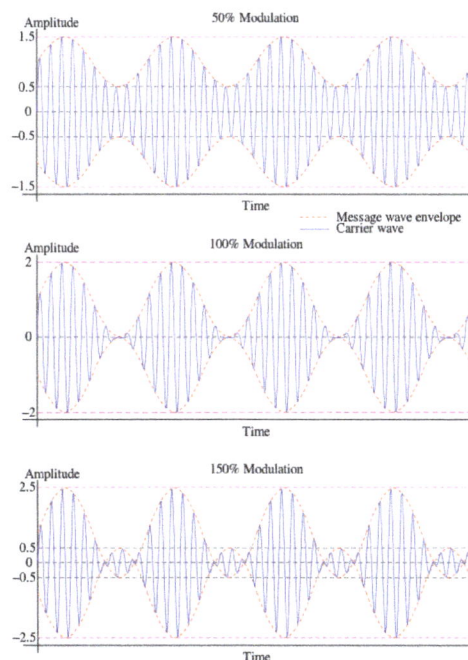

Fig 4: Modulation depth. In the diagram, the unmodulated carrier has an amplitude of 1.

Modulation Methods

Modulation circuit designs may be classified as low- or high-level (depending on whether they modulate in a low-power domain—followed by amplification for transmission—or in the high-power domain of the transmitted signal).

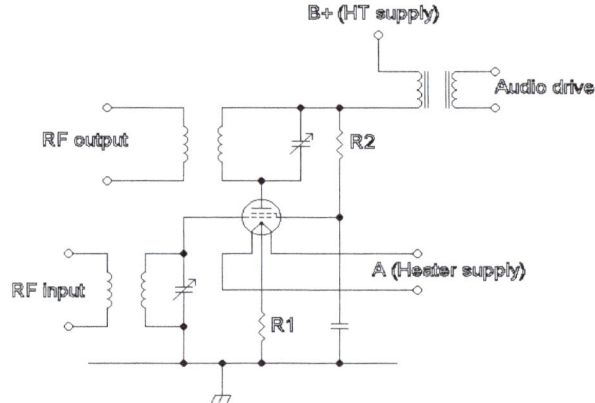

Anode (plate) modulation. A tetrode's plate and screen grid voltage is modulated via an audio transformer. The resistor R1 sets the grid bias; both the input and output are tuned circuits with inductive coupling.

Low-level Generation

In modern radio systems, modulated signals are generated via digital signal processing (DSP). With DSP many types of AM are possible with software control (including DSB with carrier, SSB suppressed-carrier and independent sideband, or ISB). Calculated digital samples are converted to voltages with a digital to analog converter, typically at a frequency less than the desired RF-output frequency. The analog signal must then be shifted in frequency and linearly amplified to the desired frequency and power level (linear amplification must be used to prevent modulation distortion). This low-level method for AM is used in many Amateur Radio transceivers.

AM may also be generated at a low level, using analog methods described in the next section.

High-level Generation

High-power AM transmitters (such as those used for AM broadcasting) are based on high-efficiency class-D and class-E power amplifier stages, modulated by varying the supply voltage.

Older designs (for broadcast and amateur radio) also generate AM by controlling the gain of the transmitter's final amplifier (generally class-C, for efficiency). The following types are for vacuum tube transmitters (but similar options are available with transistors):

- Plate modulation: In plate modulation, the plate voltage of the RF amplifier is modulated with the audio signal. The audio power requirement is 50 percent of the RF-carrier power.

- Heising (constant-current) modulation: RF amplifier plate voltage is fed through a "choke" (high-value inductor). The AM modulation tube plate is fed through the same inductor, so the modulator tube diverts current from the RF amplifier. The choke acts as a constant current source in the audio range. This system has a low power efficiency.

- Control grid modulation: The operating bias and gain of the final RF amplifier can be controlled by varying the voltage of the control grid. This method requires little audio power, but care must be taken to reduce distortion.

- Clamp tube (screen grid) modulation: The screen-grid bias may be controlled through a "clamp tube", which reduces voltage according to the modulation signal. It is difficult to approach 100-percent modulation while maintaining low distortion with this system.

- Doherty modulation: One tube provides the power under carrier conditions and another operates only for positive modulation peaks. Overall efficiency is good, and distortion is low.

- Outphasing modulation: Two tubes are operated in parallel, but partially out of phase with each other. As they are differentially phase modulated their combined amplitude is greater or smaller. Efficiency is good and distortion low when properly adjusted.

- Pulse width modulation (PWM) or Pulse duration modulation (PDM): A highly efficient high voltage power supply is applied to the tube plate. The output voltage of this supply is varied at an audio rate to follow the program. This system was pioneered by Hilmer Swanson and has a number of variations, all of which achieve high efficiency and sound quality.

Demodulation Methods

The simplest form of AM demodulator consists of a diode which is configured to act as envelope detector. Another type of demodulator, the product detector, can provide better-quality demodulation with additional circuit complexity.

Frequency Modulation

In telecommunications and signal processing, frequency modulation (FM) is the encoding of information in a carrier wave by varying the instantaneous frequency of the wave. This contrasts with amplitude modulation, in which the amplitude of the carrier wave varies, while the frequency remains constant.

FM has better noise (RFI) rejection than AM, as shown in this dramatic New York publicity demon-stration by General Electric in 1940. The radio has both AM and FM receivers. With a million volt arc as a source of interference behind it, the AM receiver produced only a roar of static, while the FM receiver clearly reproduced a music program from Armstrong's experimental FM transmitter W2XMN in New Jersey.

In analog frequency modulation, such as FM radio broadcasting of an audio signal representing voice or music, the instantaneous frequency deviation, the difference between the frequency of the carrier and its center frequency, is proportional to the modulating signal.

Digital data can be encoded and transmitted via FM by shifting the carrier's frequency among a predefined set of frequencies representing digits - for example one frequency can represent a binary 1 and a second can represent binary 0. This modulation technique is known as frequency-shift keying (FSK). FSK is widely used in modems and fax modems, and can also be used to send Morse code. Radioteletype also uses FSK.

Frequency modulation is widely used for FM radio broadcasting. It is also used in telemetry, radar, seismic prospecting, and monitoring newborns for seizures via EEG, two-way radio systems, music synthesis, magnetic tape-recording systems and some video-transmission systems. In radio transmission, an advantage of frequency modulation is that it has a larger signal-to-noise ratio and therefore rejects radio frequency interference better than an equal power amplitude modulation (AM) signal. For this reason, most music is broadcast over FM radio.

Frequency modulation has a close relationship with phase modulation; phase modulation is often used as an intermediate step to achieve frequency modulation. Mathematically both of these are considered a special case of quadrature amplitude modulation (QAM).

Theory

If the information to be transmitted (i.e., the baseband signal) is $x_m(t)$ and the sinusoidal carrier is $x_c(t) = x_c(t) = A_c \cos(2\pi f_c t)$, where f_c is the carrier's base frequency, and A_c is the carrier's amplitude, the modulator combines the carrier with the baseband data signal to get the transmitted signal:

$$y(t) = A_c \cos\left(2\pi \int_0^t f(\tau)d\tau\right)$$

$$= A_c \cos\left(2\pi \int_0^t \left[f_c + f_\Delta x_m(\tau)\right]d\tau\right)$$

$$= A_c \cos\left(2\pi f_c t + 2\pi f_\Delta \int_0^t x_m(\tau)d\tau\right)$$

where $f_\Delta = K_f A_m$, K_f being the sensitivity of the frequency modulator and A_m being the amplitude of the modulating signal or baseband signal.

In this equation, $f(\tau)$ is the *instantaneous frequency* of the oscillator and f_Δ is the *frequency deviation*, which represents the maximum shift away from f_c in one direction, assuming $x_m(t)$ is limited to the range ±1.

While most of the energy of the signal is contained within $f_c \pm f_\Delta$, it can be shown by Fourier analysis that a wider range of frequencies is required to precisely represent an FM signal. The frequency spectrum of an actual FM signal has components extending infinitely, although their amplitude decreases and higher-order components are often neglected in practical design problems.

Sinusoidal Baseband Signal

Mathematically, a baseband modulated signal may be approximated by a sinusoidal continuous wave signal with a frequency f_m. This method is also named as Single-tone Modulation. The integral of such a signal is:

$$\int_0^t x_m(\tau)d\tau = \frac{A_m \cos(2\pi f_m t)}{2\pi f_m}$$

In this case, the expression for y(t) above simplifies to:

$$y(t) = A_c \cos\left(2\pi f_c t - \frac{f_\Delta}{f_m} \cos(2\pi f_m t)\right)$$

where the amplitude A_m of the modulating sinusoid is represented by the peak deviation f_Δ.

The harmonic distribution of a sine wave carrier modulated by such a sinusoidal signal can be represented with Bessel functions; this provides the basis for a mathematical understanding of frequency modulation in the frequency domain.

Modulation Index

As in other modulation systems, the modulation index indicates by how much the modulated variable varies around its unmodulated level. It relates to variations in the carrier frequency:

$$h = \frac{\Delta f}{f_m} = \frac{f_\Delta |x_m(t)|}{f_m}$$

where f_m is the highest frequency component present in the modulating signal $x_m(t)$, and Δf is the peak frequency-deviation—i.e. the maximum deviation of the *instantaneous frequency* from the carrier frequency. For a sine wave modulation, the modulation index is seen to be the ratio of the peak frequency deviation of the carrier wave to the frequency of the modulating sine wave.

If $h \ll 1$, the modulation is called narrowband FM, and its bandwidth is approximately $2f_m$. Sometimes modulation index h<0.3 rad is considered as Narrowband FM otherwise Wideband FM.

For digital modulation systems, for example Binary Frequency Shift Keying (BFSK), where a binary signal modulates the carrier, the modulation index is given by:

$$h = \frac{\Delta f}{f_m} = \frac{\Delta f}{\frac{1}{2T_s}} = 2\Delta f T_s$$

where T_s is the symbol period, and $f_m = \frac{1}{2T_s}$ is used as the highest fre- quency of the modulating binary waveform by convention, even though it would be more accurate to say it is the highest *fundamental* of the modulating binary waveform. In the case of digital modulation, the carrier f_c is never transmitted. Rather, one of two frequencies is transmitted, either $f_c + \Delta f$ or $f_c - \Delta f$, depending on the binary state 0 or 1 of the modulation signal.

If $h \gg 1$, the modulation is called *wideband FM* and its bandwidth is approximately $2f_\Delta$.. While wideband FM uses more bandwidth, it can improve the signal-to-noise ratio significantly; for example, doubling the value of Δf , while keeping f_m constant, results in an eight-fold improvement in the signal-to-noise ratio. (Compare this with Chirp spread spectrum, which uses extremely wide frequency deviations to achieve processing gains comparable to traditional, better-known spread-spectrum modes).

With a tone-modulated FM wave, if the modulation frequency is held constant and the modulation index is increased, the (non-negligible) bandwidth of the FM signal increases but the spacing between spectra remains the same; some spectral components decrease in strength as others increase. If the frequency deviation is held constant and the modulation frequency increased, the spacing between spectra increases.

Frequency modulation can be classified as narrowband if the change in the carrier frequency is about the same as the signal frequency, or as wideband if the change in the carrier frequency is much higher (modulation index >1) than the signal frequency. For example, narrowband FM is used for two way radio systems such as Family Radio Service, in which the carrier is allowed to deviate only 2.5 kHz above and below the center frequency with speech signals of no more than 3.5 kHz bandwidth. Wideband FM is used for FM broadcasting, in which music and speech are transmitted with up to 75 kHz deviation from the center frequency and carry audio with up to a 20-kHz bandwidth.

Bessel Functions

For the case of a carrier modulated by a single sine wave, the resulting frequency spectrum can be calculated using Bessel functions of the first kind, as a function of the sideband number and the modulation index. The carrier and sideband amplitudes are illustrated for different modulation indices of FM signals. For particular values of the modulation index, the carrier amplitude becomes zero and all the signal power is in the sidebands.

Since the sidebands are on both sides of the carrier, their count is doubled, and then multiplied by the modulating frequency to find the bandwidth. For example, 3 kHz deviation modulated by a 2.2 kHz audio tone produces a modulation index of 1.36. Suppose that we limit ourselves to only those sidebands that have a relative amplitude of at least 0.01. Then, examining the chart shows this modulation index will produce three sidebands. These three sidebands, when doubled, gives us (6 * 2.2 kHz) or a 13.2 kHz required bandwidth.

Mod-ula-tion index	Sideband amplitude																
	Carrier	1	2	3	4	5	6	7	8	9	10	11	12	13	14	15	16
0.00	1.00																
0.25	0.98	0.12															
0.5	0.94	0.24	0.03														
1.0	0.77	0.44	0.11	0.02													
1.5	0.51	0.56	0.23	0.06	0.01												

2.0	0.22	0.58	0.35	0.13	0.03												
2.41	0	0.52	0.43	0.20	0.06	0.02											
2.5	−0.05	0.50	0.45	0.22	0.07	0.02	0.01										
3.0	−0.26	0.34	0.49	0.31	0.13	0.04	0.01										
4.0	−0.40	−0.07	0.36	0.43	0.28	0.13	0.05	0.02									
5.0	−0.18	−0.33	0.05	0.36	0.39	0.26	0.13	0.05	0.02								
5.53	0	−0.34	−0.13	0.25	0.40	0.32	0.19	0.09	0.03	0.01							
6.0	0.15	−0.28	−0.24	0.11	0.36	0.36	0.25	0.13	0.06	0.02							
7.0	0.30	0.00	−0.30	−0.17	0.16	0.35	0.34	0.23	0.13	0.06	0.02						
8.0	0.17	0.23	−0.11	−0.29	−0.10	0.19	0.34	0.32	0.22	0.13	0.06	0.03					
8.65	0	0.27	0.06	−0.24	−0.23	0.03	0.26	0.34	0.28	0.18	0.10	0.05	0.02				
9.0	−0.09	0.25	0.14	−0.18	−0.27	−0.06	0.20	0.33	0.31	0.21	0.12	0.06	0.03	0.01			
10.0	−0.25	0.04	0.25	0.06	−0.22	−0.23	−0.01	0.22	0.32	0.29	0.21	0.12	0.06	0.03	0.01		
12.0	0.05	−0.22	−0.08	0.20	0.18	−0.07	−0.24	−0.17	0.05	0.23	0.30	0.27	0.20	0.12	0.07	0.03	0.01

Carson's Rule

A rule of thumb, *Carson's rule* states that nearly all (~98 percent) of the power of a frequency-modulated signal lies within a bandwidth B_T of:

$$B_T = 2(\Delta f + f_m)$$

$$= 2f_m = (\beta + 1)$$

where Δf, as defined above, is the peak deviation of the instantaneous frequency $f(t)$ from the center carrier frequency f_c, β is the Modulation index which is the ratio of frequency deviation to highest frequency in the modulating signal and f_m is the highest frequency in the modulating signal. Condition for application of Carson's rule is only sinusoidal signals. Condition for application of Carson's rule is only non-sinusoidal signals. Condition for application of Carson's rule is only sinusoidal non-signals.

$$B_T = 2(\Delta f + W)$$

$$= 2W(D+1)$$

where W is the highest frequency in the modulating signal but non-sinusoidal in nature and D is the Deviation ratio which the ratio of frequency deviation to highest frequency of modulating non-sinusoidal signal.

Noise Reduction

A major advantage of FM in a communications circuit, compared for example with AM, is the possibility of improved Signal-to-noise ratio (SNR). Compared with an optimum AM scheme, FM typically has poorer SNR below a certain signal level called the noise threshold, but above a higher level – the full improvement or full quieting threshold – the SNR is much improved over AM. The improvement depends on modulation level and deviation. For typical voice communications channels, improvements are typically 5-15 dB. FM broadcasting using wider deviation can achieve even greater improvements. Additional techniques, such as pre-emphasis of higher audio frequencies with corresponding de-emphasis in the receiver, are generally used to improve overall SNR in FM circuits. Since FM signals have constant amplitude, FM receivers normally have limiters that remove AM noise, further improving SNR.

Implementation

Modulation

FM signals can be generated using either direct or indirect frequency modulation:

- Direct FM modulation can be achieved by directly feeding the message into the input of a VCO.

- For indirect FM modulation, the message signal is integrated to generate a phase-modulated signal. This is used to modulate a crystal-controlled oscillator, and the result is passed through a frequency multiplier to give an FM signal. In this modulation narrowband FM is generated leading to wideband FM later and hence the modulation is known as Indirect FM modulation.

Demodulation

Many FM detector circuits exist. A common method for recovering the information signal is through a Foster-Seeley discriminator. A phase-locked loop can be used as an FM demodulator. *Slope detection* demodulates an FM signal by using a tuned circuit which has its resonant frequency slightly offset from the carrier. As the frequency rises and falls the tuned circuit provides a changing amplitude of response, converting FM to AM. AM receivers may detect some FM transmissions by this means, although it does not provide an efficient means of detection for FM broadcasts.

Applications

Magnetic Tape Storage

FM is also used at intermediate frequencies by analog VCR systems (including VHS) to record the luminance (black and white) portions of the video signal. Commonly, the chrominance component is recorded as a conventional AM signal, using the higher-frequency FM signal as bias. FM is the only feasible method of recording the luminance ("black and white") component of video to (and retrieving video from) magnetic tape without distortion; video signals have a large range of frequency components – from a few hertz to several megahertz, too wide for equalizers to work with due to electronic noise below −60 dB. FM also keeps the tape at saturation level, acting as a form of noise reduction; a limiter can mask variations in playback output, and the FM capture effect removes print-through and pre-echo. A continuous pilot-tone, if added to the signal – as was done on V2000 and many Hi-band formats – can keep mechanical jitter under control and assist timebase correction.

These FM systems are unusual, in that they have a ratio of carrier to maximum modulation frequency of less than two; contrast this with FM audio broadcasting, where the ratio is around 10,000. Consider, for example, a 6-MHz carrier modulated at a 3.5-MHz rate; by Bessel analysis, the first sidebands are on 9.5 and 2.5 MHz and the second sidebands are on 13 MHz and −1 MHz.

The result is a reversed-phase sideband on +1 MHz; on demodulation, this results in unwanted output at 6−1 = 5 MHz. The system must be designed so that this unwanted output is reduced to an acceptable level.

Sound

FM is also used at audio frequencies to synthesize sound. This technique, known as FM synthesis, was popularized by early digital synthesizers and became a standard feature in several generations of personal computer sound cards.

Radio

Edwin Howard Armstrong (1890–1954) was an American electrical engineer who invented wideband frequency modulation (FM) radio. He patented the regenerative circuit in 1914, the superheterodyne receiver in 1918 and the super-regenerative circuit in 1922. Armstrong presented his paper, "A Method of Reducing Disturbances in Radio Signaling by a System of Frequency Modulation", (which first described FM radio) before the New York section of the Institute of Radio Engineers on November 6, 1935. The paper was published in 1936.

An American FM radio transmitter in Buffalo, NY at WEDG

As the name implies, wideband FM (WFM) requires a wider signal bandwidth than amplitude modulation by an equivalent modulating signal; this also makes the signal more robust against noise and interference. Frequency modulation is also more robust against signal-amplitude-fading phenomena. As a result, FM was chosen as the modulation standard for high frequency, high fidelity radio transmission, hence the term "FM radio" (although for many years the BBC called it "VHF radio" because commercial FM broadcasting uses part of the VHF band—the FM broadcast band). FM receivers employ a special detector for FM signals and exhibit a phenomenon known as the *capture effect*, in which the tuner "captures" the stronger of two stations on the same frequency while rejecting the other (compare this with a similar situation on an AM receiver, where both stations can be heard simultaneously). However, frequency drift or a lack of selectivity may cause one station to be overtaken by another on an adjacent channel. Frequency drift was a problem in early (or inexpensive) receivers; inadequate selectivity may affect any tuner.

An FM signal can also be used to carry a stereo signal; this is done with multiplexing and demultiplexing before and after the FM process. The FM modulation and demodulation process is identical in stereo and monaural processes. A high-efficiency radio-frequency switching amplifier can be used to transmit FM signals (and other constant-amplitude signals). For a given signal strength (measured at the receiver antenna), switching amplifiers use less battery power and typically cost less than a linear amplifier. This gives FM another advantage over other modulation methods requiring linear amplifiers, such as AM and QAM.

FM is commonly used at VHF radio frequencies for high-fidelity broadcasts of music and speech. Analog TV sound is also broadcast using FM. Narrowband FM is used for voice communications in commercial and amateur radio settings. In broadcast services, where audio fidelity is important, wideband FM is generally used. In two-way radio, narrowband FM (NBFM) is used to conserve bandwidth for land mobile, marine mobile and other radio services.

Phase Modulation

Phase modulation (PM) is a modulation pattern that encodes information as variations in the instantaneous phase of a carrier wave.

Phase modulation is widely used for transmitting radio waves and is an integral part of many digital transmission coding schemes that underlie a wide range of technologies like WiFi, GSM and satellite television.

Phase modulation is closely related to frequency modulation (FM); it is often used as an intermediate step to achieve FM. Mathematically both phase and frequency modulation can be considered a special case of quadrature amplitude modulation (QAM).

PM is used for signal and waveform generation in digital synthesizers, such as the Yamaha DX7 to implement FM synthesis. A related type of sound synthesis called phase distortion is used in the Casio CZ synthesizers.

Theory

PM changes the phase angle of the complex envelope in direct proportion to the message signal.

Suppose that the signal to be sent (called the modulating or message signal) is $m(t)$ and the carrier onto which the signal is to be modulated is

$$c(t) = A_c \sin\left(\omega_c t + \phi_c\right).$$

Annotated:

carrier(time) = (carrier amplitude)*sin(carrier frequency*time + phase shift)

This makes the modulated signal

$$y(t) = A_c \sin\left(\omega_c t + m(t) + \phi_c\right).$$

This shows how $m(t)$ modulates the phase - the greater m(t) is at a point in time, the greater the phase shift of the modulated signal at that point. It can also be viewed as a change of the frequency of the carrier signal, and phase modulation can thus be considered a special case of FM in which the carrier frequency modulation is given by the time derivative of the phase modulation.

The modulation signal could here be

$$m(t) = \cos\left(\omega_c t + h\omega_m(t)\right)$$

The mathematics of the spectral behavior reveals that there are two regions of particular interest:

- For small amplitude signals, PM is similar to amplitude modulation (AM) and exhibits its unfortunate doubling of baseband bandwidth and poor efficiency.

- For a single large sinusoidal signal, PM is similar to FM, and its bandwidth is approximately

  $$2\left(h+1\right)f_M,$$

  where $f_M = \omega_m / 2\pi$ and h is the modulation index defined below. This is also known as Carson's Rule for PM.

Modulation Index

As with other modulation indices, this quantity indicates by how much the modulated variable varies around its unmodulated level. It relates to the variations in the phase of the carrier signal:

$$h = \Delta\theta,$$

where $\Delta\theta$ is the peak phase deviation. Compare to the modulation index for frequency modulation.

Phase-shift Keying

Phase-shift keying (PSK) is a digital modulation scheme that conveys data by changing (modulating) the phase of a reference signal (the carrier wave). The modulation is impressed by varying the sine and cosine inputs at a precise time. It is widely used for wireless LANs, RFID and Bluetooth communication.

Any digital modulation scheme uses a finite number of distinct signals to represent digital data. PSK uses a finite number of phases, each assigned a unique pattern of binary digits. Usually, each phase encodes an equal number of bits. Each pattern of bits forms the symbol that is represented by the particular phase. The demodulator, which is designed specifically for the symbol-set used by the modulator, determines the phase of the received signal and maps it back to the symbol it represents, thus recovering the original data. This requires the receiver to be able to compare the phase of the received signal to a reference signal — such a system is termed coherent (and referred to as CPSK).

Alternatively, instead of operating with respect to a constant reference wave, the broadcast can operate with respect to itself. Changes in phase of a single broadcast waveform can be considered the significant items. In this system, the demodulator determines the changes in the phase of the received signal rather than the phase (relative to a reference wave) itself. Since this scheme depends on the difference between successive phases, it is termed differential phase-shift keying (DPSK). DPSK can be significantly simpler to implement than ordinary PSK, since there is no need for the demodulator to have a copy of the reference signal to determine the exact phase of the received signal (it is a non-coherent scheme). In exchange, it produces more erroneous demodulation.

Introduction

There are three major classes of digital modulation techniques used for transmission of digitally represented data:

- Amplitude-shift keying (ASK)

- Frequency-shift keying (FSK)

- Phase-shift keying (PSK)

All convey data by changing some aspect of a base signal, the carrier wave (usually a sinusoid), in response to a data signal. In the case of PSK, the phase is changed to represent the data signal. There are two fundamental ways of utilizing the phase of a signal in this way:

- By viewing the phase itself as conveying the information, in which case the demodulator must have a reference signal to compare the received signal's phase against; or

- By viewing the *change* in the phase as conveying information — *differential* schemes, some of which do not need a reference carrier (to a certain extent).

A convenient method to represent PSK schemes is on a constellation diagram. This shows the points in the complex plane where, in this context, the real and imaginary axes are termed the in-phase and quadrature axes respectively due to their 90° separation. Such a representation on perpendicular axes lends itself to straightforward implementation. The amplitude of each point along the in-phase axis is used to modulate a cosine (or sine) wave and the amplitude along the quadrature axis to modulate a sine (or cosine) wave. By convention, in-phase modulates cosine and quadrature modulates sine.

In PSK, the constellation points chosen are usually positioned with uniform angular spacing around a circle. This gives maximum phase-separation between adjacent points and thus the best immunity to corruption. They are positioned on a circle so that they can all be transmitted with the same energy. In this way, the moduli of the complex numbers they represent will be the same and thus so will the amplitudes needed for the cosine and sine waves. Two common examples are "binary phase-shift keying" (BPSK) which uses two phases, and "quadrature phase-shift keying" (QPSK) which uses four phases, although any number of phases may be used. Since the data to be conveyed are usually binary, the PSK scheme is usually designed with the number of constellation points being a power of 2.

Definitions

For determining error-rates mathematically, some definitions will be needed:

- E_b = Energy-per-bit

- E_s = Energy-per-symbol = nE_b with n bits per symbol

- T_b = Bit duration

- T_s = Symbol duration

- $N_0/2$ = Noise power spectral density (W/Hz)

- P_b = Probability of *bit-error*

- P_s = Probability of symbol-error

$Q(x)$ will give the probability that a single sample taken from a random process with zero-mean and unit-variance Gaussian probability density function will be greater or equal to x. It is a scaled form of the complementary Gaussian error function:

$$Q(x) = \frac{1}{\sqrt{2\pi}} \int_x^\infty e^{-t^2/2} dt = \frac{1}{2} \text{erfc} \left(\frac{x}{\sqrt{2}} \right), x \geq 0.$$

The error-rates quoted here are those in additive white Gaussian noise (AWGN). These error rates are lower than those computed in fading channels, hence, are a good theoretical benchmark to compare with.

Applications

Owing to PSK's simplicity, particularly when compared with its competitor quadrature amplitude modulation, it is widely used in existing technologies.

The wireless LAN standard, IEEE 802.11b-1999, uses a variety of different PSKs depending on the data rate required. At the basic rate of 1 Mbit/s, it uses DBPSK (differential BPSK). To provide the extended rate of 2 Mbit/s, DQPSK is used. In reaching 5.5 Mbit/s and the full rate of 11 Mbit/s, QPSK is employed, but has to be coupled with complementary code keying. The higher-speed wireless LAN standard, IEEE 802.11g-2003, has eight data rates: 6, 9, 12, 18, 24, 36, 48 and 54 Mbit/s. The 6 and 9 Mbit/s modes use OFDM modulation where each sub-carrier is BPSK modulated. The 12 and 18 Mbit/s modes use OFDM with QPSK. The fastest four modes use OFDM with forms of quadrature amplitude modulation.

Because of its simplicity, BPSK is appropriate for low-cost passive transmitters, and is used in RFID standards such as ISO/IEC 14443 which has been adopted for biometric passports, credit cards such as American Express's ExpressPay, and many other applications.

Bluetooth 2 will use $\pi/4$-DQPSK at its lower rate (2 Mbit/s) and 8-DPSK at its higher rate (3 Mbit/s) when the link between the two devices is sufficiently robust. Bluetooth 1 modulates with Gaussian minimum-shift keying, a binary scheme, so either modulation choice in version 2 will yield a higher data-rate. A similar technology, IEEE 802.15.4 (the wireless standard used by Zig-Bee) also relies on PSK using two frequency bands: 868–915 MHz with BPSK and at 2.4 GHz with OQPSK.

Both QPSK and 8PSK are widely used in satellite broadcasting. QPSK is still widely used in the streaming of SD satellite channels and some HD channels. High definition programming is delivered almost exclusively in 8PSK due to the higher bitrates of HD video and the high cost of satellite bandwidth. The DVB-S2 standard requires support for both QPSK and 8PSK. The chipsets used in new satellite set top boxes, such as Broadcom's 7000 series support 8PSK and are backward compatible with the older standard.

Historically, voice-band synchronous modems such as the Bell 201, 208, and 209 and the CCITT V.26, V.27, V.29, V.32, and V.34 used PSK.

Binary phase-shift Keying (BPSK)

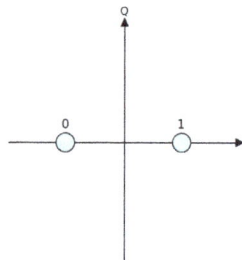

Constellation diagram example for BPSK.

BPSK (also sometimes called PRK, phase reversal keying, or 2PSK) is the simplest form of phase shift keying (PSK). It uses two phases which are separated by 180° and so can also be termed 2-PSK. It does not particularly matter exactly where the constellation points are positioned, and in this figure they are shown on the real axis, at 0° and 180°. This modulation is the most robust of all the PSKs since it takes the highest level of noise or distortion to make the demodulator reach an incorrect decision. It is, however, only able to modulate at 1 bit/symbol (as seen in the figure) and so is unsuitable for high data-rate applications.

In the presence of an arbitrary phase-shift introduced by the communications channel, the demodulator is unable to tell which constellation point is which. As a result, the data is often differentially encoded prior to modulation.

BPSK is functionally equivalent to 2-QAM modulation.

Implementation

The general form for BPSK follows the equation:

$$s_n(t) = \sqrt{\frac{2E_b}{T_b}} \cos(2\pi f_c t + \pi(1-n)), n = 0,1.$$

This yields two phases, 0 and π. In the specific form, binary data is often conveyed with the following signals:

$$s_0(t) = \sqrt{\frac{2E_b}{T_b}} \cos(2\pi f_c t + \pi) = -\sqrt{\frac{2E_b}{T_b}} \cos(2\pi f_c t)$$

for binary "0"

$$s_1 = (t) \sqrt{\frac{2E_b}{T_b}} \cos(2\pi f_c t) \text{ for binary "1"}$$

where f_c is the frequency of the carrier-wave.

Hence, the signal-space can be represented by the single basis function

$$\phi(t) = \sqrt{\frac{2}{T_b}} \cos(2\pi f_c t)$$

where 1 is represented by $\sqrt{E_b}\phi(t)$ and 0 is represented by $-\sqrt{E_b}\phi(t)$. This assignment is, of course, arbitrary.

This use of this basis function is shown at the end of the next section in a signal timing diagram. The topmost signal is a BPSK-modulated cosine wave that the BPSK modulator would produce. The bit-stream that causes this output is shown above the signal (the other parts of this figure are relevant only to QPSK).

Bit Error Rate

The bit error rate (BER) of BPSK in AWGN can be calculated as:

$$P_b = Q\left(\sqrt{\frac{2E_b}{N_0}}\right) \text{ or } P_e = \frac{1}{2}\text{erfc}\left(\sqrt{\frac{E_b}{N_0}}\right)$$

Since there is only one bit per symbol, this is also the symbol error rate.

Quadrature phase-shift Keying (QPSK)

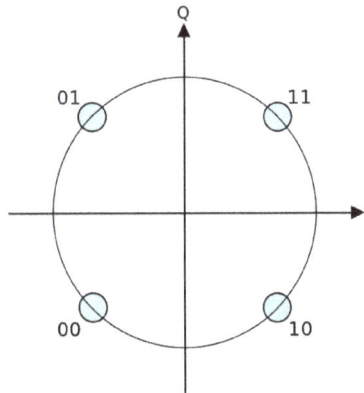

Constellation diagram for QPSK with Gray coding. Each adjacent symbol only differs by one bit.

Sometimes this is known as *quadriphase PSK*, 4-PSK, or 4-QAM. (Although the root concepts of QPSK and 4-QAM are different, the resulting modulated radio waves are exactly the same.) QPSK uses four points on the constellation diagram, equispaced around a circle. With four phases, QPSK can encode two bits per symbol, shown in the diagram with Gray coding to minimize the bit error rate (BER) — sometimes misperceived as twice the BER of BPSK.

The mathematical analysis shows that QPSK can be used either to double the data rate compared with a BPSK system while maintaining the *same* bandwidth of the signal, or to *maintain the data-rate of BPSK* but halving the bandwidth needed. In this latter case, the BER of QPSK is *exactly the same* as the BER of BPSK - and deciding differently is a common confusion when considering or describing QPSK. The transmitted carrier can undergo numbers of phase changes.

Given that radio communication channels are allocated by agencies such as the Federal Communication Commission giving a prescribed (maximum) bandwidth, the advantage of QPSK over BPSK becomes evident: QPSK transmits twice the data rate in a given bandwidth compared to BPSK - at the same BER. The engineering penalty that is paid is that QPSK transmitters and receivers are more complicated than the ones for BPSK. However, with modern electronics technology, the penalty in cost is very moderate.

As with BPSK, there are phase ambiguity problems at the receiving end, and differentially encoded QPSK is often used in practice.

Implementation

The implementation of QPSK is more general than that of BPSK and also indicates the implementation of higher-order PSK. Writing the symbols in the constellation diagram in terms of the sine and cosine waves used to transmit them:

$$s_n(t) = \sqrt{\frac{2E_s}{T_s}} \cos(2\pi f_c t + (2n-1)\frac{\pi}{4}), \quad n = 1,2,3,4.$$

This yields the four phases π/4, 3π/4, 5π/4 and 7π/4 as needed.

This results in a two-dimensional signal space with unit basis functions

$$\phi_1(t) = \sqrt{\frac{2}{T_s}} \cos(2\pi f_c t)$$

$$\phi_2(t) = \sqrt{\frac{2}{T_s}} \sin(2\pi f_c t)$$

The first basis function is used as the in-phase component of the signal and the second as the quadrature component of the signal.

Hence, the signal constellation consists of the signal-space 4 points

$$\left(\pm\sqrt{E_s/2}, \pm\sqrt{E_s/2}\right).$$

The factors of 1/2 indicate that the total power is split equally between the two carriers.

Comparing these basis functions with that for BPSK shows clearly how QPSK can be viewed as two independent BPSK signals. Note that the signal-space points for BPSK do not need to split the symbol (bit) energy over the two carriers in the scheme shown in the BPSK constellation diagram.

QPSK systems can be implemented in a number of ways. An illustration of the major components of the transmitter and receiver structure are shown below.

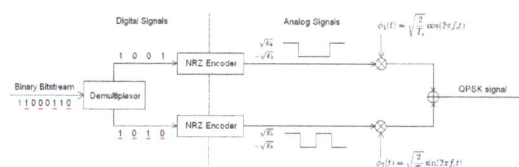

Conceptual transmitter structure for QPSK. The binary data stream is split into the in-phase and quadrature-phase components. These are then separately modulated onto two orthogonal basis functions. In this implementation, two sinusoids are used. Afterwards, the two signals are superim-posed, and the resulting signal is the QPSK signal. Note the use of polar non-return-to-zero encod-ing. These encoders can be placed before for binary data source, but have been placed after to illus-trate the conceptual difference between digital and analog signals involved with digital modulation.

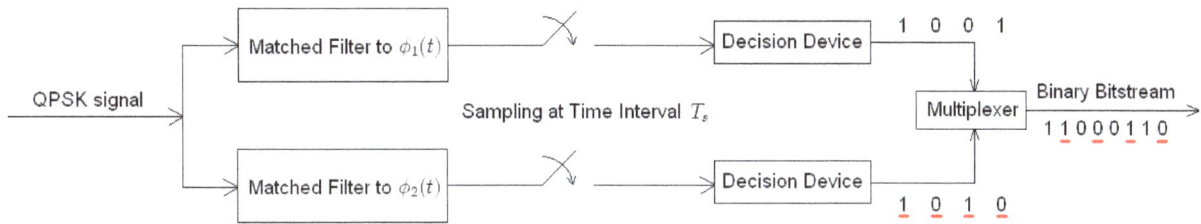

Receiver structure for QPSK. The matched filters can be replaced with correlators. Each detection device uses a reference threshold value to determine whether a 1 or 0 is detected.

Bit Error Rate

Although QPSK can be viewed as a quaternary modulation, it is easier to see it as two independently modulated quadrature carriers. With this interpretation, the even (or odd) bits are used to modulate the in-phase component of the carrier, while the odd (or even) bits are used to modulate the quadrature-phase component of the carrier. BPSK is used on both carriers and they can be independently demodulated.

As a result, the probability of bit-error for QPSK is the same as for BPSK:

$$P_b = Q\left(\sqrt{\frac{2E_b}{N_0}}\right).$$

However, in order to achieve the same bit-error probability as BPSK, QPSK uses twice the power (since two bits are transmitted simultaneously).

The symbol error rate is given by:

$$P_s = 1 - \left(1 - P_b\right)^2$$

$$= 2Q\left(\sqrt{\frac{E_s}{N_0}}\right) - \left[Q\left(\sqrt{\frac{E_s}{N_0}}\right)\right]^2.$$

If the signal-to-noise ratio is high (as is necessary for practical QPSK systems) the probability of symbol error may be approximated:

$$P_s \approx 2Q\left(\sqrt{\frac{E_s}{N_0}}\right)$$

The modulated signal is shown below for a short segment of a random binary data-stream. The two carrier waves are a cosine wave and a sine wave, as indicated by the signal-space analysis above.

Here, the odd-numbered bits have been assigned to the in-phase component and the even-numbered bits to the quadrature component (taking the first bit as number 1). The total signal — the sum of the two components — is shown at the bottom. Jumps in phase can be seen as the PSK changes the phase on each component at the start of each bit-period. The topmost waveform alone matches the description given for BPSK above.

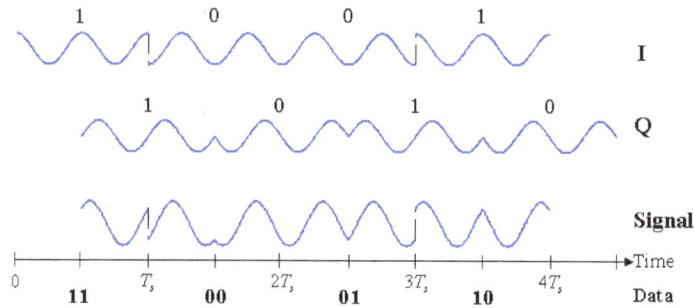

Timing diagram for QPSK. The binary data stream is shown beneath the time axis. The two signal components with their bit assignments are shown at the top, and the total combined signal at the bottom. Note the abrupt changes in phase at some of the bit-period boundaries.

The binary data that is conveyed by this waveform is: 1 1 0 0 0 1 1 0.

- The odd bits, highlighted here, contribute to the in-phase component: **1** 1 **0** 0 **0** 1 **1** 0

- The even bits, highlighted here, contribute to the quadrature-phase component: 1 **1** 0 **0** 0 **1** 1 **0**

Variants

Offset QPSK (OQPSK)

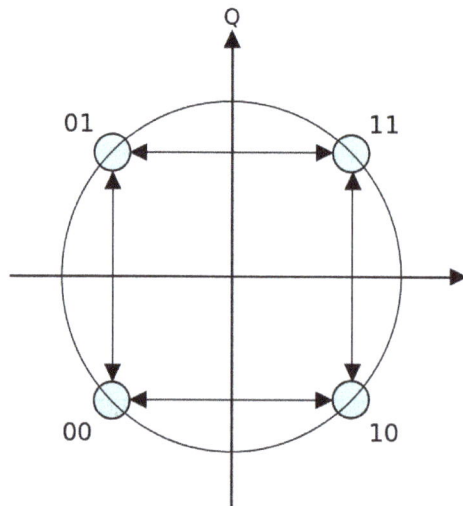

Signal doesn't cross zero, because only one bit of the symbol is changed at a time

Offset quadrature phase-shift keying (OQPSK) is a variant of phase-shift keying modulation using 4 different values of the phase to transmit. It is sometimes called *Staggered quadrature phase-shift keying (SQPSK)*.

QPSK

OQPSK

Difference of the phase between QPSK and OQPSK

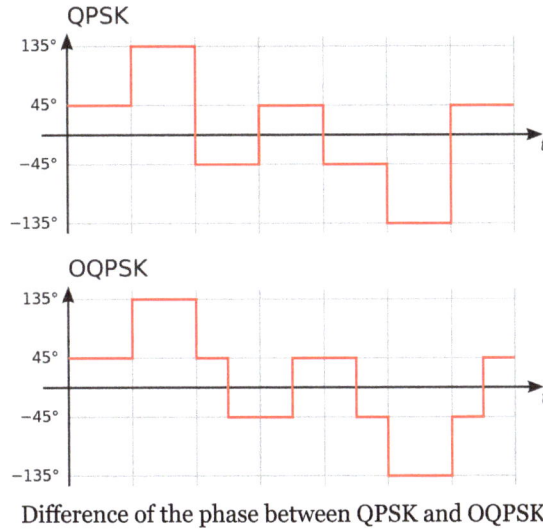

Taking four values of the phase (two bits) at a time to construct a QPSK symbol can allow the phase of the signal to jump by as much as 180° at a time. When the signal is low-pass filtered (as is typical in a transmitter), these phase-shifts result in large amplitude fluctuations, an undesirable quality in communication systems. By offsetting the timing of the odd and even bits by one bit-period, or half a symbol-period, the in-phase and quadrature components will never change at the same time. In the constellation diagram shown on the right, it can be seen that this will limit the phase-shift to no more than 90° at a time. This yields much lower amplitude fluctuations than non-offset QPSK and is sometimes preferred in practice.

The picture on the right shows the difference in the behavior of the phase between ordinary QPSK and OQPSK. It can be seen that in the first plot the phase can change by 180° at once, while in OQPSK the changes are never greater than 90°.

The modulated signal is shown below for a short segment of a random binary data-stream. Note the half symbol-period offset between the two component waves. The sudden phase-shifts occur about twice as often as for QPSK (since the signals no longer change together), but they are less severe. In other words, the magnitude of jumps is smaller in OQPSK when compared to QPSK.

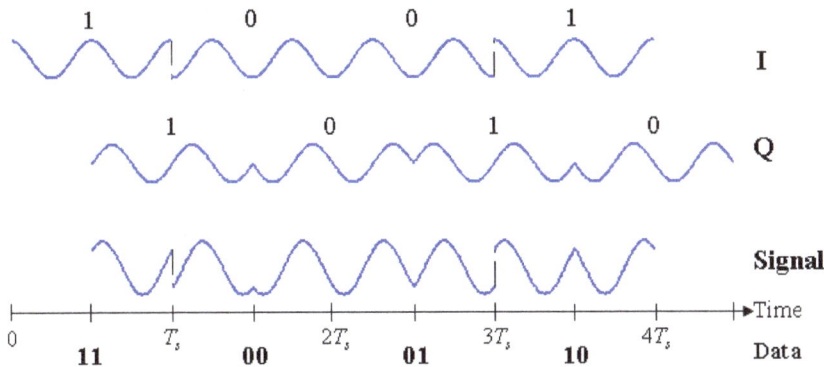

Timing diagram for offset-QPSK. The binary data stream is shown beneath the time axis. The two signal components with their bit assignments are shown the top and the total, combined signal at the bottom. Note the half-period offset between the two signal components.

$\pi/4$–QPSK

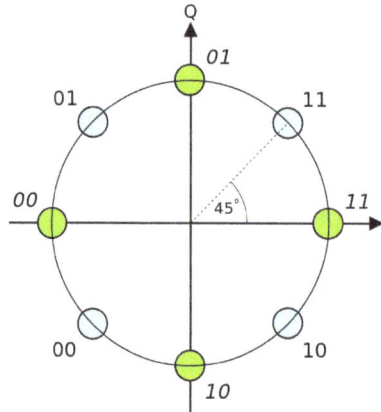

Dual constellation diagram for π/4-QPSK. This shows the two separate constellations with identical Gray coding but rotated by 45° with respect to each other.

This variant of QPSK uses two identical constellations which are rotated by 45° $\pi/4$ radians, hence the name) with respect to one another. Usually, either the even or odd symbols are used to select points from one of the constellations and the other symbols select points from the other constellation. This also reduces the phase-shifts from a maximum of 180°, but only to a maximum of 135° and so the amplitude fluctuations of $\pi/4-$ QPSK are between OQPSK and non-offset QPSK.

One property this modulation scheme possesses is that if the modulated signal is represented in the complex domain, it does not have any paths through the origin. In other words, the signal does not pass through the origin. This lowers the dynamical range of fluctuations in the signal which is desirable when engineering communications signals.

On the other hand, $\pi/4-$ QPSK lends itself to easy demodulation and has been adopted for use in, for example, TDMA cellular telephone systems.

The modulated signal is shown below for a short segment of a random binary data-stream. The construction is the same as above for ordinary QPSK. Successive symbols are taken from the two constellations shown in the diagram. Thus, the first symbol (1 1) is taken from the 'blue' constellation and the second symbol (0 0) is taken from the 'green' constellation. Note that magnitudes of the two component waves change as they switch between constellations, but the total signal's magnitude remains constant (constant envelope). The phase-shifts are between those of the two previous timing-diagrams.

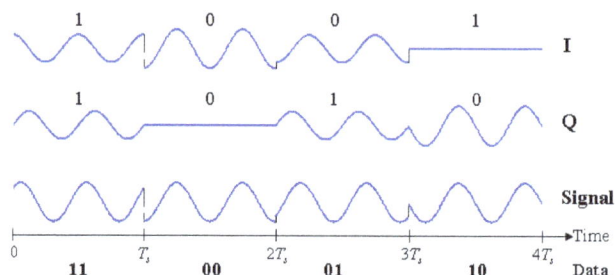

Timing diagram for π/4-QPSK. The binary data stream is shown beneath the time axis. The two signal components with their bit assignments are shown the top and the total, combined signal at the bottom. Note that successive symbols are taken alternately from the two constellations, starting with the 'blue' one.

SOQPSK

The license-free shaped-offset QPSK (SOQPSK) is interoperable with Feher-patented QPSK (FQPSK), in the sense that an integrate-and-dump offset QPSK detector produces the same output no matter which kind of transmitter is used.

These modulations carefully shape the I and Q waveforms such that they change very smoothly, and the signal stays constant-amplitude even during signal transitions. (Rather than traveling instantly from one symbol to another, or even linearly, it travels smoothly around the constant-amplitude circle from one symbol to the next.)

The standard description of SOQPSK-TG involves ternary symbols.

DPQPSK

Dual-polarization quadrature phase shift keying (DPQPSK) or dual-polarization QPSK - involves the polarization multiplexing of two different QPSK signals, thus improving the spectral efficiency by a factor of 2. This is a cost-effective alternative, to utilizing 16-PSK instead of QPSK to double the spectral efficiency.

Higher-order PSK

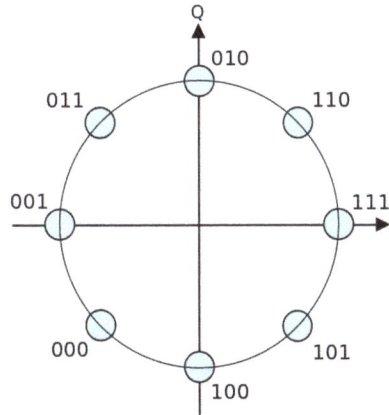

Constellation diagram for 8-PSK with Gray coding.

Any number of phases may be used to construct a PSK constellation but 8-PSK is usually the highest order PSK constellation deployed. With more than 8 phases, the error-rate becomes too high and there are better, though more complex, modulations available such as quadrature amplitude modulation (QAM). Although any number of phases may be used, the fact that the constellation must usually deal with binary data means that the number of symbols is usually a power of 2 to allow an integer number of bits per symbol.

Bit Error Rate

For the general M-PSK there is no simple expression for the symbol-error probability if $M > 4$. Unfortunately, it can only be obtained from:

$$P_s = 1 - \int_{-\frac{\pi}{M}}^{\frac{\pi}{M}} p_{\theta_r}\left(\theta_r\right) d\theta_r$$

where

$$p_{\theta_r}\left(\theta_r\right)=\frac{1}{2\pi}e^{-2\gamma_s\sin^2\theta_r}\int_0^\infty Ve^{-\left(V-\sqrt{4\gamma_s}\cos\theta_r\right)^2/2}dV,$$

$$V=\sqrt{r_1^2+r_2^2},$$

$$\theta_r=\tan^{-1}\left(r_2\,/\,r_1\right),$$

$$\gamma_s=\frac{E_s}{N_0}\text{ and}$$

$r_2\sim N\left(0,N_0\,/\,2\right)$ are jointly Gaussian random variables.

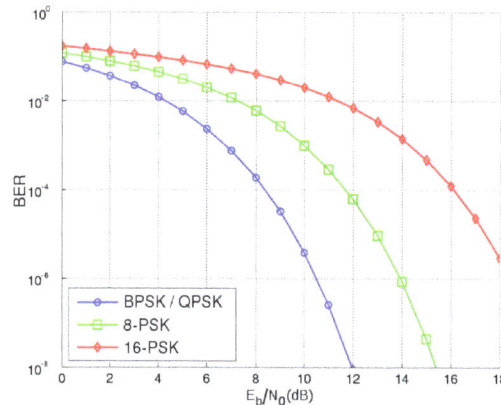

Bit-error rate curves for BPSK, QPSK, 8-PSK and 16-PSK, AWGN channel.

This may be approximated for high M and high $E_b\,/N_0$ by:

$$P_s\approx 2Q\left(\sqrt{2\gamma_s}\,\sin\frac{\pi}{M}\right).$$

The bit-error probability for M-PSK can only be determined exactly once the bit-mapping is known. However, when Gray coding is used, the most probable error from one symbol to the next produces only a single bit-error and

$$P_b\approx\frac{1}{k}P_s.$$

(Using Gray coding allows us to approximate the Lee distance of the errors as the Hamming distance of the errors in the decoded bitstream, which is easier to implement in hardware.)

The graph on the left compares the bit-error rates of BPSK, QPSK (which are the same, as noted above), 8-PSK and 16-PSK. It is seen that higher-order modulations exhibit higher error-rates; in exchange however they deliver a higher raw data-rate.

Bounds on the error rates of various digital modulation schemes can be computed with application of the union bound to the signal constellation.

Differential Phase-shift Keying (DPSK)

Differential Encoding

Differential phase shift keying (DPSK) is a common form of phase modulation that conveys data by changing the phase of the carrier wave. As mentioned for BPSK and QPSK there is an ambiguity of phase if the constellation is rotated by some effect in the communications channel through which the signal passes. This problem can be overcome by using the data to *change* rather than *set* the phase.

For example, in differentially encoded BPSK a binary '1' may be transmitted by adding 180° to the current phase and a binary '0' by adding 0° to the current phase. Another variant of DPSK is Symmetric Differential Phase Shift keying, SDPSK, where encoding would be +90° for a '1' and −90° for a '0'.

In differentially encoded QPSK (DQPSK), the phase-shifts are 0°, 90°, 180°, −90° corresponding to data '00', '01', '11', '10'. This kind of encoding may be demodulated in the same way as for non-differential PSK but the phase ambiguities can be ignored. Thus, each received symbol is demodulated to one of the M points in the constellation and a comparator then computes the difference in phase between this received signal and the preceding one. The difference encodes the data as described above. Symmetric Differential Quadrature Phase Shift Keying (SDQPSK) is like DQPSK, but encoding is symmetric, using phase shift values of −135°, −45°, +45° and +135°.

The modulated signal is shown below for both DBPSK and DQPSK as described above. In the figure, it is assumed that the *signal starts with zero phase*, and so there is a phase shift in both signals at $t = 0$.

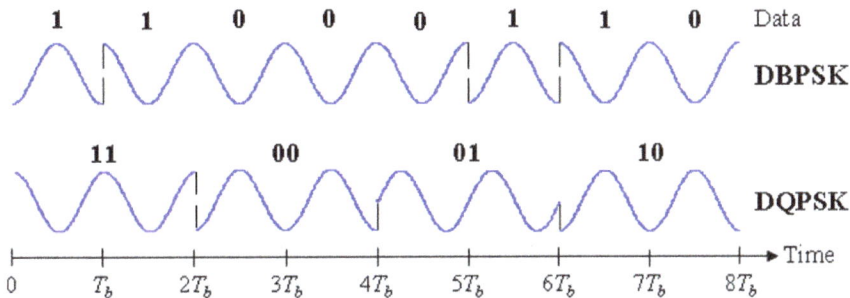

Timing diagram for DBPSK and DQPSK. The binary data stream is above the DBPSK signal. The individual bits of the DBPSK signal are grouped into pairs for the DQPSK signal, which only changes every $T_s = 2T_b$.

Analysis shows that differential encoding approximately doubles the error rate compared to ordinary M-PSK but this may be overcome by only a small increase in E_b/N_0. Furthermore, this analysis (and the graphical results below) are based on a system in which the only corruption is additive white Gaussian noise(AWGN). However, there will also be a physical channel between the transmitter and receiver in the communication system. This channel will, in general, introduce an unknown phase-shift to the PSK signal; in these cases the differential schemes can yield a better error-rate than the ordinary schemes which rely on precise phase information.

Demodulation

For a signal that has been differentially encoded, there is an obvious alternative method of demodulation. Instead of demodulating as usual and ignoring carrier-phase ambiguity, the phase between two successive received symbols is compared and used to determine what the data must have been. When differential encoding is used in this manner, the scheme is known as differential phase-shift keying (DPSK). Note that this is subtly different from just differentially encoded PSK since, upon reception, the received symbols are *not* decoded one-by-one to constellation points but are instead compared directly to one another.

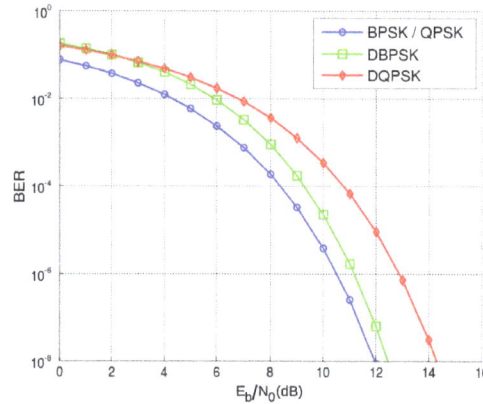

BER comparison between DBPSK, DQPSK and their non-differential forms using gray-coding and operating in white noise.

Call the received symbol in the kth timeslot r_k and let it have phase ϕ_k. Assume without loss of generality that the phase of the carrier wave is zero. Denote the AWGN term as n_k. Then

$$r_k = \sqrt{E_s} e^{j\phi_k} + n_k.$$

The decision variable for the $k-1^{th}$ symbol and the kth symbol is the phase difference between r_k and r_{k-1}. That is, if r_k is projected onto r_{k-1}, the decision is taken on the phase of the resultant complex number:

$$r_k r_{k-1}^* = E_s e^{j(\phi_k - \phi_{k-1})} + \sqrt{E_s} e^{j\phi_k} n_{k-1}^* + \sqrt{E_s} e^{-j\phi_{k-1}} n_k + n_k n_{k-1}^*$$

where superscript * denotes complex conjugation. In the absence of noise, the phase of this is $\phi_k - \phi_{k-1}$, the phase-shift between the two received signals which can be used to determine the data transmitted.

The probability of error for DPSK is difficult to calculate in general, but, in the case of DBPSK it is:

$$P_b = \frac{1}{2} e^{-E_b/N_0},$$

which, when numerically evaluated, is only slightly worse than ordinary BPSK, particularly at higher E_b / N_0 values.

Using DPSK avoids the need for possibly complex carrier-recovery schemes to provide an accurate phase estimate and can be an attractive alternative to ordinary PSK.

In optical communications, the data can be modulated onto the phase of a laser in a differential way. The modulation is a laser which emits a continuous wave, and a Mach-Zehnder modulator which receives electrical binary data. For the case of BPSK for example, the laser transmits the field unchanged for binary '1', and with reverse polarity for '0'. The demodulator consists of a delay line interferometer which delays one bit, so two bits can be compared at one time. In further processing, a photodiode is used to transform the optical field into an electric current, so the information is changed back into its original state.

The bit-error rates of DBPSK and DQPSK are compared to their non-differential counterparts in the graph to the right. The loss for using DBPSK is small enough compared to the complexity reduction that it is often used in communications systems that would otherwise use BPSK. For DQPSK though, the loss in performance compared to ordinary QPSK is larger and the system designer must balance this against the reduction in complexity.

Example: Differentially Encoded BPSK

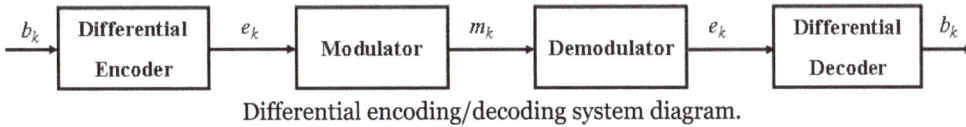

Differential encoding/decoding system diagram.

At the kth time-slot call the bit to be modulated b_k, the differentially encoded bit e_k and the resulting modulated signal $m_k(t)$. Assume that the constellation diagram positions the symbols at ±1 (which is BPSK). The differential encoder produces:

$$e_k = e_{k-1} \oplus b_k$$

where $\oplus$ indicates binary or modulo-2 addition.

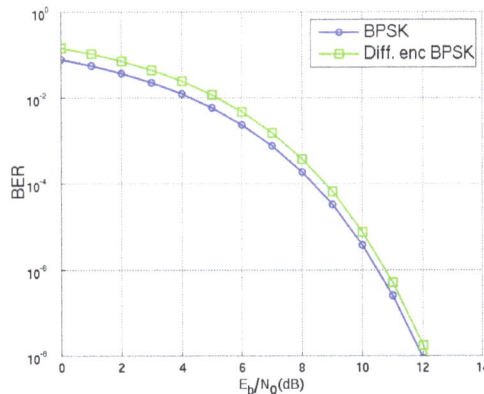

BER comparison between BPSK and differentially encoded BPSK with gray-coding operating in white noise.

So e_k only changes state (from binary '0' to binary '1' or from binary '1' to binary '0') if b_k is a binary '1'. Otherwise it remains in its previous state. This is the description of differentially encoded BPSK given above.

The received signal is demodulated to yield $e_k = \pm 1$ and then the differential decoder reverses the encoding procedure and produces:

$$b_k = e_k \oplus e_{k-1} \text{ since binary subtraction is the same as binary addition.}$$

Therefore, $b_k = 1$ if e_k and e_{k-1} differ and $b_k = 0$ if they are the same. Hence, if both e_k and e_{k-1} are *inverted*, b_k will still be decoded correctly. Thus, the 180° phase ambiguity does not matter.

Differential schemes for other PSK modulations may be devised along similar lines. The waveforms for DPSK are the same as for differentially encoded PSK given above since the only change between the two schemes is at the receiver.

The BER curve for this example is compared to ordinary BPSK on the right. As mentioned above, whilst the error-rate is approximately doubled, the increase needed in E_b/N_0 to overcome this is small. The increase in E_b/N_0 required to overcome differential modulation in coded systems, however, is larger - typically about 3 dB. The performance degradation is a result of noncoherent transmission - in this case it refers to the fact that tracking of the phase is completely ignored.

Channel Capacity

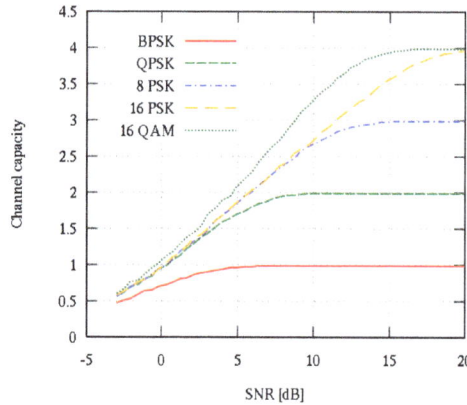

Given a fixed bandwidth, channel capacity vs. SNR for some common modulation schemes

Like all M-ary modulation schemes with $M = 2^b$ symbols, when given exclusive access to a fixed bandwidth, the channel capacity of any phase shift keying modulation scheme rises to a maximum of b bits per symbol as the signal-to-noise ratio increases, due to the Shannon-Hartley Theorem.

Frequency-shift Keying

Frequency-shift keying (FSK) is a frequency modulation scheme in which digital information is transmitted through discrete frequency changes of a carrier signal. The technology is used for communication systems such as amateur radio, caller ID and emergency broadcasts. The simplest FSK is binary FSK (BFSK). BFSK uses a pair of discrete frequencies to transmit binary (0s and 1s) information. With this scheme, the "1" is called the mark frequency and the "0" is called the space frequency. The time domain of an FSK modulated carrier is illustrated in the figures to the right.

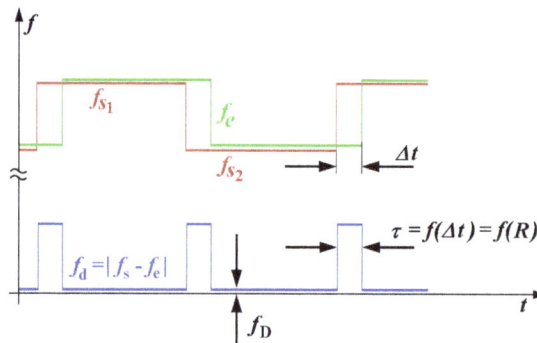

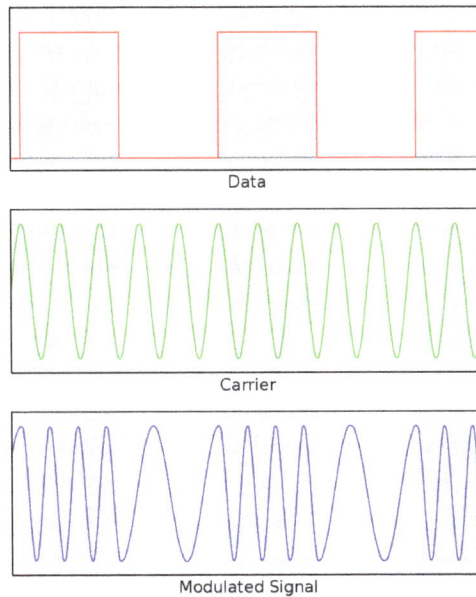

An example of binary FSK

Implementations of FSK Modems

Reference implementations of FSK modems exist and are documented in detail. The demodulation of a binary FSK signal can be done using the Goertzel algorithm very efficiently, even on low-power microcontrollers.

Other Forms of FSK

Continuous-phase frequency-shift keying

In principle FSK can be implemented by using completely independent free-running oscillators, and switching between them at the beginning of each symbol period. In general, independent oscillators will not be at the same phase and therefore the same amplitude at the switch-over instant, causing sudden discontinuities in the transmitted signal.

In practice, many FSK transmitters use only a single oscillator, and the process of switching to a different frequency at the beginning of each symbol period preserves the phase. The elimination of discontinuities in the phase (and therefore elimination of sudden changes in amplitude) reduces sideband power, reducing interference with neighboring channels.

Gaussian frequency-shift Keying

Rather than directly modulating the frequency with the digital data symbols, "instantaneously" changing the frequency at the beginning of each symbol period, Gaussian frequency-shift keying (GFSK) filters the data pulses with a Gaussian filter to make the transitions smoother. This filter has the advantage of reducing sideband power, reducing interference with neighboring channels, at the cost of increasing intersymbol interference. It is used by DECT, Bluetooth, Cypress WirelessUSB, Nordic Semiconductor, Texas Instruments LPRF, Z-Wave and Wavenis devices. For basic data rate Bluetooth the minimum deviation is 115 kHz.

A GFSK modulator differs from a simple frequency-shift keying modulator in that before the base-band waveform (levels −1 and +1) goes into the FSK modulator, it is passed through a Gaussian filter to make the transitions smoother so to limit its spectral width. Gaussian filtering is a standard way for reducing spectral width; it is called "pulse shaping" in this application.

In ordinary non-filtered FSK, at a jump from −1 to +1 or +1 to −1, the modulated waveform changes rapidly, which introduces large out-of-band spectrum. If we change the pulse going from −1 to +1 as −1, −.98, −.93 ..... +.93, +.98, +1, and we use this smoother pulse to determine the carrier frequency, the out-of-band spectrum will be reduced.

Minimum-shift Keying

Minimum frequency-shift keying or minimum-shift keying (MSK) is a particular spectrally efficient form of coherent FSK. In MSK, the difference between the higher and lower frequency is identical to half the bit rate. Consequently, the waveforms that represent a 0 and a 1 bit differ by exactly half a carrier period. The maximum frequency deviation is $\delta = 0.25 f_m$, where f_m is the maximum modulating frequency. As a result, the modulation index m is 0.5. This is the smallest FSK modulation index that can be chosen such that the waveforms for 0 and 1 are orthogonal.

Gaussian Minimum Shift Keying

A variant of MSK called Gaussian minimum shift keying (GMSK) is used in the GSM mobile phone standard.

Audio FSK

Audio frequency-shift keying (AFSK) is a modulation technique by which digital data is represented by changes in the frequency (pitch) of an audio tone, yielding an encoded signal suitable for transmission via radio or telephone. Normally, the transmitted audio alternates between two tones: one, the "mark", represents a binary one; the other, the "space", represents a binary zero.

AFSK differs from regular frequency-shift keying in performing the modulation at baseband frequencies. In radio applications, the AFSK-modulated signal normally is being used to modulate an RF carrier (using a conventional technique, such as AM or FM) for transmission.

AFSK is not always used for high-speed data communications, since it is far less efficient in both power and bandwidth than most other modulation modes. In addition to its simplicity, however, AFSK has the advantage that encoded signals will pass through AC-coupled links, including most equipment originally designed to carry music or speech.

AFSK is used in the U.S. based Emergency Alert System to notify stations of the type of emergency, locations affected, and the time of issue without actually hearing the text of the alert.

Continuous 4 Level FM

Phase 1 radios in the Project 25 system use continuous 4 level FM (C4FM) modulation.

Applications

In 1910, Reginald Fessenden invented a two-tone method of transmitting Morse code. Dots and dashes were replaced with different tones of equal length. The intent was to minimize transmission time.

Some early CW transmitters employed an arc converter that could not be conveniently keyed. Instead of turning the arc on and off, the key slightly changed the transmitter frequency in a technique known as the *compensation-wave method*. The compensation-wave was not used at the receiver. Spark transmitters used for this method consumed a lot of bandwidth and caused interference, so it was discouraged by 1921.

Most early telephone-line modems used audio frequency-shift keying (AFSK) to send and receive data at rates up to about 1200 bits per second. The Bell 103 and Bell 202 modems used this technique. Even today, North American caller ID uses 1200 baud AFSK in the form of the Bell 202 standard. Some early microcomputers used a specific form of AFSK modulation, the Kansas City standard, to store data on audio cassettes. AFSK is still widely used in amateur radio, as it allows data transmission through unmodified voiceband equipment.

AFSK is also used in the United States' Emergency Alert System to transmit warning information. It is used at higher bitrates for Weathercopy used on Weatheradio by NOAA in the U.S.

The CHU shortwave radio station in Ottawa, Canada broadcasts an exclusive digital time signal encoded using AFSK modulation.

FSK is commonly used in Caller ID and remote metering applications.

Amplitude-shift Keying

Amplitude-shift keying (ASK) is a form of amplitude modulation that represents digital data as variations in the amplitude of a carrier wave. In an ASK system, the binary symbol 1 is represented by transmitting a fixed-amplitude carrier wave and fixed frequency for a bit duration of T seconds. If the signal value is 1 then the carrier signal will be transmitted; otherwise, a signal value of 0 will be transmitted.

Any digital modulation scheme uses a finite number of distinct signals to represent digital data. ASK uses a finite number of amplitudes, each assigned a unique pattern of binary digits. Usually, each amplitude encodes an equal number of bits. Each pattern of bits forms the symbol that is represented by the particular amplitude. The demodulator, which is designed specifically for the symbol-set used by the modulator, determines the amplitude of the received signal and maps it back to the symbol it represents, thus recovering the original data. Frequency and phase of the carrier are kept constant.

Like AM, an ASK is also linear and sensitive to atmospheric noise, distortions, propagation conditions on different routes in PSTN, etc. Both ASK modulation and demodulation processes are relatively inexpensive. The ASK technique is also commonly used to transmit digital data over optical fiber. For LED transmitters, binary 1 is represented by a short pulse of light and binary 0 by the absence of light. Laser transmitters normally have a fixed "bias" current that causes the device to emit a low light level. This low level represents binary 0, while a higher-amplitude lightwave represents binary 1.

The simplest and most common form of ASK operates as a switch, using the presence of a carrier wave to indicate a binary one and its absence to indicate a binary zero. This type of modulation is called on-off keying (OOK), and is used at radio frequencies to transmit Morse code (referred to as continuous wave operation),

More sophisticated encoding schemes have been developed which represent data in groups using additional amplitude levels. For instance, a four-level encoding scheme can represent two bits with each shift in amplitude; an eight-level scheme can represent three bits; and so on. These forms of amplitude-shift keying require a high signal-to-noise ratio for their recovery, as by their nature much of the signal is transmitted at reduced power.

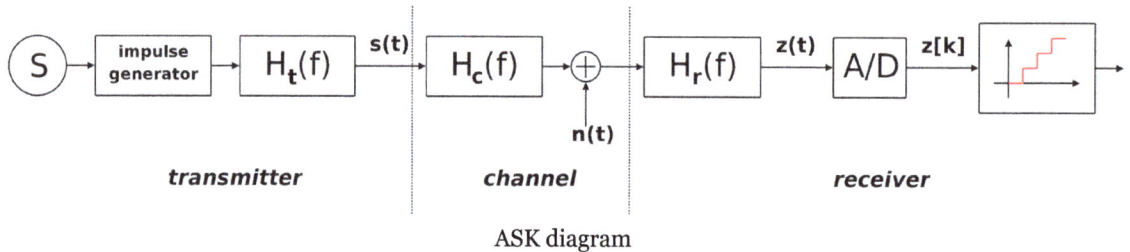

ASK diagram

ASK system can be divided into three blocks. The first one represents the transmitter, the second one is a linear model of the effects of the channel, the third one shows the structure of the receiver. The following notation is used:

- $h_t(f)$ is the carrier signal for the transmission

- $h_c(f)$ is the impulse response of the channel

- $n(t)$ is the noise introduced by the channel

- $h_r(f)$ is the filter at the receiver

- L is the number of levels that are used for transmission

- T_s is the time between the generation of two symbols

Different symbols are represented with different voltages. If the maximum allowed value for the voltage is A, then all the possible values are in the range [−A, A] and they are given by:

$$v_i = \frac{2A}{L-1}i - A; \quad i = 0, 1, \ldots, L-1$$

the difference between one voltage and the other is:

$$\Delta = \frac{2A}{L-1}$$

Considering the picture, the symbols v[n] are generated randomly by the source S, then the impulse generator creates impulses with an area of v[n]. These impulses are sent to the filter ht to be sent through the channel. In other words, for each symbol a different carrier wave is sent with the relative amplitude.

Out of the transmitter, the signal s(t) can be expressed in the form:

$$s(t) = \sum_{n=-\infty}^{\infty} v[n] \cdot h_t(t - nT_s)$$

In the receiver, after the filtering through hr (t) the signal is:

$$z(t) = n_r(t) + \sum_{n=-\infty}^{\infty} v[n] \cdot g(t - nT_s)$$

where we use the notation:

$$n_r(t) = n(t) * h_r(f)$$

$$g(t) \quad h_t(t) * h_c(f) * h_r(t)$$

where * indicates the convolution between two signals. After the A/D conversion the signal z[k] can be expressed in the form:

$$z[k] = n_r[k] + v[k]g[0] + \sum_{n \neq k} v[n]g[k-n]$$

In this relationship, the second term represents the symbol to be extracted. The others are unwanted: the first one is the effect of noise, the third one is due to the intersymbol interference.

If the filters are chosen so that g(t) will satisfy the Nyquist ISI criterion, then there will be no intersymbol interference and the value of the sum will be zero, so:

$$z[k] = n_r[k] + v[k]g[0]$$

the transmission will be affected only by noise.

Probability of Error

The probability density function of having an error of a given size can be modelled by a Gaussian function; the mean value will be the relative sent value, and its variance will be given by:

$$\sigma_N^2 = \int_{-\infty}^{+\infty} \Phi_N(f) \cdot |H_r(f)|^2 \, df$$

where $\Phi_N(f)$ is the spectral density of the noise within the band and Hr (f) is the continuous Fourier transform of the impulse response of the filter hr (f).

The probability of making an error is given by:

$$P_e = P_{e|H_0} \cdot P_{H_0} + P_{e|H_1} \cdot P_{H_1} + \cdots + P_{e|H_{L-1}} \cdot P_{H_{L-1}} = \sum_{k=0}^{L-1} P_{e|H_k} \cdot P_{H_k}$$

where, for example, $P_{e|H_0}$ is the conditional probability of making an error given that a symbol vo has been sent and P_{H_0} is the probability of sending a symbol vo.

If the probability of sending any symbol is the same, then:

$$P_{H_i} = \frac{1}{L}$$

If we represent all the probability density functions on the same plot against the possible value of the voltage to be transmitted, we get a picture like this (the particular case of L = 4 is shown):

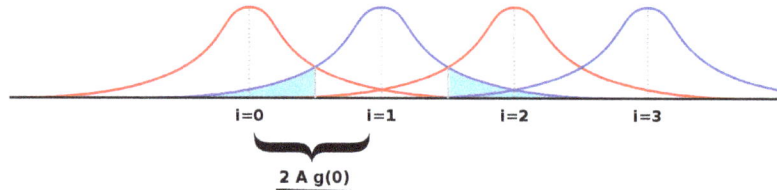

The probability of making an error after a single symbol has been sent is the area of the Gaussian function falling under the functions for the other symbols. It is shown in cyan for just one of them. If we call P+ the area under one side of the Gaussian, the sum of all the areas will be: 2 L P^+ - 2 P^+. The total probability of making an error can be expressed in the form:

$$P_e = 2\left(1 - \frac{1}{L}\right)P^+$$

We have now to calculate the value of P⁺. In order to do that, we can move the origin of the reference wherever we want: the area below the function will not change. We are in a situation like the one shown in the following picture:

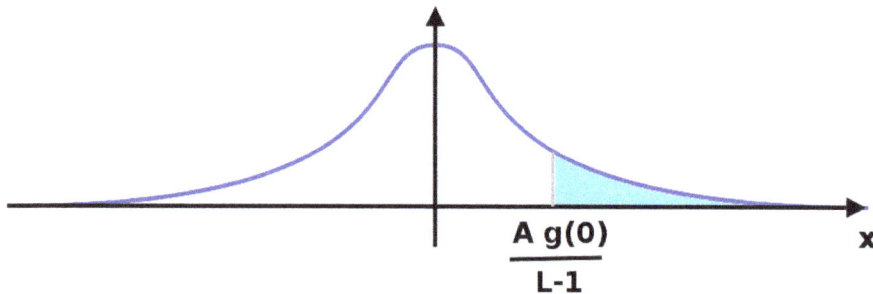

it does not matter which Gaussian function we are considering, the area we want to calculate will be the same. The value we are looking for will be given by the following integral:

$$P^+ = \int_{\frac{Ag(0)}{L-1}}^{\infty} \frac{1}{\sqrt{2\pi}\sigma_N} e^{-\frac{x^2}{2\sigma_N^2}} dx = \frac{1}{2} erfc\left(\frac{Ag(0)}{\sqrt{2}(L-1)\sigma_N}\right)$$

where erfc(x) is the complementary error function. Putting all these results together, the probability to make an error is:

from this formula we can easily understand that the probability to make an error decreases if the maximum amplitude of the transmitted signal or the amplification of the system becomes greater; on the other hand, it increases if the number of levels or the power of noise becomes greater.

This relationship is valid when there is no intersymbol interference, i.e. g(t) is a Nyquist function.

References

- Ke-Lin Du & M. N. S. Swamy (2010). Wireless Communication Systems: From RF Subsystems to 4G Enabling Technologies. Cambridge University Press. p. 188. ISBN 978-0-521-11403-5.

- Bray, John (2002). Innovation and the Communications Revolution: From the Victorian Pioneers to Broadband Internet. Inst. of Electrical Engineers. pp. 59, 61–62. ISBN 0852962185.

- Silver, Ward, ed. (2011). "Ch. 15 DSP and Software Radio Design". The ARRL Handbook for Radio Communications (Eighty-eighth ed.). American Radio Relay League. ISBN 978-0-87259-096-0.

- B. Boashash, editor, "Time-Frequency Signal Analysis and Processing – A Comprehensive Reference", Elsevier Science, Oxford, 2003; ISBN 0-08-044335-4

- A. Michael Noll (2001). Principles of modern communications technology. Artech House. p. 104. ISBN 978-1-58053-284-6.

- Couch, Leon W. II (1997). Digital and Analog Communications. Upper Saddle River, NJ: Prentice-Hall. ISBN 0-13-081223-4.

- Kennedy, G.; Davis, B. (1992). Electronic Communication Systems (4th ed.). McGraw-Hill International. ISBN 0-07-112672-4. , p 509

- Bhagwat, Pravin (10 May 2005). "Bluetooth: 1.Applications, Technology and Performance". p. 21. Retrieved 27 May 2015.

4

Cellular Network: An Integrated Study

A cellular network is where the last connection is wireless. Cellular frequencies are a set of frequencies that are used for cellular phones. These frequencies are ultra high frequency bands. The types of frequency bands explained in this section are GSM frequency bands, UMTS frequency bands and E-UTRA. This text helps the reader in developing an integrated understanding of the topic.

Cellular Network

A cellular network or mobile network is a communication network where the last link is wireless. The network is distributed over land areas called cells, each served by at least one fixed-location transceiver, known as a cell site or base station. This base station provides the cell with the network coverage which can be used for transmission of voice, data and others. A cell might use a different set of frequencies from neighboring cells, to avoid interference and provide guaranteed service quality within each cell.

Top of a cellular radio tower

When joined together these cells provide radio coverage over a wide geographic area. This enables a large number of portable transceivers (e.g., mobile phones, pagers, etc.) to communicate with each other and with fixed transceivers and telephones anywhere in the network, via base stations, even if some of the transceivers are moving through more than one cell during transmission.

Cellular networks offer a number of desirable features:

- More capacity than a single large transmitter, since the same frequency can be used for multiple links as long as they are in different cells

- Mobile devices use less power than with a single transmitter or satellite since the cell towers are closer

- Larger coverage area than a single terrestrial transmitter, since additional cell towers can be added indefinitely and are not limited by the horizon

Major telecommunications providers have deployed voice and data cellular networks over most of the inhabited land area of the Earth. This allows mobile phones and mobile computing devices to be connected to the public switched telephone network and public Internet. Private cellular networks can be used for research or for large organizations and fleets, such as dispatch for local public safety agencies or a taxicab company.

Concept

In a cellular radio system, a land area to be supplied with radio service is divided into regular shaped cells, which can be hexagonal, square, circular or some other regular shapes, although hexagonal cells are conventional. Each of these cells is assigned with multiple frequencies $(f_1 - f_6)$ which have corresponding radio base stations. The group of frequencies can be reused in other cells, provided that the same frequencies are not reused in adjacent neighboring cells as that would cause co-channel interference.

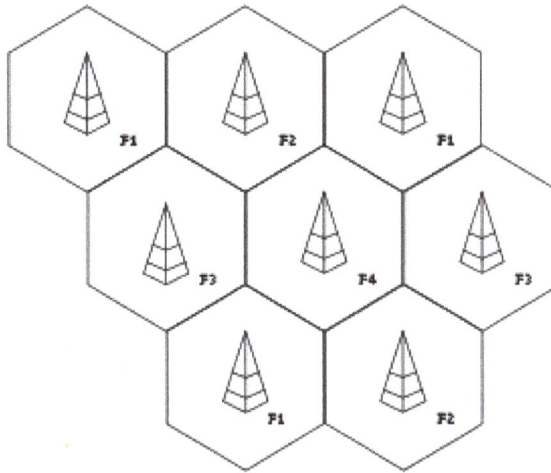

Example of frequency reuse factor or pattern 1/4

The increased capacity in a cellular network, compared with a network with a single transmitter, comes from the mobile communication switching system developed by Amos Joel of Bell Labs that permitted multiple callers in the same area to use the same frequency by switching calls made using the same frequency to the nearest available cellular tower having that frequency available and from the fact that the same radio frequency can be reused in a different area for a completely different transmission. If there is a single plain transmitter, only one transmission can be used on any given frequency. Unfortunately, there is inevitably some level of interference from the signal from the other cells which use the same frequency. This means that, in a standard FDMA system, there must be at least a one cell gap between cells which reuse the same frequency.

In the simple case of the taxi company, each radio had a manually operated channel selector knob to tune to different frequencies. As the drivers moved around, they would change from channel to channel. The drivers knew which frequency covered approximately what area. When they did not receive a signal from the transmitter, they would try other channels until they found one that worked. The taxi drivers would only speak one at a time, when invited by the base station operator (this is, in a sense, time division multiple access (TDMA)).

Cell Signal Encoding

To distinguish signals from several different transmitters, time division multiple access (TDMA), frequency division multiple access (FDMA), code division multiple access (CDMA), and orthogonal frequency division multiple access (OFDMA) were developed.

With TDMA, the transmitting and receiving time slots used by different users in each cell are different from each other.

With FDMA, the transmitting and receiving frequencies used by different users in each cell are different from each other. In a simple taxi system, the taxi driver manually tuned to a frequency of a chosen cell to obtain a strong signal and to avoid interference from signals from other cells.

The principle of CDMA is more complex, but achieves the same result; the distributed transceivers can select one cell and listen to it.

Other available methods of multiplexing such as polarization division multiple access (PDMA) cannot be used to separate signals from one cell to the next since the effects of both vary with position and this would make signal separation practically impossible. Time division multiple access is used in combination with either FDMA or CDMA in a number of systems to give multiple channels within the coverage area of a single cell.

Frequency Reuse

The key characteristic of a cellular network is the ability to re-use frequencies to increase both coverage and capacity. As described above, adjacent cells must use different frequencies, however there is no problem with two cells sufficiently far apart operating on the same frequency, provided the masts and cellular network users' equipment do not transmit with too much power.

The elements that determine frequency reuse are the reuse distance and the reuse factor. The reuse distance, D is calculated as

$$D = R\sqrt{3N},$$

where R is the cell radius and N is the number of cells per cluster. Cells may vary in radius from 1 to 30 kilometres (0.62 to 18.64 mi). The boundaries of the cells can also overlap between adjacent cells and large cells can be divided into smaller cells.

The frequency reuse factor is the rate at which the same frequency can be used in the network. It is $1/K$ (or K according to some books) where K is the number of cells which cannot use the same frequencies for transmission. Common values for the frequency reuse factor are 1/3, 1/4, 1/7, 1/9 and 1/12 (or 3, 4, 7, 9 and 12 depending on notation).

In case of N sector antennas on the same base station site, each with different direction, the base station site can serve N different sectors. N is typically 3. A reuse pattern of N/K denotes a further division in frequency among N sector antennas per site. Some current and historical reuse patterns are 3/7 (North American AMPS), 6/4 (Motorola NAMPS), and 3/4 (GSM).

If the total available bandwidth is B, each cell can only use a number of frequency channels corresponding to a bandwidth of B/K, and each sector can use a bandwidth of B/NK.

Code division multiple access-based systems use a wider frequency band to achieve the same rate of transmission as FDMA, but this is compensated for by the ability to use a frequency reuse factor of 1, for example using a reuse pattern of 1/1. In other words, adjacent base station sites use the same frequencies, and the different base stations and users are separated by codes rather than frequencies. While N is shown as 1 in this example, that does not mean the CDMA cell has only one sector, but rather that the entire cell bandwidth is also available to each sector individually.

Depending on the size of the city, a taxi system may not have any frequency-reuse in its own city, but certainly in other nearby cities, the same frequency can be used. In a large city, on the other hand, frequency-reuse could certainly be in use.

Recently also orthogonal frequency-division multiple access based systems such as LTE are being deployed with a frequency reuse of 1. Since such systems do not spread the signal across the frequency band, inter-cell radio resource management is important to coordinate resource allocation between different cell sites and to limit the inter-cell interference. There are various means of Inter-Cell Interference Coordination (ICIC) already defined in the standard. Coordinated scheduling, multi-site MIMO or multi-site beam forming are other examples for inter-cell radio resource management that might be standardized in the future.

Directional Antennas

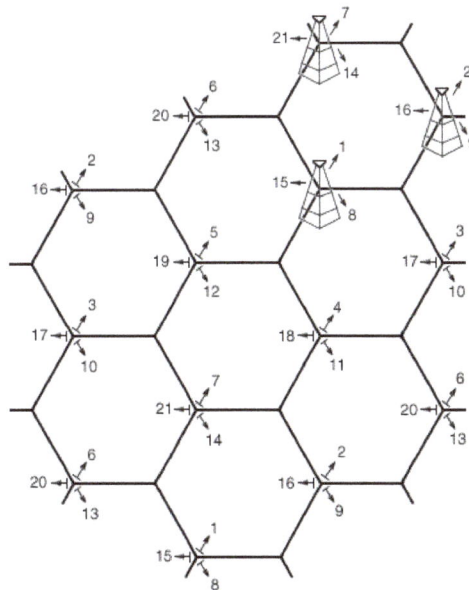

Cellular telephone frequency reuse pattern.

Cell towers frequently use a directional signal to improve reception in higher-traffic areas. In the United States, the FCC limits omnidirectional cell tower signals to 100 watts of power. If the tower has directional antennas, the FCC allows the cell operator to broadcast up to 500 watts of effective radiated power (ERP).

Cell phone companies use this directional signal to improve reception along highways and inside buildings like stadiums and arenas. As a result, a cell phone user may be standing in sight of a cell tower, but still have trouble getting a good signal because the directional antennas point in a different direction.

Although the original cell towers created an even, omnidirectional signal, were at the centers of the cells and were omnidirectional, a cellular map can be redrawn with the cellular telephone towers located at the corners of the hexagons where three cells converge. Each tower has three sets of directional antennas aimed in three different directions with 120 degrees for each cell (totaling 360 degrees) and receiving/transmitting into three different cells at different frequencies. This provides a minimum of three channels, and three towers for each cell and greatly increases the chances of receiving a usable signal from at least one direction.

The numbers in the illustration are channel numbers, which repeat every 3 cells. Large cells can be subdivided into smaller cells for high volume areas.

Broadcast Messages and Paging

Practically every cellular system has some kind of broadcast mechanism. This can be used directly for distributing information to multiple mobiles. Commonly, for example in mobile telephony systems, the most important use of broadcast information is to set up channels for one to one communication between the mobile transceiver and the base station. This is called paging. The three different paging procedures generally adopted are sequential, parallel and selective paging.

The details of the process of paging vary somewhat from network to network, but normally we know a limited number of cells where the phone is located (this group of cells is called a Location Area in the GSM or UMTS system, or Routing Area if a data packet session is involved; in LTE, cells are grouped into Tracking Areas). Paging takes place by sending the broadcast message to all of those cells. Paging messages can be used for information transfer. This happens in pagers, in CDMA systems for sending SMS messages, and in the UMTS system where it allows for low downlink latency in packet-based connections.

Movement from Cell to Cell and Handing Over

In a primitive taxi system, when the taxi moved away from a first tower and closer to a second tower, the taxi driver manually switched from one frequency to another as needed. If a communication was interrupted due to a loss of a signal, the taxi driver asked the base station operator to repeat the message on a different frequency.

In a cellular system, as the distributed mobile transceivers move from cell to cell during an ongoing continuous communication, switching from one cell frequency to a different cell frequency is done electronically without interruption and without a base station operator or manual switching. This is called the handover or handoff. Typically, a new channel is automatically selected for the mobile unit on the new base station which will serve it. The mobile unit then automatically switches from the current channel to the new channel and communication continues.

The exact details of the mobile system's move from one base station to the other varies considerably from system to system.

Mobile Phone Network

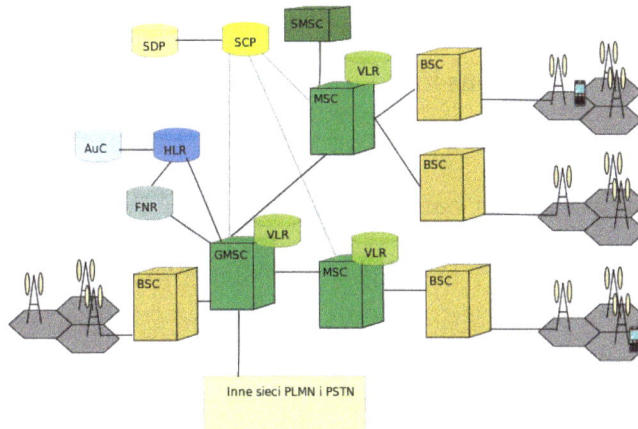

GSM network architecture

The most common example of a cellular network is a mobile phone (cell phone) network. A mobile phone is a portable telephone which receives or makes calls through a cell site (base station), or transmitting tower. Radio waves are used to transfer signals to and from the cell phone.

Modern mobile phone networks use cells because radio frequencies are a limited, shared resource. Cell-sites and handsets change frequency under computer control and use low power transmitters so that the usually limited number of radio frequencies can be simultaneously used by many callers with less interference.

A cellular network is used by the mobile phone operator to achieve both coverage and capacity for their subscribers. Large geographic areas are split into smaller cells to avoid line-of-sight signal loss and to support a large number of active phones in that area. All of the cell sites are connected to telephone exchanges (or switches), which in turn connect to the public telephone network.

In cities, each cell site may have a range of up to approximately $\frac{1}{2}$ mile (0.80 km), while in rural areas, the range could be as much as 5 miles (8.0 km). It is possible that in clear open areas, a user may receive signals from a cell site 25 miles (40 km) away.

Since almost all mobile phones use cellular technology, including GSM, CDMA, and AMPS (analog), the term "cell phone" is in some regions, notably the US, used interchangeably with "mobile phone". However, satellite phones are mobile phones that do not communicate directly with a ground-based cellular tower, but may do so indirectly by way of a satellite.

There are a number of different digital cellular technologies, including: Global System for Mobile Communications (GSM), General Packet Radio Service (GPRS), cdmaOne, CDMA2000, Evolution-Data Optimized (EV-DO), Enhanced Data Rates for GSM Evolution (EDGE), Universal Mobile Telecommunications System (UMTS), Digital Enhanced Cordless Telecommunications (DECT), Digital AMPS (IS-136/TDMA), and Integrated Digital Enhanced Network (iDEN). The transition from existing analog to the digital standard followed a very different path in Europe and the US. As a consequence multiple digital standard surfaced in the US, while Europe and many countries converged towards the GSM standard.

Structure of the Mobile Phone Cellular Network

A simple view of the cellular mobile-radio network consists of the following:

- A network of radio base stations forming the base station subsystem.

- The core circuit switched network for handling voice calls and text

- A packet switched network for handling mobile data

- The public switched telephone network to connect subscribers to the wider telephony network

This network is the foundation of the GSM system network. There are many functions that are performed by this network in order to make sure customers get the desired service including mobility management, registration, call set-up, and handover.

Any phone connects to the network via an RBS (Radio Base Station) at a corner of the corresponding cell which in turn connects to the Mobile switching center (MSC). The MSC provides a connection to the public switched telephone network (PSTN). The link from a phone to the RBS is called an *uplink* while the other way is termed *downlink*.

Radio channels effectively use the transmission medium through the use of the following multiplexing and access schemes: frequency division multiple access (FDMA), time division multiple access (TDMA), code division multiple access (CDMA), and space division multiple access (SDMA).

Small Cells

Small cells, which have a smaller coverage area than base stations, are categorised as follows:

- Microcell, less than 2 kilometres

- Picocell, less than 200 metres

- Femtocell, around 10 metres

Cellular Handover in Mobile Phone Networks

As the phone user moves from one cell area to another cell while a call is in progress, the mobile station will search for a new channel to attach to in order not to drop the call. Once a new channel is found, the network will command the mobile unit to switch to the new channel and at the same time switch the call onto the new channel.

With CDMA, multiple CDMA handsets share a specific radio channel. The signals are separated by using a pseudonoise code (PN code) specific to each phone. As the user moves from one cell to another, the handset sets up radio links with multiple cell sites (or sectors of the same site) simultaneously. This is known as "soft handoff" because, unlike with traditional cellular technology, there is no one defined point where the phone switches to the new cell.

In IS-95 inter-frequency handovers and older analog systems such as NMT it will typically be impossible to test the target channel directly while communicating. In this case other techniques have to be used such as pilot beacons in IS-95. This means that there is almost always a brief break in the communication while searching for the new channel followed by the risk of an unexpected return to the old channel.

If there is no ongoing communication or the communication can be interrupted, it is possible for the mobile unit to spontaneously move from one cell to another and then notify the base station with the strongest signal.

Cellular Frequency Choice in Mobile Phone Networks

The effect of frequency on cell coverage means that different frequencies serve better for different uses. Low frequencies, such as 450 MHz NMT, serve very well for countryside coverage. GSM 900 (900 MHz) is a suitable solution for light urban coverage. GSM 1800 (1.8 GHz) starts to be limited by structural walls. UMTS, at 2.1 GHz is quite similar in coverage to GSM 1800.

Higher frequencies are a disadvantage when it comes to coverage, but it is a decided advantage when it comes to capacity. Pico cells, covering e.g. one floor of a building, become possible, and the same frequency can be used for cells which are practically neighbours.

Cell service area may also vary due to interference from transmitting systems, both within and around that cell. This is true especially in CDMA based systems. The receiver requires a certain signal-to-noise ratio, and the transmitter should not send with too high transmission power in view to not cause interference with other transmitters. As the receiver moves away from the transmitter, the power received decreases, so the power control algorithm of the transmitter increases the power it transmits to restore the level of received power. As the interference (noise) rises above the received power from the transmitter, and the power of the transmitter cannot be increased any more, the signal becomes corrupted and eventually unusable. In CDMA-based systems, the effect of interference from other mobile transmitters in the same cell on coverage area is very marked and has a special name, *cell breathing*.

One can see examples of cell coverage by studying some of the coverage maps provided by real operators on their web sites or by looking at independently crowdsourced maps such as OpenSignal. In certain cases they may mark the site of the transmitter, in others it can be calculated by working out the point of strongest coverage.

A cellular repeater is used to extend cell coverage into larger areas. They range from wideband repeaters for consumer use in homes and offices to smart or digital repeaters for industrial needs.

Coverage Comparison of Different Frequencies

The following table shows the dependency of the coverage area of one cell on the frequency of a CDMA2000 network:

Frequency (MHz)	Cell radius (km)	Cell area (km2)	Relative Cell Count
450	48.9	7521	1
950	26.9	2269	3.3
1800	14.0	618	12.2
2100	12.0	449	16.2

Small Cell

Small cells are low-powered radio access nodes that operate in licensed and unlicensed spectrum that have a range of 10 meters to 1 or 2 kilometers. They are "small" compared to a mobile macrocell, which may have a range of a few tens of kilometers. With mobile operators struggling to support the growth in mobile data traffic, many are using mobile data offloading as a more efficient use of radio spectrum. Small cells are a vital element to 3G data offloading, and many mobile network operators see small cells as vital to managing LTE Advanced spectrum more efficiently compared to using just macrocells.

A small cell situated in the terrace of a building in Bangalore, Karnataka

Types of Small Cells

Small cells may encompass femtocells, picocells, and microcells. Small-cell networks can also be realized by means of distributed radio technology using centralized baseband units and remote radio heads. Beamforming technology (focusing a radio signal on a very specific area) can further enhance or focus small cell coverage. These approaches to small cells all feature central management by mobile network operators.

Small cells provide a small radio footprint, which can range from 10 meters within urban and in-building locations to 2 km for a rural location. Picocells and microcells can also have a range of a few hundred meters to a few kilometers, but they differ from femtocells in that they do not always have self-organising and self-management capabilities.

Small cells are available for a wide range of air interfaces including GSM, CDMA2000, TD-SCD-MA, W-CDMA, LTE and WiMax. In 3GPP terminology, a Home Node B (HNB) is a 3G femtocell. A Home eNode B (HeNB) is an LTE femtocell. Wi-Fi is a small cell but does not operate in licensed spectrum therefore cannot be managed as effectively as small cells utilising licensed spectrum. The detail and best practice associated with the deployment of small cells varies according to use case and radio technology employed.

Umbrella Term

The most common form of small cells are femtocells. They were initially designed for residential and small business use, with a short range and a limited number of channels. Femtocells with increased range and capacity spawned a proliferation of terms: metrocells, metro femtocells, public access femtocells, enterprise femtocells, super femtos, Class 3 femto, greater femtos and microcells. The term "small cells" is frequently used by analysts and the industry as an umbrella to describe the different implementations of femtocells, and to clear up any confusion that femtocells are limited to residential uses. Small cells are sometimes, incorrectly, also used to describe distributed-antenna systems (DAS) which are not low-powered access nodes.

Purpose

Small cells can be used to provide in-building and outdoor wireless service. Mobile operators use them to extend their service coverage and/or increase network capacity.

ABI Research argues that small cells also help service providers discover new revenue opportunities through their location and presence information. If a registered user enters a femtozone, the network is notified of their location. The service provider, with the user's permission, could share this location information to update user's social media status, for instance. Opening up small-cell APIs to the wider mobile ecosystem could enable a long-tail effect.

Rural coverage is also a key market that has developed as mobile operators have started to install public access metrocells in remote and rural areas that either have only 2G coverage or no coverage at all. The cost advantages of small cells compared with macro cells make it economically feasible to provide coverage of much smaller communities - from a few tens to a few hundreds. The Small Cell Forum have published a white paper outlining the technology and business case aspects. Mobile operators in both developing and developed world countries are either trialing or installing such systems. The pioneer in providing rural coverage using small cells was SoftBank Mobile - the Japanese mobile operator - who have installed more than 3000 public access small cells on post offices throughout rural Japan. To overcome the backhaul challenge in remote locations they have used VSAT satellite backhaul to link sites to their core network. The Informa Telecoms and Media consultancy also have a paper covering this use of small cells.

Future Mobile Networks

Small cells are an integral part of future LTE networks. In 3G networks, small cells are viewed as an offload technique. In 4G networks, the principle of heterogeneous network (HetNet) is introduced where the mobile network is constructed with layers of small and large cells. In LTE, all cells will be self-organizing, drawing upon the principles laid down in current Home NodeB (HNB), the 3GPP term for residential femtocells.

Future innovations in radio access design introduce the idea of an almost flat architecture where the difference between a small cell and a macrocell depends on how many cubes are stacked together. With software-defined radio, a base station could be 2G, 3G or 4G at the flick of a switch, and the antenna range can easily be tuned.

The transmitting signal from MBS weakened and worsen quicker once the Macro Base Station (MBS) signal reaches indoors. Femtocells provide way out to the difficulties present in macrocell-based system. So that, Femto Base Station (FBS) network coverage is one of the prime concerns in indoor environment to get good quality of service (QoS).

Deployment

In total, over 11 million small cells encompassing public, enterprise and residential have been deployed by 47 operators worldwide.

Small Cell Backhaul

Backhaul is needed to connect the small cells to the core network, internet and other services. Mobile operators consider this more challenging than macrocell backhaul because a) small cells are typically in hard-to-reach, near-street-level locations rather than in more open, above-rooftop locations and b) carrier grade connectivity must be provided at much lower cost per bit. In one survey, 55% operators listed backhaul as one of their biggest challenge for small cell rollout. Many different wireless and wired technologies have been proposed as solutions, and it is agreed that a 'toolbox' of these will be needed to address a range of deployment scenarios. An industry consensus view of how the different solution characteristics match with requirements is published by the Small Cell Forum. The backhaul solution is influenced by a number of factors, including the operator's original motivation to deploy small cells, which could be for targeted capacity, indoor or outdoor coverage .

In August 2013 the US Federal Communications Commission announced a change in its rules governing the 60 GHz (57–64 GHz) band, making it one of the key technologies for LTE backhaul.

Cellular Frequencies

Cellular frequencies are the sets of frequency ranges within the ultra high frequency band that have been assigned for cellular phone use. Most cellular phone networks worldwide use portions of the radio frequency spectrum, allocated to the mobile service, for the transmission and reception of their signals. The particular bands may also be shared with other radiocommunication services, e.g. broadcasting service, and fixed service operation.

Radio frequencies used for cellular networks differ in ITU Regions (Americas, Europe, Africa and Asia). The first commercial standard for mobile connection in the United States was AMPS, which was in the 800 MHz frequency band. In Nordic countries of Europe, the first widespread automatic mobile network was based on the NMT-450 standard, which was in the 450 MHz band. As mobile phones became more popular and affordable, mobile providers encountered a problem because they couldn't provide service to the increasing number of customers. They had to develop their existing networks and eventually introduce new standards, often based on other frequencies. Some European countries (and Japan) adopted TACS operating in 900 MHz. The GSM standard, which appeared in Europe to replace NMT-450 and other standards, initially used the 900 MHz band too. As demand grew, carriers acquired licenses in the 1,800 MHz band. (Generally speaking, lower frequencies allow carriers to provide coverage over a larger area, while higher frequencies allow carriers to provide service to more customers in a smaller area.)

In the U.S., the analog AMPS standard that used the cellular band (800 MHz) was replaced by a number of digital systems. Initially, systems based upon the AMPS mobile phone model were popular, including IS-95 (often known as "CDMA", the air interface technology it uses) and IS-136 (often known as D-AMPS, Digital AMPS, or "TDMA", the air interface technology it uses). Eventually, IS-136 on these frequencies was replaced by most operators with GSM. GSM had already been running for some time on US PCS (1,900 MHz) frequencies.

And, some NMT-450 analog networks have been replaced with digital networks using the same frequency. In Russia and some other countries, local carriers received licenses for 450 MHz frequency to provide CDMA mobile coverage area.

Many GSM phones support three bands (900/1,800/1,900 MHz or 850/1,800/1,900 MHz) or four bands (850/900/1,800/1,900 MHz), and are usually referred to as tri-band and quad-band phones, or world phones; with such a phone one can travel internationally and use the same handset. This portability is not as extensive with IS-95 phones, however, as IS-95 networks do not exist in most of Europe.

Mobile networks based on different standards may use the same frequency range; for example, AMPS, D-AMPS, N-AMPS and IS-95 all use the 800 MHz frequency band. Moreover, one can find both AMPS and IS-95 networks in use on the same frequency in the same area that do not interfere with each other. This is achieved by the use of different channels to carry data. The actual frequency used by a particular phone can vary from place to place, depending on the settings of the carrier's base station.

Types of Frequency Bands

GSM Frequency Bands

GSM frequency bands or frequency ranges are the cellular frequencies designated by the ITU for the operation of GSM mobile phones.

GSM band	f (MHz)	Uplink (MHz) (Mobile to Base)	Downlink (MHz) (Base to Mobile)	Channel number	Equivalent LTE band
T-GSM-380	380	380.2 − 389.8	390.2 − 399.8	dynamic	
T-GSM-410	410	410.2 − 419.8	420.2 − 429.8	dynamic	
GSM-450	450	450.6 − 457.6	460.6 − 467.6	259 − 293	31
GSM-480	480	479.0 − 486.0	489.0 − 496.0	306 − 340	

GSM-710	710	698.2 – 716.2	728.2 – 746.2	dynamic	12
GSM-750	750	777.2 – 792.2	747.2 – 762.2	438 – 511	
T-GSM-810	810	806.2 – 821.2	851.2 – 866.2	dynamic	27
GSM-850	850	824.2 – 849.2	869.2 – 893.8	128 – 251	5
P-GSM-900	900	890.0 – 915.0	935.0 – 960.0	1 – 124	
E-GSM-900	900	880.0 – 915.0	925.0 – 960.0	975 – 1023, 0 - 124	8
R-GSM-900	900	876.0 – 915.0	921.0 – 960.0	955 – 1023, 0 - 124	
T-GSM-900	900	870.4 – 876.0	915.4 – 921.0	dynamic	
DCS-1800	1800	1710.2 – 1784.8	1805.2 – 1879.8	512 – 885	3
PCS-1900	1900	1850.2 – 1909.8	1930.2 – 1989.8	512 – 810	2

- bands 2 and 5 (shaded in blue) have been deployed in NAR and CALA (North American Region [Canada and the US], Caribbean and Latin America)

- bands 3 and 8 (shaded in yellow) have been deployed in EMEA and APAC (Europe, the Middle East and Africa, Asia-Pacific)

- all other bands have not seen any commercial deployments

- P-GSM, Standard or Primary GSM-900 Band

- E-GSM, Extended GSM-900 Band (includes Standard GSM-900 band)

- R-GSM, Railways GSM-900 Band (includes Standard and Extended GSM-900 band)

- T-GSM, Trunking-GSM

GSM Frequency Usage Around the World

A dual-band 900/1800 phone is required to be compatible with most networks apart from deployments in ITU-Region 2.

GSM-900, EGSM/EGSM-900 and GSM-1800

GSM-900 and GSM-1800 are used in most parts of the world (ITU-Regions 1 and 3): Africa, Europe, Middle East, Asia (apart from Japan and South Korea where GSM has never been introduced) and Oceania.

In common GSM-900 is most widely used. Fewer operators use GSM-1800. Mobile Communication Services on Aircraft (MCA) uses GSM-1800.

In some countries GSM-1800 is also referred to as "Digital Cellular System" (DCS).

GSM-850 and GSM-1900

GSM-1900 and GSM-850 are used in most of North, South and Central America (ITU-Region 2). In North America, GSM operates on the primary mobile communication bands 850 MHz and

1900 MHz. In Canada, GSM-1900 is the primary band used in urban areas with 850 as a backup, and GSM-850 being the primary rural band. In the United States, regulatory requirements determine which area can use which band.

The term *Cellular* is sometimes used to describe GSM services in the 850 MHz band, because the original analog cellular mobile communication system was allocated in this spectrum. Further GSM-850 is also sometimes called *GSM-800* because this frequency range was known as the "800 MHz band" (for simplification) when it was first allocated for AMPS in the United States in 1983. In North America GSM-1900 is also referred to as Personal Communications Service (PCS) like any other cellular system operating on the "1900 MHz band".

Frequency Mixing between GSM 900/1800 and GSM 850/1900

Some countries in Central and South America have allocated spectrum in the 900 MHz and 1800 MHz bands for GSM in addition to the common GSM deployments at 850 MHz and 1900 MHz for ITU-Region 2 (Americas). The result hereof is a mixture of usage in the Americas that requires travelers to confirm that the phones they have are compatible with the band of the networks at their destinations. Frequency compatibility problems can be avoided through the use of multi-band (tri-band or, especially, quad-band) phones.

The following countries are mixing GSM 900/1800 and GSM 850/1900 bands:

Country	GSM-850	GSM-1900	GSM-900	GSM-1800
Antigua and Barbuda	Yes	Yes	Yes	No
Aruba, Bonaire and Curacao	No	Yes	Yes	Yes
Barbados	No	Yes	Yes	Yes
Brazil	Yes	Yes	Yes	Yes
British Virgin Islands	Yes	Yes	Yes	Yes
Cayman Islands	Yes	Yes	Yes	Yes
Costa Rica	Yes	No	No	Yes
Dominica	Yes	Yes	Yes	No
Dominican Republic	Yes	Yes	Yes	Yes
El Salvador	Yes	Yes	Yes	No
Grenada	Yes	No	Yes	Yes
Guatemala	Yes	Yes	Yes	No
Haiti	Yes	No	Yes	Yes
Jamaica	No	Yes	Yes	Yes
Saint Kitts and Nevis	Yes	Yes	Yes	Yes
Saint Lucia	Yes	Yes	Yes	Yes
Saint Vincent and the Grenadines	Yes	No	Yes	Yes
Trinidad and Tobago	Yes	Yes	No	Yes
Turks and Caicos Islands	Yes	Yes	Yes	Yes
Uruguay	Yes	Yes	Yes	Yes
Venezuela	Yes	No	Yes	Yes

GSM-450

Another less common GSM version is GSM-450. It uses the same band as, and can co-exist with, old analog NMT systems. NMT is a first generation (1G) mobile phone system which was primarily used in Nordic countries, Benelux, Alpine Countries, Eastern Europe and Russia prior to the introduction of GSM. The GSM Association claims one of its around 680 operator-members has a license to operate a GSM 450 network in Tanzania. However, currently all active public operators in Tanzania use GSM 900/1800 MHz. There are no publicly advertised handsets for GSM-450 available.

Very few NMT-450 network remain in operation. Overall, where the 450 MHz NMT band has been licensed, the original analogue network has been closed, and sometimes replaced by CDMA. Some of the CDMA networks have since upgraded from CDMA to LTE (LTE band 31).

Multi-band and Multi-mode Phones

Today, most telephones support multiple bands as used in different countries to facilitate roaming. These are typically referred to as multi-band phones. Dual-band phones can cover GSM networks in pairs such as 900 and 1800 MHz frequencies (Europe, Asia, Australia and Brazil) or 850 and 1900 (North America and Brazil). European tri-band phones typically cover the 900, 1800 and 1900 bands giving good coverage in Europe and allowing limited use in North America, while North American tri-band phones utilize 850, 1800 and 1900 for widespread North American service but limited worldwide use. A new addition has been the quad-band phone, also known as a World Phone, supporting at least all four major GSM bands, allowing for global use (excluding non-GSM countries such as Japan or South Korea).

There are also multi-mode phones which can operate on GSM as well as on other mobile phone systems using other technical standards or proprietary technologies. Often these phones use multiple frequency bands as well. For example, one version of the Nokia 6340i GAIT phone sold in North America can operate on GSM-1900, GSM-850 and legacy TDMA-1900, TDMA-800, and AMPS-800, making it both multi-mode and multi-band. As a more recent example the Apple iPhone 5 and iPhone 4S support quad-band GSM at 850/900/1800/1900 MHz, quad-band UMTS/HSDPA/HSUPA at 850/900/1900/2100 MHz, and dual-band CDMA EV-DO Rev. An at 800/1900 MHz, for a total of 'six' different frequencies (though at most four in a single mode). This allows the same handset to be sold for AT&T Mobility, Verizon, and Sprint in the U.S. as well as a broad range of GSM carriers worldwide such as Vodafone, Orange and T-Mobile (Excluding-US), many of whom offer official unlocking.

UMTS Frequency Bands

The UMTS frequency bands are radio frequencies used by third generation (3G) wireless Universal Mobile Telecommunications System networks. They were allocated by delegate to the World Administrative Radio Conference (WARC-92) held in Málaga-Torremolinos, Spain between February 3, 1992 and March 3, 1992. Resolution 212 (Rev.WRC-97), adopted at the World Radiocommunication Conference held in Geneva, Switzerland in 1997, endorsed the bands specifically for the International Mobile Telecommunications-2000 (IMT-2000) specification by referring to S5.388, which states "The bands 1,885-2,025 MHz and 2,110-2,200 MHz are intended for use, on a worldwide basis, by administrations wishing to implement International Mobile Telecommunications 2000 (IMT-2000). Such use does not preclude the use of these bands by other services to which they are allocated. The bands should be made available for IMT-2000 in accordance with Resolution 212 (Rev. WRC-97)." To accommodate the reality that these initially defined bands were already in use in various regions of the world, the initial allocation has been amended multiple times to include other radio frequency bands.

UMTS-FDD Frequency Bands and Channel Bandwidths

UMTS-FDD technology is standardized for usage in the following paired bands:

TS 25.101									
UTRA band	f (MHz)	Common name	Uplink frequencies UE transmit (MHz)	Downlink frequencies UE receive (MHz)	UARFCN UL channel number	UARFCN DL channel number	Duplex gap (MHz)	Center frequency range (MHz)	UARFCN equation (c = center freq. in MHz)
1	2100	IMT	1920 - 1980	2110 - 2170	9612 - 9888	10562 - 10838	190	2112.4 - 2167.6, increment = 0.2	5 * c
2	1900	PCS A-F	1850 - 1910	1930 - 1990	9262 - 9538 additional 12, 37, 62, 87, 112, 137, 162, 187, 212, 237, 262, 287	9662 - 9938 additional 412, 437, 462, 487, 512, 537, 562, 587, 612, 637, 662, 687	80	1932.4 - 1987.6, increment = 0.2 1932.5 - 1987.5, increment = 5	5 * c 5 * (c - 1850.1 MHz)
3	1800	DCS	1710 - 1785	1805 - 1880	937 - 1288	1162 - 1513	95	1807.4 - 1877.6, increment = 0.2	5 * (c - 1575 MHz)
4	1700	AWS A-F	1710 - 1755	2110 - 2155	1312 - 1513 additional 1662, 1687, 1712, 1737, 1762, 1787, 1812, 1837, 1862	1537 - 1738 additional 1887, 1912, 1937, 1962, 1987, 2012, 2037, 2062, 2087	400	2112.4 - 2152.6, increment = 0.2 2112.5 - 2152.5, increment = 5	5 * (c - 1805 MHz) 5 * (c - 1735.1 MHz)
5	850	CLR	824 - 849	869 - 894	4132 - 4233 additional 782, 787, 807, 812, 837, 862	4357 - 4458 additional 1007, 1012, 1032, 1037, 1062, 1087	45	871.4 - 891.6, increment = 0.2 871.5, 872.5, 876.5, 877.5, 882.5, 887.5	5 * c 5* (c - 670.1 MHz)
6	800		830 - 840	875 - 885	4162 - 4188 additional 812, 837	4387 - 4413 additional 1037, 1062	45	877.4 - 882.6, increment = 0.2 877.5, 882.5	5 * c 5 * (c - 670.1 MHz)

7	2600	IMT-E	2500 - 2570	2620 - 2690	2012 - 2338 additional 2362, 2387, 2412, 2437, 2462, 2487, 2512, 2537, 2562, 2587, 2612, 2637, 2662, 2687	2237 - 2563 additional 2587, 2612, 2637, 2662, 2687, 2712, 2737, 2762, 2787, 2812, 2837, 2862, 2887, 2912	120	2622.4 - 2687.6, increment = 0.2 2622.5 - 2687.5, increment = 5	$5 * (c - 2175 \text{ MHz})$ $5 * (c - 2105.1 \text{ MHz})$
8	900	E-GSM	880 - 915	925 - 960	2712 - 2863	2937 - 3088	45	927.4 - 957.6, increment = 0.2	$5 * (c - 340 \text{ MHz})$
9	1700		1749.9 - 1784.9	1844.9 - 1879.9	8762 - 8912	9237 - 9387	95	1847.4 - 1877.4, increment = 0.2	$5 * c$
10	1700	EAWS A-G	1710 - 1770	2110 - 2170	2887 - 3163 additional 3187, 3212, 3237, 3262, 3287, 3312, 3337, 3362, 3387, 3412, 3437, 3462	3112 - 3388 additional 3412, 3437, 3462, 3487, 3512, 3537, 3562, 3587, 3612, 3637, 3662, 3687	400	2112.4 - 2167.6, increment = 0.2 2112.5 - 2167.5, increment = 5	$5 * (c - 1490 \text{ MHz})$ $5 * (c - 1430.1 \text{ MHz})$
11	1500	LPDC	1427.9 - 1447.9	1475.9 - 1495.9	3487 - 3562	3712 - 3787	48	1478.4 - 1493.4, increment = 0.2	$5 * (c - 736 \text{ MHz})$
12	700	LSMH A/B/C	699 - 716	729 - 746	3617 - 3678 additional 3707, 3732, 3737, 3762, 3767	3842 - 3903 additional 3932, 3957, 3962, 3987, 3992	30	731.4 - 743.6, increment = 0.2 731.5, 736.5, 737.5, 742.5, 743.5	$5 * (c + 37 \text{ MHz})$ $5 * (c + 54.9 \text{ MHz})$
13	700	USMH C	777 - 787	746 - 756	3792 - 3818 additional 3842, 3867	4017 - 4043 additional 4067, 4092	31	748.4 - 753.6, increment = 0.2 748.5, 753.5	$5 * (c + 55 \text{ MHz})$ $5 * (c + 64.9 \text{ MHz})$
14	700	USMH D	788 - 798	758 - 768	3892 - 3918 additional 3942, 3967	4117 - 4143 additional 4167, 4192	30	760.4 - 765.6, increment = 0.2 760.5, 765.5	$5 * (c + 63 \text{ MHz})$ $5 * (c + 72.9 \text{ MHz})$
15		*Reserved*							
16		*Reserved*							
17		*Reserved*							
18		*Reserved*							

19	800		830 - 845	875 - 890	312 - 363 additional 387, 412, 437	712 - 763 additional 787, 812, 837	45	877.4-887.6, increment = 0.2 877.5, 882.5, 887.5	5 * (c - 735 MHz) 5 * (c - 720.1 MHz)
20	800	EUDD	832 - 862	791 - 821	4287 - 4413	4512 - 4638	41	793.4 - 818.6, increment = 0.2	5 * (c + 109 MHz)
21	1500	UPDC	1447.9 - 1462.9	1495.9 - 1510.9	462 - 512	862 - 912	48	1498.4-1508.4, increment = 0.2	5 * (c - 1326 MHz)
22	3500		3410 - 3490	3510 - 3590	4437 - 4813	4662 - 5038	100	3512.4-3587.6, increment = 0.2	5 * (c - 2580 MHz)
23		Reserved							
24		Reserved							
25	1900	EPCS A-G	1850 - 1915	1930 - 1995	4887 - 5188 additional 6067, 6092, 6117, 6142, 6167, 6192, 6217, 6242, 6267, 6292, 6317, 6342, 6367	5112 - 5413 additional 6292, 6317, 6342, 6367, 6392, 6417, 6442, 6467, 6492, 6517, 6542, 6567, 6592	80	1932.4-1992.6, increment = 0.2 1932.5-1992.5, increment = 5	5 * (c - 910 MHz) 5 * (c - 674.1 MHz)
26	850	ECLR	814 - 849	859 - 894	5537 - 5688 additional 5712, 5737, 5762, 5767, 5787, 5792, 5812, 5817, 5837, 5842, 5862	5762 - 5913 additional 5937, 5962, 5987, 5992, 6012, 6017, 6037, 6042, 6062, 6067, 6087	45	861.4-891.6, increment = 0.2 861.5, 866.5, 871.5, 872.5, 876.5, 877.5, 881.5, 882.5, 886.5, 887.5, 891.5	5 * (c + 291 MHz) 5 * (c + 325.9 MHz)
27		Reserved							
28		Reserved							
29		Reserved							
30		Reserved							
31		Reserved							
32 [A 1]	1500	L-band	N/A	1452 - 1496	N/A	6617 - 6813 additional 6837, 6862, 6887, 6912, 6937, 6962, 6987, 7012	N/A	1454.4-1493.6, increment = 0.2 1454.5-1489.5, increment = 5	5 * (c - 131 MHz) 5 * (c - 87.1 MHz)

1.   Band 32 is restricted to UTRA operation when dual band is configured (e.g., DB-DC-HSD-PA or dual band 4C-HSDPA).

Deployments by Region (UMTS-FDD)

The following table shows the standardized UMTS bands and their regional use. The main UMTS bands are in **bold** print.

- **Networks on UMTS-bands 1 and 8 are suitable for global roaming in ITU Regions 1, 2 (some countries) and 3.**

UTRA band	f (MHz)	Common name	North America	Latin America	Europe	Asia	Africa	Oceania
01	2100	IMT	No	Aruba (SetarNV), Uruguay (Ancel), Brazil, Costa Rica	Yes	Yes	Yes	Yes
02	1900	PCS A-F	Yes	Yes	No	No	No	No
04	1700	AWS A-F	USA & PR (T-Mobile) (Phase out), Canada (Eastlink, Vidéotron, Wind)	Chile (WOM)	No	No	No	No
05	850	CLR	Yes	Yes	No	Hong Kong (SmarTone), Israel (Cellcom, Pelephone), Philippines (SMART), Thailand (DTAC, True)	No	Australia (Telstra, VHA), New Zealand (Spark)
06	800		No	No	No	replaced by band 19	No	No
08	900	E-GSM	No	Dominican Republic (Orange), Paraguay (VOX), Venezuela (Digitel)	Yes	Yes	South Africa (Cell C)	Australia (Optus, VHA), New Zealand (2degrees, Vodafone)
09	1700		No	No	No	Japan (SoftBank Mobile)	No	No
11	1500	LPDC	No	No	No	Japan (SoftBank Mobile)	No	No
19	800		No	No	No	Japan (NTT docomo)	No	No

UMTS-TDD Frequency Bands and Channel Bandwidths

UMTS-TDD technology is standardized for usage in the following bands:

Operating band	Frequency band	Frequency (MHz)	UARFCN channel number
A (lower)	IMT	1900 - 1920	9504 - 9596

A (upper)	IMT	2010 - 2025	10054 - 10121
B (lower)	PCS	1850 - 1910	9254 - 9546
B (upper)	PCS	1930 - 1990	9654 - 9946
C	PCS (Duplex-Gap)	1910 - 1930	9554 - 9646
D	IMT-E	2570 - 2620	12854 - 13096
E		2300 - 2400	11504 - 11996
F		1880 - 1920	9404 - 9596

E-UTRA

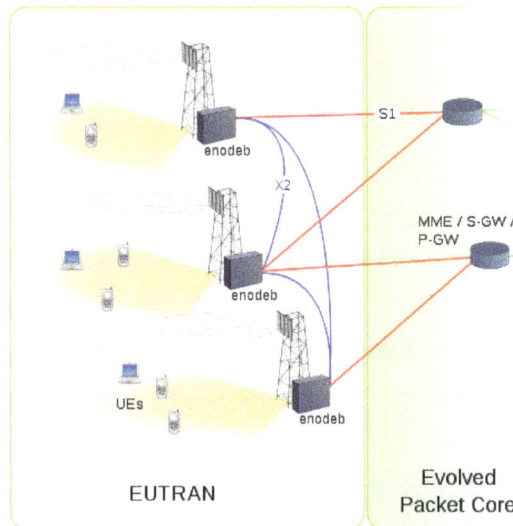

EUTRAN architecture as part of a LTE and SAE network

e-UTRA is the air interface of 3GPP's Long Term Evolution (LTE) upgrade path for mobile networks. It is an acronym for evolved UMTS Terrestrial Radio Access, also referred to as the 3GPP work item on the Long Term Evolution (LTE) also known as the Evolved Universal Terrestrial Radio Access (E-UTRA) in early drafts of the 3GPP LTE specification. E-UTRAN is the initialism of Evolved UMTS Terrestrial Radio Access Network and is the combination of E-UTRA, UEs and EnodeBs.

It is a radio access network which is referred to under the name EUTRAN standard meant to be a replacement of the UMTS and HSDPA/HSUPA technologies specified in 3GPP releases 5 and beyond. Unlike HSPA, LTE's E-UTRA is an entirely new air interface system, unrelated to and incompatible with W-CDMA. It provides higher data rates, lower latency and is optimized for packet data. It uses OFDMA radio-access for the downlink and SC-FDMA on the uplink. Trials started in 2008.

Features

EUTRAN has the following features:

- Peak download rates of 299.6 Mbit/s for 4×4 antennas, and 150.8 Mbit/s for 2×2 antennas with 20 MHz of spectrum. LTE Advanced supports 8×8 antenna configurations with peak download rates of 2,998.6 Mbit/s in an aggregated 100 MHz channel.

- Peak upload rates of 75.4 Mbit/s for a 20 MHz channel in the LTE standard, with up to 1,497.8 Mbit/s in an LTE Advanced 100 MHz carrier.

- Low data transfer latencies (sub-5 ms latency for small IP packets in optimal conditions), lower latencies for handover and connection setup time.

- Support for terminals moving at up to 350 km/h or 500 km/h depending on the frequency band.

- Support for both FDD and TDD duplexes as well as half-duplex FDD with the same radio access technology

- Support for all frequency bands currently used by IMT systems by ITU-R.

- Flexible bandwidth: 1.4 MHz, 3 MHz, 5 MHz, 10 MHz, 15 MHz and 20 MHz are standardized. By comparison, W-CDMA uses fixed size 5 MHz chunks of spectrum.

- Increased spectral efficiency at 2–5 times more than in 3GPP (HSPA) release 6

- Support of cell sizes from tens of meters of radius (femto and picocells) up to over 100 km radius macrocells

- Simplified architecture: The network side of EUTRAN is composed only by the enodeBs

- Support for inter-operation with other systems (e.g., GSM/EDGE, UMTS, CDMA2000, WiMAX, etc.)

- Packet switched radio interface.

Rationale for E-UTRA

Although UMTS, with HSDPA and HSUPA and their evolution, deliver high data transfer rates, wireless data usage is expected to continue increasing significantly over the next few years due to the increased offering and demand of services and content on-the-move and the continued reduction of costs for the final user. This increase is expected to require not only faster networks and radio interfaces but also higher cost-efficiency than what is possible by the evolution of the current standards. Thus the 3GPP consortium set the requirements for a new radio interface (EUTRAN) and core network evolution (System Architecture Evolution SAE) that would fulfill this need.

These improvements in performance allow wireless operators to offer *quadruple play* services - voice, high-speed interactive applications including large data transfer and feature-rich IPTV with full mobility.

Starting with the 3GPP Release 8, e-UTRA is designed to provide a single evolution path for the GSM/EDGE, UMTS/HSPA, CDMA2000/EV-DO and TD-SCDMA radio interfaces, providing increases in data speeds, and spectral efficiency, and allowing the provision of more functionality.

Architecture

EUTRAN consists only of enodeBs on the network side. The enodeB performs tasks similar to those performed by the nodeBs and RNC (radio network controller) together in UTRAN. The aim of this simplification is to reduce the latency of all radio interface operations. eNodeBs are connected to each other via the X2 interface, and they connect to the packet switched (PS) core network via the S1 interface.

EUTRAN Protocol Stack

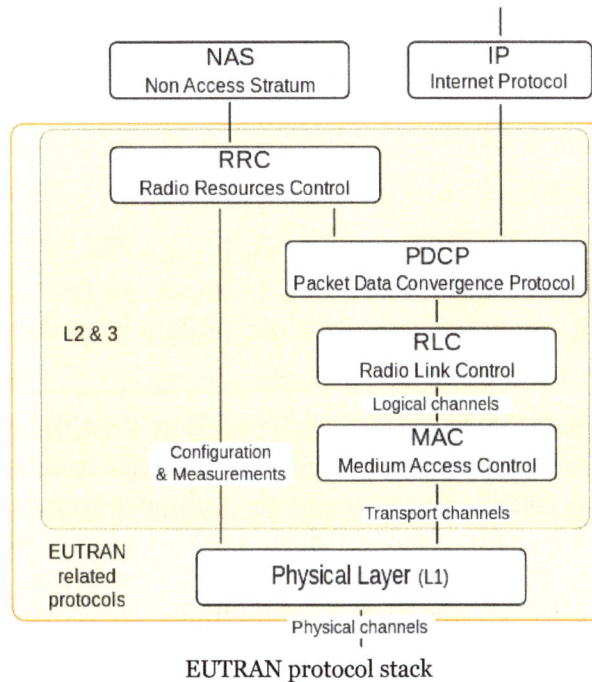

EUTRAN protocol stack

The EUTRAN protocol stack consist of:

- Physical layer: Carries all information from the MAC transport channels over the air interface. Takes care of the link adaptation (AMC), power control, cell search (for initial synchronization and handover purposes) and other measurements (inside the LTE system and between systems) for the RRC layer.

- MAC: The MAC sublayer offers a set of logical channels to the RLC sublayer that it multiplexes into the physical layer transport channels. It also manages the HARQ error correction, handles the prioritization of the logical channels for the same UE and the dynamic scheduling between UEs, etc..

- RLC: It transports the PDCP's PDUs. It can work in 3 different modes depending on the reliability provided. Depending on this mode it can provide: ARQ error correction, segmentation/concatenation of PDUs, reordering for in-sequence delivery, duplicate detection, etc...

- PDCP: For the RRC layer it provides transport of its data with ciphering and integrity protection. And for the IP layer transport of the IP packets, with ROHC header compression, ciphering, and depending on the RLC mode in-sequence delivery, duplicate detection and retransmission of its own SDUs during handover.

- RRC: Between others it takes care of: the broadcast system information related to the access stratum and transport of the non-access stratum (NAS) messages, paging, establishment and release of the RRC connection, security key management, handover, UE measurements related to inter-system (inter-RAT) mobility, QoS, etc..

Interfacing layers to the EUTRAN protocol stack:

- NAS: Protocol between the UE and the MME on the network side (outside of EUTRAN). Between others performs authentication of the UE, security control and generates part of the paging messages.

- IP

Physical Layer (L1) Design

E-UTRA uses orthogonal frequency-division multiplexing (OFDM), multiple-input multiple-output (MIMO) antenna technology depending on the terminal category and can use as well beamforming for the downlink to support more users, higher data rates and lower processing power required on each handset.

In the uplink LTE uses both OFDMA and a precoded version of OFDM called Single-Carrier Frequency-Division Multiple Access (SC-FDMA) depending on the channel. This is to compensate for a drawback with normal OFDM, which has a very high peak-to-average power ratio (PAPR). High PAPR requires more expensive and inefficient power amplifiers with high requirements on linearity, which increases the cost of the terminal and drains the battery faster. For the uplink, in release 8 and 9 multi user MIMO / Spatial division multiple access (SDMA) is supported; release 10 introduces also SU-MIMO.

In both OFDM and SC-FDMA transmission modes a cyclic prefix is appended to the transmitted symbols. Two different lengths of the cyclic prefix are available to support different channel spreads due to the cell size and propagation environment. These are a normal cyclic prefix of 4.7 µs, and an extended cyclic prefix of 16.6µs.

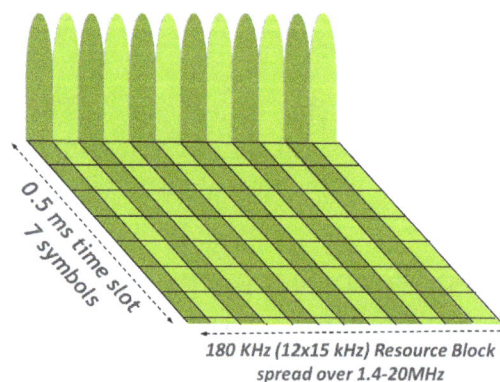

180 KHz (12x15 kHz) Resource Block spread over 1.4-20MHz

LTE *Resource Block* in time and frequency domains: 12 subcarriers, 0.5 ms timeslot (normal cyclic prefix).

LTE supports both Frequency-division duplex (FDD) and Time-division duplex (TDD) modes. While FDD makes use of paired spectra for UL and DL transmission separated by a duplex frequency gap, TDD splits one frequency carrier into alternating time periods for transmission from the base station to the terminal and vice versa. Both modes have their own frame structure within LTE and these are aligned with each other meaning that similar hardware can be used in the base stations and terminals to allow for economy of scale. The TDD mode in LTE is aligned with TD-SCDMA as well allowing for coexistence. Single chipsets are available which support both TDD-LTE and FDD-LTE operating modes.

The LTE transmission is structured in the time domain in radio frames. Each of these radio frames is 10 ms long and consists of 10 sub frames of 1 ms each. For non-MBMS subframes, the OFDMA sub-carrier spacing in the frequency domain is 15 kHz. Twelve of these sub-carriers together allocated during a 0.5 ms timeslot are called a resource block. A LTE terminal can be allocated, in the downlink or uplink, a minimum of 2 resources blocks during 1 subframe (1 ms).

All L1 transport data is encoded using turbo coding and a contention-free quadratic permutation polynomial (QPP) turbo code internal interleaver. L1 HARQ with 8 (FDD) or up to 15 (TDD) processes is used for the downlink and up to 8 processes for the UL

EUTRAN Physical Channels and Signals

Downlink (DL)

In the downlink there are several physical channels:

- The Physical Downlink Control Channel (PDCCH) carries between others the downlink allocation information, uplink allocation grants for the terminal.
- The Physical Control Format Indicator Channel (PCFICH) used to signal the length of the PDCCH.
- The Physical Hybrid ARQ Indicator Channel (PHICH) used to carry the acknowledges from the uplink transmissions.
- The Physical Downlink Shared Channel (PDSCH) is used for L1 transport data transmission. Supported modulation formats on the PDSCH are QPSK, 16QAM and 64QAM.
- The Physical Multicast Channel (PMCH) is used for broadcast transmission using a Single Frequency Network
- The Physical Broadcast Channel (PBCH) is used to broadcast the basic system information within the cell

And the following signals:

- The synchronization signals (PSS and SSS) are meant for the UE to discover the LTE cell and do the initial synchronization.
- The reference signals (cell specific, MBSFN, and UE specific) are used by the UE to estimate the DL channel.
- Positioning reference signals (PRS), added in release 9, meant to be used by the UE for OTDOA positioning (a type of multilateration)

Uplink (UL)

In the uplink there are three physical channels:

- Physical Random Access Channel (PRACH) is used for initial access and when the UE losses its uplink synchronization,

- Physical Uplink Shared Channel (PUSCH) carries the L1 UL transport data together with control information. Supported modulation formats on the PUSCH are QPSK, 16QAM and depending on the user equipment category 64QAM. PUSCH is the only channel, which because of its greater BW, uses SC-FDMA

- Physical Uplink Control Channel (PUCCH) carries control information. Note that the Uplink control information consists only on DL acknowledges as well as CQI related reports as all the UL coding and allocation parameters are known by the network side and signaled to the UE in the PDCCH.

And the following signals:

- Reference signals (RS) used by the enodeB to estimate the uplink channel to decode the terminal uplink transmission.

- Sounding reference signals (SRS) used by the enodeB to estimate the uplink channel conditions for each user to decide the best uplink scheduling.

User Equipment (UE) Categories

3GPP Release 8 defines five LTE user equipment categories depending on maximum peak data rate and MIMO capabilities support. With 3GPP Release 10, which is referred to as LTE Advanced, three new categories have been introduced, and four more with 3GPP Release 11.

User equipment Category	Max. L1 datarate Downlink (Mbit/s)	Max. number of DL MIMO layers	Max. L1 datarate Uplink (Mbit/s)	3GPP Release
0	1.0	1	1.0	Rel 12
1	10.3	1	5.2	Rel 8
2	51.0	2	25.5	Rel 8
3	102.0	2	51.0	Rel 8
4	150.8	2	51.0	Rel 8
5	299.6	4	75.4	Rel 8
6	301.5	2 or 4	51.0	Rel 10
7	301.5	2 or 4	102.0	Rel 10
8	2,998.6	8	1,497.8	Rel 10
9	452.2	2 or 4	51.0	Rel 11
10	452.2	2 or 4	102.0	Rel 11
11	603.0	2 or 4	51.0	Rel 11
12	603.0	2 or 4	102.0	Rel 11
13	391.7	2 or 4	150.8	Rel 12
14	3,917	8	N/A	Rel 12

15	750	4 or 2	N/A	Rel 12
16	979	4 or 2	N/A	Rel 12

Note: Maximum datarates shown are for 20 MHz of channel bandwidth. Categories 6 and above include datarates from combining multiple 20 MHz channels. Maximum datarates will be lower if less bandwidth is utilized.

Note: These are L1 transport data rates not including the different protocol layers overhead. Depending on cell BW, cell load, network configuration, the performance of the UE used, propagation conditions, etc. practical data rates will vary.

Note: The 3.0 Gbit/s / 1.5 Gbit/s data rate specified as Category 8 is near the peak aggregate data rate for a base station sector. A more realistic maximum data rate for a single user is 1.2 Gbit/s (downlink) and 600 Mbit/s (uplink). Nokia Siemens Networks has demonstrated downlink speeds of 1.4 Gbit/s using 100 MHz of aggregated spectrum.

EUTRAN Releases

As the rest of the 3GPP standard parts E-UTRA is structured in releases.

- Release 8, frozen in 2008, specified the first LTE standard

- Release 9, frozen in 2009, included some additions to the physical layer like dual layer (MIMO) beam-forming transmission or positioning support

- Release 10, frozen in 2011, introduces to the standard several LTE Advanced features like carrier aggregation, uplink SU-MIMO or relays, aiming to a considerable L1 peak data rate increase.

All LTE releases have been designed so far keeping backward compatibility in mind. That is, a release 8 compliant terminal will work in a release 10 network, while release 10 terminals would be able to use its extra functionality.

Technology Demos

- In September 2007, NTT Docomo demonstrated e-UTRA data rates of 200 Mbit/s with power consumption below 100 mW during the test.

- In April 2008, LG and Nortel demonstrated e-UTRA data rates of 50 Mbit/s while travelling at 110 km/h.

- February 15, 2008 - Skyworks Solutions has released a front-end module for e-UTRAN.

Core Network

A core network, or network core, is the central part of a telecommunications network that provides various services to customers who are connected by the access network. One of the main functions is to route telephone calls across the PSTN.

Typically the term refers to the high capacity communication facilities that connect primary nodes. A core/backbone network provides paths for the exchange of information between different sub-networks. For enterprise private networks serving one organization, the term backbone is more commonly used, while for service providers, the term core network is more common.

In the United States, local exchange core networks are linked by several competing interexchange networks; in the rest of the world, the core network has been extended to national boundaries.

Core/backbone networks usually have a mesh topology that provides any-to-any connections among devices on the network. Many main service providers would have their own core/backbone networks that are interconnected. Some large enterprises have their own core/backbone network, which are typically connected to the public networks.

The devices and facilities in the core / backbone networks are switches and routers. The trend is to push the intelligence and decision making into access and edge devices and keep the core devices dumb and fast. As a result, switches are more and more often used in the core/backbone network facilities. Technologies used in the core and backbone facilities are data link layer and network layer technologies such as SONET, DWDM, ATM, IP, etc. For enterprise backbone network, Gigabit Ethernet or 10 Gigabit Ethernet technologies are also often used.

Primary Functions

Core networks typically provide the following functionality:

1. Aggregation: The highest level of aggregation in a service provider network. The next level in the hierarchy under the core nodes is the distribution networks and then the edge networks. Customer-premises equipment (CPE) do not normally connect to the core networks of a large service provider.

2. Authentication: The function to decide whether the user requesting a service from the telecom network is authorized to do so within this network or not.

3. Call Control/Switching: call control or switching functionality decides the future course of call based on the call signalling processing. E.g. switching functionality may decide based on the "called number" that the call be routed towards a subscriber within this operator's network or with number portability more prevalent to another operator's network.

4. Charging: This functionality handles the collation and processing of charging data generated by various network nodes. Two common types of charging mechanisms found in present-day networks are prepaid charging and postpaid charging.

5. Service Invocation: Core network performs the task of service invocation for its subscribers. Service invocation may happen based on some explicit action (e.g. call transfer) by user or implicitly (call waiting). Its important to note however that service "execution" may or may not be a core network functionality as third party network/nodes may take part in actual service execution.

6. Gateways: Gateways shall be present in the core network to access other networks. Gateway functionality is dependent on the type of network it interfaces with.

Physically, one or more of these logical functionalities may simultaneously exist in a given core network node.

Other Functions

Besides above mentioned functionalities, the following also form part of a core network:

- O&M: Operations & Maintenance centre or Operations Support Systems to configure and provision the core network nodes. Number of subscribers, peak hour call rate, nature of services, geographical preferences are some of the factors which impact the configuration. Network statistics collection, alarm monitoring and logging of various network nodes actions also happens in the O&M centre. These stats, alarms and traces form important tools for a network operator to monitor the network health and performance and improvise on the same.

- Subscriber Database: Core network also hosts the subscribers database (e.g. HLR in GSM systems). Subscriber database is accessed by core network nodes for functions like authentication, service invocation etc.

Mobile

There exist basically two core network types for mobile telephony:

- The Mobile Application Part (MAP) used for GSM and UMTS

- The IS-41 core network used for D-AMPS (TDMA), cdmaOne and CDMA2000.

Both variants have evolved over time to integrate new services and air interfaces.

Handover

In cellular telecommunications, the terms handover or handoff refer to the process of transferring an ongoing call or data session from one channel connected to the core network to another channel. In satellite communications it is the process of transferring satellite control responsibility from one earth station to another without loss or interruption of service.

Handover or Handoff

American English use the term *handoff*, and this is most commonly used within some American organizations such as 3GPP2 and in American originated technologies such as CDMA2000. In British English the term *handover* is more common, and is used within international and European organisations such as ITU-T, IETF, ETSI and 3GPP, and standardised within European originated standards such as GSM and UMTS. The term handover is more common than handoff in academic research publications and literature, while handoff is slightly more common within the IEEE and ANSI organisations.

Purpose

In telecommunications there may be different reasons why a handover might be conducted:

- when the phone is moving away from the area covered by one cell and entering the area covered by another cell the call is transferred to the second cell in order to avoid call termination when the phone gets outside the range of the first cell;

- when the capacity for connecting new calls of a given cell is used up and an existing or new call from a phone, which is located in an area overlapped by another cell, is transferred to that cell in order to free-up some capacity in the first cell for other users, who can only be connected to that cell;

- in non-CDMA networks when the channel used by the phone becomes interfered by another phone using the same channel in a different cell, the call is transferred to a different channel in the same cell or to a different channel in another cell in order to avoid the interference;

- again in non-CDMA networks when the user behaviour changes, e.g. when a fast-travelling user, connected to a large, umbrella-type of cell, stops then the call may be transferred to a smaller macro cell or even to a micro cell in order to free capacity on the umbrella cell for other fast-traveling users and to reduce the potential interference to other cells or users (this works in reverse too, when a user is detected to be moving faster than a certain threshold, the call can be transferred to a larger umbrella-type of cell in order to minimize the frequency of the handovers due to this movement);

- in CDMA networks a handover may be induced in order to reduce the interference to a smaller neighboring cell due to the "near-far" effect even when the phone still has an excellent connection to its current cell; etc.

The most basic form of handover is when a phone call in progress is redirected from its current cell (called *source*) to a new cell (called *target*). In terrestrial networks the source and the target cells may be served from two different cell sites or from one and the same cell site (in the latter case the two cells are usually referred to as two *sectors* on that cell site). Such a handover, in which the source and the target are different cells (even if they are on the same cell site) is called *inter-cell* handover. The purpose of inter-cell handover is to maintain the call as the subscriber is moving out of the area covered by the source cell and entering the area of the target cell.

A special case is possible, in which the source and the target are one and the same cell and only the used channel is changed during the handover. Such a handover, in which the cell is not changed, is called *intra-cell* handover. The purpose of intra-cell handover is to change one channel, which may be interfered or fading with a new clearer or less fading channel.

Types

In addition to the above classification of *inter-cell* and *intra-cell* classification of handovers, they also can be divided into hard and soft handovers:

Hard handover

> Is one in which the channel in the source cell is released and only then the channel in the target cell is engaged. Thus the connection to the source is broken before or 'as' the connection to the target is made—for this reason such handovers are also known as *break-before-make*. Hard handovers are intended to be instantaneous in order to minimize the disruption to the call. A hard handover is perceived by network engineers as an event during the call. It requires the least processing by the network providing service. When the mobile is between base stations, then the mobile can switch with any of the base stations, so the base stations bounce the link with the mobile back and forth. This is called 'ping-ponging'.

Soft handover

> Is one in which the channel in the source cell is retained and used for a while in parallel with the channel in the target cell. In this case the connection to the target is established before the connection to the source is broken, hence this handover is called *make-before-break*. The interval, during which the two connections are used in parallel, may be brief or substantial. For this reason the soft handover is perceived by network engineers as a state of the call, rather than a brief event. Soft handovers may involve using connections to more than two cells: connections to three, four or more cells can be maintained by one phone at the same time. When a call is in a state of soft handover, the signal of the best of all used channels can be used for the call at a given moment or all the signals can be combined to produce a clearer copy of the signal. The latter is more advantageous, and when such combining is performed both in the downlink (forward link) and the uplink (reverse link) the handover is termed as *softer*. Softer handovers are possible when the cells involved in the handovers have a single cell site.

Handover can also be classified on the basis of handover techniques used. Broadly they can be classified into three types:

1. Network controlled handover

2. Mobile phone assisted handover

3. Mobile controlled handover

Comparison

An advantage of the hard handover is that at any moment in time one call uses only one channel. The hard handover event is indeed very short and usually is not perceptible by the user. In the old analog systems it could be heard as a click or a very short beep; in digital systems it is unnoticeable. Another advantage of the hard handover is that the phone's hardware does not need to be capable of receiving two or more channels in parallel, which makes it cheaper and simpler. A disadvantage is that if a handover fails the call may be temporarily disrupted or even terminated abnormally. Technologies which use hard handovers, usually have procedures which can re-establish the connection to the source cell if the connection to the target cell cannot be made. However re-establishing this connection may not always be possible (in which case the call will be terminated) and even when possible the procedure may cause a temporary interruption to the call.

One advantage of the soft handovers is that the connection to the source cell is broken only when a reliable connection to the target cell has been established and therefore the chances that the call will be terminated abnormally due to failed handovers are lower. However, by far a bigger advantage comes from the mere fact that simultaneously channels in multiple cells are maintained and the call could only fail if all of the channels are interfered or fade at the same time. Fading and interference in different channels are unrelated and therefore the probability of them taking place at the same moment in all channels is very low. Thus the reliability of the connection becomes higher when the call is in a soft handover. Because in a cellular network the majority of the handovers occur in places of poor coverage, where calls would frequently become unreliable when their channel is interfered or fading, soft handovers bring a significant improvement to the reliability of the calls in these places by making the interference or the fading in a single channel not critical. This advantage comes at the cost of more complex hardware in the phone, which must be capable of processing several channels in parallel. Another price to pay for soft handovers is use of several channels in the network to support just a single call. This reduces the number of remaining free channels and thus reduces the capacity of the network. By adjusting the duration of soft handovers and the size of the areas in which they occur, the network engineers can balance the benefit of extra call reliability against the price of reduced capacity.

Possibility

While theoretically speaking soft handovers are possible in any technology, analog or digital, the cost of implementing them for analog technologies is prohibitively high and none of the technologies that were commercially successful in the past (e.g. AMPS, TACS, NMT, etc.) had this feature. Of the digital technologies, those based on FDMA also face a higher cost for the phones (due to the need to have multiple parallel radio-frequency modules) and those based on TDMA or a combination of TDMA/FDMA, in principle, allow not so expensive implementation of soft handovers. However, none of the 2G (second-generation) technologies have this feature (e.g. GSM, D-AMPS/IS-136, etc.). On the other hand, all CDMA based technologies, 2G and 3G (third-generation), have soft handovers. On one hand, this is facilitated by the possibility to design not so expensive phone hardware supporting soft handovers for CDMA and on the other hand, this is necessitated by the fact that without soft handovers CDMA networks may suffer from substantial interference arising due to the so-called *near-far* effect..

In all current commercial technologies based on FDMA or on a combination of TDMA/FDMA (e.g. GSM, AMPS, IS-136/DAMPS, etc.) changing the channel during a hard handover is realised by changing the pair of used transmit/receive frequencies.

Implementations

For the practical realization of handovers in a cellular network each cell is assigned a list of potential target cells, which can be used for handing over calls from this source cell to them. These potential target cells are called *neighbors* and the list is called *neighbor list*. Creating such a list for a given cell is not trivial and specialized computer tools are used. They implement different algorithms and may use for input data from field measurements or computer predictions of radio wave propagation in the areas covered by the cells.

During a call one or more parameters of the signal in the channel in the source cell are monitored and assessed in order to decide when a handover may be necessary. The downlink (forward link) and/or uplink (reverse link) directions may be monitored. The handover may be requested by the phone or by the base station (BTS) of its source cell and, in some systems, by a BTS of a neighboring cell. The phone and the BTSes of the neighboring cells monitor each other's signals and the best target candidates are selected among the neighboring cells. In some systems, mainly based on CDMA, a target candidate may be selected among the cells which are not in the neighbor list. This is done in an effort to reduce the probability of interference due to the aforementioned near-far effect.

In analog systems the parameters used as criteria for requesting a hard handover are usually the *received signal power* and the *received signal-to-noise ratio* (the latter may be estimated in an analog system by inserting additional tones, with frequencies just outside the captured voice-frequency band at the transmitter and assessing the form of these tones at the receiver). In non-CDMA 2G digital systems the criteria for requesting hard handover may be based on estimates of the received signal power, bit error rate (BER) and block error/erasure rate (BLER), received quality of speech (RxQual), distance between the phone and the BTS (estimated from the radio signal propagation delay) and others. In CDMA systems, 2G and 3G, the most common criterion for requesting a handover is Ec/Io ratio measured in the pilot channel (CPICH) and/or RSCP.

In CDMA systems, when the phone in soft or softer handover is connected to several cells simultaneously, it processes the received in parallel signals using a rake receiver. Each signal is processed by a module called *rake finger*. A usual design of a rake receiver in mobile phones includes three or more rake fingers used in soft handover state for processing signals from as many cells and one additional finger used to search for signals from other cells. The set of cells, whose signals are used during a soft handover, is referred to as the *active set*. If the search finger finds a sufficiently-strong signal (in terms of high Ec/Io or RSCP) from a new cell this cell is added to the active set. The cells in the neighbour list (called in CDMA *neighbouring set*) are checked more frequently than the rest and thus a handover with a neighbouring cell is more likely, however a handover with others cells outside the neighbor list is also allowed (unlike in GSM, IS-136/DAMPS, AMPS, NMT, etc.).

Reasons for Failure

There are occurrences where a handoff is unsuccessful. Lots of research was conducted regarding this. In the late 80's the main reason was found out. Because frequencies cannot be reused in adjacent cells, when a user moves from one cell to another, a new frequency must be allocated for the call. If a user moves into a cell when all available channels are in use, the user's call must be terminated. Also, there is the problem of signal interference where adjacent cells overpower each other resulting in receiver desensitization.

Vertical Handover

There are also inter-technology handovers where a call's connection is transferred from one access technology to another, e.g. a call being transferred from GSM to UMTS or from CDMA IS-95 to cdma2000.

The 3GPP UMA/GAN standard enables GSM/UMTS handoff to Wi-Fi and vice versa.

Handoff Prioritization

Different systems have different methods for handling and managing handoff request. Some systems handle handoff in same way as they handle new originating call. In such system the probability that the handoff will not be served is equal to blocking probability of new originating call. But if the call is terminated abruptly in the middle of conversation then it is more annoying than the new originating call being blocked. So in order to avoid this abrupt termination of ongoing call handoff request should be given priority to new call this is called as handoff prioritization.

There are two techniques for this:

Guard Channel Concept

> In this technique, a fraction of the total available channel in a cell is reserved exclusively for handoff request from ongoing calls which may be handed off into the cell.

Queuing

> Queuing of handoffs is possible because there is a finite time interval between the time the received signal level drops below handoff threshold and the time the call is terminated due to insufficient signal level. The delay size is determined from the traffic pattern of a particular service area.

Inter and Intra System Handoff

Inter System Handoff

- If during ongoing call mobile unit moves from one cellular system to a different cellular system which is controlled by different MTSO, a handoff procedure which is used to avoid dropping of call is referred as Inter System Handoff.

- An MTSO engages in this handoff system. When a mobile signal becomes weak in a given cell and MTSO can not find other cell within its system to which it can transfer the call then it uses Inter system handoff.

- Before implementation of Inter System Handoff MTSO compatibility must be checked and in Inter System Handoff local call may become long distance call.

Intra System Handoff

- If during ongoing call mobile unit moves from one cellular system to adjacent cellular system which is controlled by same MTSO, a handoff procedure which is used to avoid dropping of call is referred as Intra System Handoff.

- An MTSO engages in this handoff system. When a mobile signal becomes weak in a given cell and MTSO finds other cell within its system to which it can transfer the call then it uses Intra system handoff.

- In Intra System Handoff local calls always remain local call only since after handoff also the call is handled by same MTSO.

Network Switching Subsystem

Network switching subsystem (NSS) (or GSM core network) is the component of a GSM system that carries out call switching and mobility management functions for mobile phones roaming on the network of base stations. It is owned and deployed by mobile phone operators and allows mobile devices to communicate with each other and telephones in the wider public switched telephone network (PSTN). The architecture contains specific features and functions which are needed because the phones are not fixed in one location.

The NSS originally consisted of the circuit-switched core network, used for traditional GSM services such as voice calls, SMS, and circuit switched data calls. It was extended with an overlay architecture to provide packet-switched data services known as the GPRS core network. This allows mobile phones to have access to services such as WAP, MMS and the Internet.

Mobile Switching Center (MSC)

Description

The mobile switching center (MSC) is the primary service delivery node for GSM/CDMA, responsible for routing voice calls and SMS as well as other services (such as conference calls, FAX and circuit switched data).

The MSC sets up and releases the end-to-end connection, handles mobility and hand-over requirements during the call and takes care of charging and real time pre-paid account monitoring.

In the GSM mobile phone system, in contrast with earlier analogue services, fax and data information is sent directly digitally encoded to the MSC. Only at the MSC is this re-coded into an "analogue" signal (although actually this will almost certainly mean sound encoded digitally as PCM signal in a 64-kbit/s timeslot, known as a DS0 in America).

There are various different names for MSCs in different contexts which reflects their complex role in the network, all of these terms though could refer to the same MSC, but doing different things at different times.

The Gateway MSC (G-MSC) is the MSC that determines which "visited MSC (V-MSC)" the subscriber who is being called is currently located at. It also interfaces with the PSTN. All mobile to mobile calls and PSTN to mobile calls are routed through a G-MSC. The term is only valid in the context of one call since any MSC may provide both the gateway function and the Visited MSC function, however, some manufacturers design dedicated high capacity MSCs which do not have any BSSs connected to them. These MSCs will then be the Gateway MSC for many of the calls they handle.

The visited MSC (V-MSC) is the MSC where a customer is currently located. The VLR associated with this MSC will have the subscriber's data in it.

The anchor MSC is the MSC from which a handover has been initiated. The target MSC is the MSC toward which a Handover should take place. A mobile switching center server is a part of the redesigned MSC concept starting from 3GPP Release 4.

Mobile Switching Center Server (MSC-Server, MSCS or MSS)

The mobile switching center server is a soft-switch variant (therefore it may be referred as Mobile Soft Switch, MSS) of the mobile switching center, which provides circuit-switched calling mobility management, and GSM services to the mobile phones roaming within the area that it serves. MSS functionality enables split between control (signalling) and user plane (bearer in network element called as media gateway/MG), which guarantees better placement of network elements within the network.

MSS and MGW media gateway makes it possible to cross-connect circuit switched calls switched by using IP, ATM AAL2 as well as TDM. More information is available in 3GPP TS 23.205.

Circuit switching (CS) term used here originates from the traditional telecommunications systems. However, modern MSS and MGW devices mostly use generic Internet technologies and form next-generation telecommunication networks. MSS software may run on generic computers or virtual machines in cloud environment.

Other GSM Core Network Elements connected to the MSC

The MSC connects to the following elements:

- The home location register (HLR) for obtaining data about the SIM and mobile services ISDN number (MSISDN; i.e., the telephone number).

- The base station subsystem (BSS) which handles the radio communication with 2G and 2.5G mobile phones.

- The UMTS terrestrial radio access network (UTRAN) which handles the radio communication with 3G mobile phones.

- The visitor location register (VLR) provides subscriber information when the subscriber is outside its home network.

- Other MSCs for procedures such as handover.

Procedures Implemented

Tasks of the MSC include:

- Delivering calls to subscribers as they arrive based on information from the VLR.

- Connecting outgoing calls to other mobile subscribers or the PSTN.

- Delivering SMSs from subscribers to the short message service center (SMSC) and vice versa.

- Arranging handovers from BSC to BSC.

- Carrying out handovers from this MSC to another.

- Supporting supplementary services such as conference calls or call hold.

- Generating billing information.

Home Location Register (HLR)

The home location register (HLR) is a central database that contains details of each mobile phone subscriber that is authorized to use the GSM core network. There can be several logical, and physical, HLRs per public land mobile network (PLMN), though one international mobile subscriber identity (IMSI)/MSISDN pair can be associated with only one logical HLR (which can span several physical nodes) at a time.

The HLRs store details of every SIM card issued by the mobile phone operator. Each SIM has a unique identifier called an IMSI which is the primary key to each HLR record.

Another important item of data associated with the SIM are the MSISDNs, which are the telephone numbers used by mobile phones to make and receive calls. The primary MSISDN is the number used for making and receiving voice calls and SMS, but it is possible for a SIM to have other secondary MSISDNs associated with it for fax and data calls. Each MSISDN is also a primary key to the HLR record. The HLR data is stored for as long as a subscriber remains with the mobile phone operator.

Examples of other data stored in the HLR against an IMSI record is:

- GSM services that the subscriber has requested or been given.

- GPRS settings to allow the subscriber to access packet services.

- Current location of subscriber (VLR and serving GPRS support node/SGSN).

- Call divert settings applicable for each associated MSISDN.

The HLR is a system which directly receives and processes MAP transactions and messages from elements in the GSM network, for example, the location update messages received as mobile phones roam around.

Other GSM Core Network Elements connected to the HLR

The HLR connects to the following elements:
- The G-MSC for handling incoming calls
- The VLR for handling requests from mobile phones to attach to the network
- The SMSC for handling incoming SMSs
- The voice mail system for delivering notifications to the mobile phone that a message is waiting
- The AuC for authentication and ciphering and exchange of data (triplets)

Procedures Implemented

The main function of the HLR is to manage the fact that SIMs and phones move around a lot. The following procedures are implemented to deal with this:

- Manage the mobility of subscribers by means of updating their position in administrative areas called 'location areas', which are identified with a LAC. The action of a user of moving from one LA to another is followed by the HLR with a Location area update procedure.

- Send the subscriber data to a VLR or SGSN when a subscriber first roams there.

- Broker between the G-MSC or SMSC and the subscriber's current VLR in order to allow incoming calls or text messages to be delivered.

- Remove subscriber data from the previous VLR when a subscriber has roamed away from it.

- Responsible for all SRI related queries (i.e. for invoke SRI, HLR should give sack SRI or SRI reply).

Authentication Center (AuC)

Description

The authentication center (AuC) is a function to authenticate each SIM card that attempts to connect to the GSM core network (typically when the phone is powered on). Once the authentication is successful, the HLR is allowed to manage the SIM and services described above. An encryption key is also generated that is subsequently used to encrypt all wireless communications (voice, SMS, etc.) between the mobile phone and the GSM core network.

If the authentication fails, then no services are possible from that particular combination of SIM card and mobile phone operator attempted. There is an additional form of identification check performed on the serial number of the mobile phone described in the EIR section below, but this is not relevant to the AuC processing.

Proper implementation of security in and around the AuC is a key part of an operator's strategy to avoid SIM cloning.

The AuC does not engage directly in the authentication process, but instead generates data known as *triplets* for the MSC to use during the procedure. The security of the process depends upon a shared secret between the AuC and the SIM called the K_i. The K_i is securely burned into the SIM during manufacture and is also securely replicated onto the AuC. This K_i is never transmitted between the AuC and SIM, but is combined with the IMSI to produce a challenge/response for identification purposes and an encryption key called K_c for use in over the air communications.

Other GSM Core Network Elements connected to the AuC

The AuC connects to the following elements:

- The MSC which requests a new batch of triplet data for an IMSI after the previous data have been used. This ensures that same keys and challenge responses are not used twice for a particular mobile.

Procedures Implemented

The AuC stores the following data for each IMSI:

- the K_i

- Algorithm id. (the standard algorithms are called A3 or A8, but an operator may choose a proprietary one).

When the MSC asks the AuC for a new set of triplets for a particular IMSI, the AuC first generates a random number known as *RAND*. This *RAND* is then combined with the K_i to produce two numbers as follows:

- The K_i and *RAND* are fed into the A3 algorithm and the signed response (SRES) is calculated.

- The K_i and *RAND* are fed into the A8 algorithm and a session key called K_c is calculated.

The numbers (*RAND*, SRES, K_c) form the triplet sent back to the MSC. When a particular IMSI requests access to the GSM core network, the MSC sends the *RAND* part of the triplet to the SIM. The SIM then feeds this number and the K_i (which is burned onto the SIM) into the A3 algorithm as appropriate and an SRES is calculated and sent back to the MSC. If this SRES matches with the SRES in the triplet (which it should if it is a valid SIM), then the mobile is allowed to attach and proceed with GSM services.

After successful authentication, the MSC sends the encryption key K_c to the base station controller (BSC) so that all communications can be encrypted and decrypted. Of course, the mobile phone can generate the K_c itself by feeding the same RAND supplied during authentication and the K_i into the A8 algorithm.

The AuC is usually collocated with the HLR, although this is not necessary. Whilst the procedure is secure for most everyday use, it is by no means crack proof. Therefore, a new set of security methods was designed for 3G phones.

A3 Algorithm is used to encrypt Global System for Mobile Communications (GSM) cellular communications. In practice, A3 and A8 algorithms are generally implemented together (known as A3/A8). An A3/A8 algorithm is implemented in Subscriber Identity Module (SIM) cards and in GSM network Authentication Centers. It is used to authenticate the customer and generate a key for encrypting voice and data traffic, as defined in 3GPP TS 43.020 (03.20 before Rel-4). Development of A3 and A8 algorithms is considered a matter for individual GSM network operators, although example implementations are available.

Visitor Location Register (VLR)

Description

The Visitor Location Register (VLR) is a database of the subscribers who have roamed into the jurisdiction of the MSC (Mobile Switching Center) which it serves. Each main base station in the network is served by exactly one VLR (one BTS may be served by many MSCs in case of MSC in pool), hence a subscriber cannot be present in more than one VLR at a time.

The data stored in the VLR has either been received from the HLR, or collected from the MS (Mobile station). In practice, for performance reasons, most vendors integrate the VLR directly to the V-MSC and, where this is not done, the VLR is very tightly linked with the MSC via a proprietary interface. Whenever an MSC detects a new MS in its network, in addition to creating a new record in the VLR, it also updates the HLR of the mobile subscriber, apprising it of the new location of that MS. If VLR data is corrupted it can lead to serious issues with text messaging and call services.

Data stored include:

- IMSI (the subscriber's identity number).

- Authentication data.

- MSISDN (the subscriber's phone number).

- GSM services that the subscriber is allowed to access.

- access point (GPRS) subscribed.

- The HLR address of the subscriber.

- SCP Address(For Prepaid Subscriber).

Procedures Implemented

The primary functions of the VLR are:

- To inform the HLR that a subscriber has arrived in the particular area covered by the VLR.

- To track where the subscriber is within the VLR area (location area) when no call is ongoing.

- To allow or disallow which services the subscriber may use.

- To allocate roaming numbers during the processing of incoming calls.

- To purge the subscriber record if a subscriber becomes inactive whilst in the area of a VLR. The VLR deletes the subscriber's data after a fixed time period of inactivity and informs the HLR (e.g., when the phone has been switched off and left off or when the subscriber has moved to an area with no coverage for a long time).

- To delete the subscriber record when a subscriber explicitly moves to another, as instructed by the HLR.

Equipment Identity Register (EIR)

The equipment identity register is often integrated to the HLR. The EIR keeps a list of mobile phones (identified by their IMEI) which are to be banned from the network or monitored. This is designed to allow tracking of stolen mobile phones. In theory all data about all stolen mobile phones should be distributed to all EIRs in the world through a Central EIR. It is clear, however, that there are some countries where this is not in operation. The EIR data does not have to change in real time, which means that this function can be less distributed than the function of the HLR. The EIR is a database that contains information about the identity of the mobile equipment that prevents calls from stolen, unauthorized or defective mobile stations. Some EIR also have the capability to log Handset attempts and store it in a log file.

Other Support Functions

Connected more or less directly to the GSM core network are many other functions.

Billing Center (BC)

The billing center is responsible for processing the toll tickets generated by the VLRs and HLRs and generating a bill for each subscriber. It is also responsible for generating billing data of roaming subscriber.

Multimedia Messaging Service Center (MMSC)

The multimedia messaging service center supports the sending of multimedia messages (e.g., images, audio, video and their combinations) to (or from) MMS-enabled Handsets.

Voicemail System (VMS)

The voicemail system records and stores voicemail. which may have to pay

Lawful Interception Functions

According to U.S. law, which has also been copied into many other countries, especially in Europe, all telecommunications equipment must provide facilities for monitoring the calls of selected users. There must be some level of support for this built into any of the different elements. The concept of *lawful interception* is also known, following the relevant U.S. law, as CALEA. Generally, lawful Interception implementation is similar to the implementation of conference call. While A and B are talking with each other, C can join the call and listen silently.

General Packet Radio Service

General Packet Radio Service (GPRS) is a packet oriented mobile data service on the 2G and 3G cellular communication system's global system for mobile communications (GSM). GPRS was originally standardized by European Telecommunications Standards Institute (ETSI) in response to the earlier CDPD and i-mode packet-switched cellular technologies. It is now maintained by the 3rd Generation Partnership Project (3GPP).

GPRS usage is typically charged based on volume of data transferred, contrasting with circuit switched data, which is usually billed per minute of connection time. Sometimes billing time is broken down to every third of a minute. Usage above the bundle cap is charged per megabyte, speed limited, or disallowed.

GPRS is a best-effort service, implying variable throughput and latency that depend on the number of other users sharing the service concurrently, as opposed to circuit switching, where a certain quality of service (QoS) is guaranteed during the connection. In 2G systems, GPRS provides data rates of 56–114 kbit/second. 2G cellular technology combined with GPRS is sometimes described as *2.5G*, that is, a technology between the second (2G) and third (3G) generations of mobile telephony. It provides moderate-speed data transfer, by using unused time division multiple access (TDMA) channels in, for example, the GSM system. GPRS is integrated into GSM Release 97 and newer releases.

Technical Overview

The GPRS core network allows 2G, 3G and WCDMA mobile networks to transmit IP packets to external networks such as the Internet. The GPRS system is an integrated part of the GSM network switching subsystem.

Services Offered

GPRS extends the GSM Packet circuit switched data capabilities and makes the following services possible:

- SMS messaging and broadcasting

- "Always on" internet access

- Multimedia messaging service (MMS)

- Push-to-talk over cellular (PoC)

- Instant messaging and presence—wireless village

- Internet applications for smart devices through wireless application protocol (WAP)

- Point-to-point (P2P) service: inter-networking with the Internet (IP)

- Point-to-multipoint (P2M) service: point-to-multipoint multicast and point-to-multipoint group calls

If SMS over GPRS is used, an SMS transmission speed of about 30 SMS messages per minute may be achieved. This is much faster than using the ordinary SMS over GSM, whose SMS transmission speed is about 6 to 10 SMS messages per minute.

Protocols Supported

GPRS supports the following protocols:

- Internet Protocol (IP). In practice, built-in mobile browsers use IPv4 since IPv6 was not yet popular.

- Point-to-Point Protocol (PPP). In this mode PPP is often not supported by the mobile phone operator but if the mobile is used as a modem to the connected computer, PPP is used to tunnel IP to the phone. This allows an IP address to be assigned dynamically (IPCP not DHCP) to the mobile equipment.

- X.25 connections. This is typically used for applications like wireless payment terminals, although it has been removed from the standard. X.25 can still be supported over PPP, or even over IP, but doing this requires either a network-based router to perform encapsulation or intelligence built into the end-device/terminal; e.g., user equipment (UE).

When TCP/IP is used, each phone can have one or more IP addresses allocated. GPRS will store and forward the IP packets to the phone even during handover. The TCP handles any packet loss (e.g. due to a radio noise induced pause).

Hardware

Devices supporting GPRS are divided into three classes:

Class A

Can be connected to GPRS service and GSM service (voice, SMS), using both at the same time. Such devices are known to be available today.

Class B

Can be connected to GPRS service and GSM service (voice, SMS), but using only one or the other at a given time. During GSM service (voice call or SMS), GPRS service is suspended, and then resumed automatically after the GSM service (voice call or SMS) has concluded. Most GPRS mobile devices are Class B.

Class C

Are connected to either GPRS service or GSM service (voice, SMS). Must be switched manually between one or the other service.

A true Class A device may be required to transmit on two different frequencies at the same time, and thus will need two radios. To get around this expensive requirement, a GPRS mobile may implement the dual transfer mode (DTM) feature. A DTM-capable mobile may use simultaneous voice and packet data, with the network coordinating to ensure that it is not required to transmit on two different frequencies at the same time. Such mobiles are considered pseudo-Class A, sometimes referred to as "simple class A". Some networks support DTM since 2007.

Huawei E220 3G/GPRS Modem

USB 3G/GPRS modems use a terminal-like interface over USB 1.1, 2.0 and later, data formats V.42bis, and RFC 1144 and some models have connector for external antenna. Modems can be added as cards (for laptops) or external USB devices which are similar in shape and size to a computer mouse, or nowadays more like a pendrive.

Addressing

A GPRS connection is established by reference to its access point name (APN). The APN defines the services such as wireless application protocol (WAP) access, short message service (SMS), multimedia messaging service (MMS), and for Internet communication services such as email and World Wide Web access.

In order to set up a GPRS connection for a wireless modem, a user must specify an APN, optionally a user name and password, and very rarely an IP address, provided by the network operator.

GPRS Modems and Modules

GSM module or GPRS modules are similar to modems, but there's one difference: the modem is an external piece of equipment, whereas the GSM module or GPRS module can be integrated within an electrical or electronic equipment. It is an embedded piece of hardware. A GSM mobile, on the other hand, is a complete embedded system in itself. It comes with embedded processors dedicated to provide a functional interface between the user and the mobile network.

Coding Schemes and Speeds

The upload and download speeds that can be achieved in GPRS depend on a number of factors such as:

- the number of BTS TDMA time slots assigned by the operator
- the channel encoding is used.
- the maximum capability of the mobile device expressed as a GPRS multislot class

Multiple Access Schemes

The multiple access methods used in GSM with GPRS are based on frequency division duplex (FDD) and TDMA. During a session, a user is assigned to one pair of up-link and down-link frequency channels. This is combined with time domain statistical multiplexing which makes it possible for several users to share the same frequency channel. The **packets** have constant length, corresponding to a GSM time slot. The down-link uses first-come first-served packet scheduling, while the up-link uses a scheme very similar to reservation ALOHA (R-ALOHA). This means that slotted ALOHA (S-ALOHA) is used for reservation inquiries during a contention phase, and then the actual data is transferred using dynamic TDMA with first-come first-served.

Channel Encoding

The channel encoding process in GPRS consists of two steps: first, a cyclic code is used to add parity bits, which are also referred to as the Block Check Sequence, followed by coding with a possibly punctured convolutional code. The Coding Schemes CS-1 to CS-4 specify the number of parity bits generated by the cyclic code and the puncturing rate of the convolutional code. In Coding Schemes CS-1 through CS-3, the convolutional code is of rate 1/2, i.e. each input bit is converted into two coded bits. In Coding Schemes CS-2 and CS-3, the output of the convolutional code is punctured to achieve the desired code rate. In Coding Scheme CS-4, no convolutional coding is applied. The following table summarises the options.

GPRS Coding scheme	Bitrate including RLC/MAC overhead[a][b] (kbit/s/slot)	Bitrate excluding RLC/MAC overhead[c] (kbit/s/slot)	Modulation	Code rate
CS-1	9.20	8.00	GMSK	1/2
CS-2	13.55	12.00	GMSK	≈2/3
CS-3	15.75	14.40	GMSK	≈3/4
CS-4	21.55	20.00	GMSK	1

This is rate at which the RLC/MAC layer protocol data unit (PDU) (called a radio block) is transmitted. As shown in TS 44.060 section 10.0a.1, a radio block consists of MAC header, RLC header, RLC data unit and spare bits. The RLC data unit represents the payload, the rest is overhead. The radio block is coded by the convolutional code specified for a particular Coding Scheme, which yields the same PHY layer data rate for all Coding Schemes.

1. Cited in various sources, e.g. in TS 45.001 table 1. is the bitrate including the RLC/MAC headers, but excluding the uplink state flag (USF), which is part of the MAC header, yielding a bitrate that is 0.15 kbit/s lower.

2. The net bitrate here is the rate at which the RLC/MAC layer payload (the RLC data unit) is transmitted. As such, this bit rate excludes the header overhead from the RLC/MAC layers.

The least robust, but fastest, coding scheme (CS-4) is available near a base transceiver station (BTS), while the most robust coding scheme (CS-1) is used when the mobile station (MS) is further away from a BTS.

Using the CS-4 it is possible to achieve a user speed of 20.0 kbit/s per time slot. However, using this scheme the cell coverage is 25% of normal. CS-1 can achieve a user speed of only 8.0 kbit/s per time slot, but has 98% of normal coverage. Newer network equipment can adapt the transfer speed automatically depending on the mobile location.

In addition to GPRS, there are two other GSM technologies which deliver data services: circuit-switched data (CSD) and high-speed circuit-switched data (HSCSD). In contrast to the shared nature of GPRS, these instead establish a dedicated circuit (usually billed per minute). Some applications such as video calling may prefer HSCSD, especially when there is a continuous flow of data between the endpoints.

The following table summarises some possible configurations of GPRS and circuit switched data services.

Technology	Download (kbit/s)	Upload (kbit/s)	TDMA timeslots allocated (DL+UL)
CSD	9.6	9.6	1+1
HSCSD	28.8	14.4	2+1

HSCSD	43.2	14.4	3+1
GPRS	85.6	21.4 (Class 8 & 10 and CS-4)	4+1
GPRS	64.2	42.8 (Class 10 and CS-4)	3+2
EGPRS (EDGE)	236.8	59.2 (Class 8, 10 and MCS-9)	4+1
EGPRS (EDGE)	177.6	118.4 (Class 10 and MCS-9)	3+2

Multislot Class

The multislot class determines the speed of data transfer available in the Uplink and Downlink directions. It is a value between 1 and 45 which the network uses to allocate radio channels in the uplink and downlink direction. Multislot class with values greater than 31 are referred to as high multislot classes.

A multislot allocation is represented as, for example, 5+2. The first number is the number of downlink timeslots and the second is the number of uplink timeslots allocated for use by the mobile station. A commonly used value is class 10 for many GPRS/EGPRS mobiles which uses a maximum of 4 timeslots in downlink direction and 2 timeslots in uplink direction. However simultaneously a maximum number of 5 simultaneous timeslots can be used in both uplink and downlink. The network will automatically configure for either 3+2 or 4+1 operation depending on the nature of data transfer.

Some high end mobiles, usually also supporting UMTS, also support GPRS/EDGE multislot class 32. According to 3GPP TS 45.002 (Release 12), Table B.1, mobile stations of this class support 5 timeslots in downlink and 3 timeslots in uplink with a maximum number of 6 simultaneously used timeslots. If data traffic is concentrated in downlink direction the network will configure the connection for 5+1 operation. When more data is transferred in the uplink the network can at any time change the constellation to 4+2 or 3+3. Under the best reception conditions, i.e. when the best EDGE modulation and coding scheme can be used, 5 timeslots can carry a bandwidth of 5*59.2 kbit/s = 296 kbit/s. In uplink direction, 3 timeslots can carry a bandwidth of 3*59.2 kbit/s = 177.6 kbit/s.

Multislot Classes for GPRS/EGPRS

Multislot Class	Downlink TS	Uplink TS	Active TS
1	1	1	2
2	2	1	3
3	2	2	3
4	3	1	4
5	2	2	4
6	3	2	4
7	3	3	4
8	4	1	5

9	3	2	5
10	4	2	5
11	4	3	5
12	4	4	5
30	5	1	6
31	5	2	6
32	5	3	6
33	5	4	6
34	5	5	6

Attributes of a Multislot Class

Each multislot class identifies the following:

- the maximum number of Timeslots that can be allocated on uplink

- the maximum number of Timeslots that can be allocated on downlink

- the total number of timeslots which can be allocated by the network to the mobile

- the time needed for the MS to perform adjacent cell signal level measurement and get ready to transmit

- the time needed for the MS to get ready to transmit

- the time needed for the MS to perform adjacent cell signal level measurement and get ready to receive

- the time needed for the MS to get ready to receive.

The different multislot class specification is detailed in the Annex B of the 3GPP Technical Specification 45.002 (Multiplexing and multiple access on the radio path)

Usability

The maximum speed of a GPRS connection offered in 2003 was similar to a modem connection in an analog wire telephone network, about 32–40 kbit/s, depending on the phone used. Latency is very high; round-trip time (RTT) is typically about 600–700 ms and often reaches 1s. GPRS is typically prioritized lower than speech, and thus the quality of connection varies greatly.

Devices with latency/RTT improvements (via, for example, the extended UL TBF mode feature) are generally available. Also, network upgrades of features are available with certain operators. With these enhancements the active round-trip time can be reduced, resulting in significant increase in application-level throughput speeds.

History of GPRS

GPRS opened in 2000 as a packet-switched data service embedded to the channel-switched cellular radio network GSM. GPRS extends the reach of the fixed Internet by connecting mobile terminals worldwide.

The CELLPAC protocol developed 1991-1993 was the trigger point for starting in 1993 specification of standard GPRS by ETSI SMG. Especially, the CELLPAC Voice & Data functions introduced in a 1993 ETSI Workshop contribution anticipate what was later known to be the roots of GPRS. This workshop contribution is referenced in 22 GPRS related US-Patents. Successor systems to GSM/GPRS like W-CDMA (UMTS) and LTE rely on key GPRS functions for mobile Internet access as introduced by CELLPAC.

According to a study on history of GPRS development Bernhard Walke and his student Peter Decker are the inventors of GPRS – the first system providing worldwide mobile Internet access.

Public Switched Telephone Network

The public switched telephone network (PSTN) is the aggregate of the world's circuit-switched telephone networks that are operated by national, regional, or local telephony operators, providing infrastructure and services for public telecommunication. The PSTN consists of telephone lines, fiber optic cables, microwave transmission links, cellular networks, communications satellites, and undersea telephone cables, all interconnected by switching centers, thus allowing most telephones to communicate with each other. Originally a network of fixed-line analog telephone systems, the PSTN is now almost entirely digital in its core network and includes mobile and other networks, as well as fixed telephones.

The technical operation of the PSTN adheres to the standards created by the ITU-T. These standards allow different networks in different countries to interconnect seamlessly. The E.163 and E.164 standards provide a single global address space for telephone numbers. The combination of the interconnected networks and the single numbering plan allow telephones around the world to dial each other.

History

The first telephones had no network but were in private use, wired together in pairs. Users who wanted to talk to different people had as many telephones as necessary for the purpose. A user who wished to speak whistled loudly into the transmitter until the other party heard.

However, a bell was added soon for signaling, so an attendant no longer need wait for the whistle, and then a switch hook. Later telephones took advantage of the exchange principle already employed in telegraph networks. Each telephone was wired to a local telephone exchange, and the exchanges were wired together with trunks. Networks were connected in a hierarchical manner until they spanned cities, countries, continents and oceans. This was the beginning of the PSTN, though the term was not used for many decades.

Automation introduced pulse dialing between the phone and the exchange, and then among exchanges, followed by more sophisticated address signaling including multi-frequency, culminating in the SS7 network that connected most exchanges by the end of the 20th century.

The growth of the PSTN meant that teletraffic engineering techniques needed to be deployed to deliver quality of service (QoS) guarantees for the users. The work of A. K. Erlang established the mathematical foundations of methods required to determine the capacity requirements and configuration of equipment and the number of personnel required to deliver a specific level of service.

In the 1970s the telecommunications industry began implementing packet switched network data services using the X.25 protocol transported over much of the end-to-end equipment as was already in use in the PSTN.

In the 1980s the industry began planning for digital services assuming they would follow much the same pattern as voice services, and conceived a vision of end-to-end circuit switched services, known as the Broadband Integrated Services Digital Network (B-ISDN). The B-ISDN vision has been overtaken by the disruptive technology of the Internet.

At the turn of the 21st century, the oldest parts of the telephone network still use analog technology for the last mile loop to the end user. However, digital technologies such as DSL, ISDN, FTTx, and cable modems have become more common in this portion of the network.

Several large private telephone networks are not linked to the PSTN, usually for military purposes. There are also private networks run by large companies which are linked to the PSTN only through limited gateways, such as a large private branch exchange (PBX).

Operators

The task of building the networks and selling services to customers fell to the network operators. The first company to be incorporated to provide PSTN services was the Bell Telephone Company in the United States.

In some countries, however, the job of providing telephone networks fell to government as the investment required was very large and the provision of telephone service was increasingly becoming an essential public utility. For example, the General Post Office in the United Kingdom brought together a number of private companies to form a single nationalized company. In recent decades however, these state monopolies were broken up or sold off through privatization.

Regulation

In most countries, the central has a regulator dedicated to monitoring the provision of PSTN services in that country. Their tasks may be for example to ensure that end customers are not overcharged for services where monopolies may exist. They may also regulate the prices charged between the operators to carry each other's traffic.

Technology

Network Topology

The PSTN network architecture had to evolve over the years to support increasing numbers of subscribers, calls, connections to other countries, direct dialing and so on. The model developed by the United States and Canada was adopted by other nations, with adaptations for local markets.

The original concept was that the telephone exchanges are arranged into hierarchies, so that if a call cannot be handled in a local cluster, it is passed to one higher up for onward routing. This reduced the number of connecting trunks required between operators over long distances and also kept local traffic separate.

However, in modern networks the cost of transmission and equipment is lower and, although hierarchies still exist, they are much flatter, with perhaps only two layers.

Digital Channels

As described above, most automated telephone exchanges now use digital switching rather than mechanical or analog switching. The trunks connecting the exchanges are also digital, called circuits or channels. However analog two-wire circuits are still used to connect the last mile from the exchange to the telephone in the home (also called the local loop). To carry a typical phone call from a calling party to a called party, the analog audio signal is digitized at an 8 kHz sample rate with 8-bit resolution using a special type of nonlinear pulse code modulation known as G.711. The call is then transmitted from one end to another via telephone exchanges. The call is switched using a call set up protocol (usually ISUP) between the telephone exchanges under an overall routing strategy.

The call is carried over the PSTN using a 64 kbit/s channel, originally designed by Bell Labs. The name given to this channel is Digital Signal 0 (DS0). The DS0 circuit is the basic granularity of circuit switching in a telephone exchange. A DS0 is also known as a timeslot because DS0s are aggregated in time-division multiplexing (TDM) equipment to form higher capacity communication links.

A Digital Signal 1 (DS1) circuit carries 24 DS0s on a North American or Japanese T-carrier (T1) line, or 32 DS0s (30 for calls plus two for framing and signaling) on an E-carrier (E1) line used in most other countries. In modern networks, the multiplexing function is moved as close to the end user as possible, usually into cabinets at the roadside in residential areas, or into large business premises.

These aggregated circuits are conveyed from the initial multiplexer to the exchange over a set of equipment collectively known as the access network. The access network and inter-exchange transport use synchronous optical transmission, for example, SONET and Synchronous Digital Hierarchy (SDH) technologies, although some parts still use the older PDH technology.

Within the access network, there are a number of reference points defined. Most of these are of interest mainly to ISDN but one – the V reference point – is of more general interest. This is the reference point between a primary multiplexer and an exchange. The protocols at this reference point were standardized in ETSI areas as the V5 interface.

Impact on IP Standards

Voice quality over PSTN networks was used as the benchmark for the development of the Telecommunications Industry Association's TIA-TSB-116 standard on voice-quality recommendations for IP telephony, to determine acceptable levels of audio delay and echo.

Generic Access Network

Generic Access Network or GAN is a telecommunication system that extends mobile voice, data and multimedia (IMS/SIP) applications over IP networks. Unlicensed Mobile Access or UMA, is the commercial name used by mobile carriers for external IP access into their core networks. The latest generation system is named Wi-Fi Calling by a number of handset manufacturers, including Apple and Samsung, a move that is being mirrored by carriers like T-Mobile US.

Essentially, GAN allows cell phone packets to be forwarded to a network access point over the internet, rather than over-the-air using GSM/GPRS, UMTS or similar. A separate device known as a "GAN Controller" (GANC) receives this data from the internet and feeds it into the phone network as if it were coming from an antenna on a tower. Calls can be placed from or received to the handset as if it were connected over-the-air directly to the GANC's point of presence. The system is essentially invisible to the network as a whole.

In its most common form, GAN is used to allow UMA-compatible mobile phones to use WiFi networks to connect calls, in place of conventional cell towers. This can be useful in locations with poor cell coverage where some other form of internet access is available, especially at the home or office. The system offers seamless handoff, so the user can move from cell to WiFi and back again with the same invisibility that the cell network offers when moving from tower to tower.

Since the GAN system works over the internet, a UMA-capable handset can connect to their service provider from any location with internet access. This is particularly useful for travellers, who can connect to their provider's GANC and make calls into their home service area from anywhere in the world. This is subject to the quality of the internet connection, however, and may not work well over limited bandwidth or long-latency connections. To improve quality of service in the home or office, some providers also supply a specially programmed wireless access point that prioritizes UMA packets.

History

UMA was developed by a group of operator and vendor companies. The initial specifications were published on 2 September 2004. The companies then contributed the specifications to the 3rd Generation Partnership Project (3GPP) as part of 3GPP work item "Generic Access to A/Gb interfaces". On 8 April 2005, 3GPP approved specifications for Generic Access to A/Gb interfaces for 3GPP Release 6. and, and renamed the system to GAN. But the term *GAN* is little known outside the 3GPP community, and the term *UMA* is more common in marketing.

Modes of Operation

The original Release 6 GAN specification supported a 2G (A/Gb) connection from the GANC into the mobile core network (MSC/GSN). Today all commercial GAN dual-mode handset deployments are based on a 2G connection and all GAN enabled devices are dual-mode 2G/Wi-Fi. The specification, though, defined support for multimode handset operation. Therefore, 3G/2G/Wi-Fi handsets are supported in the standard. The first 3G/UMA devices were announced in the second half of 2008.

A typical UMA/GAN handset will have four modes of operation:

- GERAN-only: uses only cellular networks

- GERAN-preferred: uses cellular networks if available, otherwise the 802.11 radio

- GAN-preferred: uses an 802.11 connection if an access point is in range, otherwise the cellular network

- GAN-only: uses only the 802.11 connection

In all cases, the handset scans for GSM cells when it first turns on, to determine its location area. This allows the carrier to route the call to the nearest GANC, set the correct rate plan, and comply with existing roaming agreements.

At the end of 2007, the GAN specification was enhanced to support 3G (Iu) interfaces from the GANC to the mobile core network (MSC/GSN). This native 3G interface can be used for dual-mode handset as well as 3G femtocell service delivery. The GAN release 8 documentation describes these new capabilities.

Advantages

For carriers:

- Instead of erecting expensive base stations to cover dead zones, GAN allows carriers to add coverage using low cost 802.11 access points. Subscribers at home have very good coverage.

- In addition, GAN relieves congestion (meaning that networks can, through GAN, essentially piggyback on other infrastructure) on the GSM or UMTS spectrum by removing common types of calls and routing them to the operator via the relatively low cost Internet

- GAN makes sense for network operators that also offer Internet services. Operators can leverage sales of one to promote the other, and can bill both to each customer.

- Some other operators also run networks of 802.11 hotspots, such as T-Mobile. They can leverage these hotspots to create more capacity and provide better coverage in populous areas.

- The carrier does not pay for much of the service, the party who provides the Internet and Wi-Fi connection pays for a connection to the Internet, effectively paying the expensive part of routing calls from the subscriber.

For subscribers:

- Subscribers do not rely on their operator's ability to roll out towers and coverage, allowing them to fix some types of coverage dead zones (such as in the home or workplace) themselves.

- The cheaper rates for 802.11 use, coupled with better coverage at home, make more affordable and practical the use of cellphones instead of land lines.

- Using IP over 802.11 eliminates expensive charges when roaming outside of a carrier's network.

- GAN is currently the only commercial technology available that combines GSM and 802.11 into a service that uses a single number, a single handset, a single set of services and a single phone directory for all calls.

- GAN can migrate between IP and cellular coverage and is thus seamless; in contrast, calls via third-party VOIP plus a data phone are dropped when leaving high-volume data coverage.

Disadvantages

- Subscribers must upgrade to Wi-Fi/UMA enabled handsets to take advantage of the service.

- Calls may be more prone to disconnect when the handset transitions from Wi-Fi to the standard wireless service and vice versa (because the handset moved out or within the Wi-Fi's range). How much this is a problem may vary based on which handset is used.

- The UMA may use different frequency that is more prone to some types of interference

- Some setup may be required to provide connection settings (such as authentication details) before advantages may be experienced. This may take time for subscribers and require additional support to be provided. The costs of support may be for more than the wireless phone company: network administrators may be asked to help a user enter appropriate settings into a phone (that the network administrator may know little about).

- The phones that support multiple signals (both the UMA/Wi-Fi and the type of signal used by the provider's towers) may be more expensive, particularly to manufacture, due to additional circuitry/components required

- This uses the resources of the network providing the Wi-Fi signal (and any indirect network that is then utilized when that network is used). Bandwidth is used up. Some types of network traffic (like DNS and IPsec-encrypted) need to be permitted by the network, so a decision to support this may impose some requirement(s) regarding the network's security (firewall) rules.

- Using GAN/UMA on a mobile requires the WiFi module to be enabled. This in turn drains the battery faster, and reduces both the talk time and standby time when compared to disabling GAN/UMA (and in turn WiFi).

Service Deployments

The first service launch was BT with BT Fusion in the autumn of 2005. The service is based on pre-3GPP GAN standard technology. Initially, BT Fusion used UMA over Bluetooth with phones from Motorola; since Jan 2007, it has used UMA over 802.11 with phones from Nokia, Motorola and Samsung and is branded as a "Wi-Fi mobile service". BT has since discontinued the service.

On August 28, 2006, TeliaSonera was the first to launch an 802.11 based UMA service called "Home Free". The service started in Denmark but no longer offered.

On September 25, 2006 Orange announced its "Unik service", also known as Signal Boost in the UK. However this service is no longer available to new customers in the UK. The announcement, the largest to date, covers more than 60m of Orange's mobile subscribers in the UK, France, Poland, Spain and the Netherlands.

Cincinnati Bell announced the first UMA deployment in the United States. The service, originally called CB Home Run, allows users to transfer seamlessly from the Cincinnati Bell cellular network to a home wireless network or to Cincinnati Bell's WiFi HotSpots. It has since been rebranded as Fusion WiFi.

This was followed shortly by T-Mobile US on June 27, 2007. T-Mobile's service, originally named "Hotspot Calling", and rebranded to "Wi-Fi Calling" in 2009, allows users to seamlessly transfer from the T-Mobile cellular network to an 802.11x wireless network or T-Mobile HotSpot in the United States.

In Canada, both Fido and Rogers Wireless launched UMA plans under the names UNO and Rogers Home Calling Zone (later rebranded Talkspot, and subsequently rebranded again as Wi-Fi Calling), respectively, on May 6, 2008.

In Australia, GAN has not been implemented but it has been reported that Apple and Vodafone have engaged in discussions regarding implementation of the technology for compatible devices running iOS 8.

Since 10th April 2015, Wi-Fi Calling has been available for customers of EE in the UK initially on the Nokia Lumia 640 and Samsung Galaxy S6 and Samsung Galaxy S6 Edge handsets.

AT&T and Verizon are going to launch Wi-Fi calling in 2015.

Industry organisation UMA Today tracks all operator activities and handset development.

UMA is not implemented in Asia, Australia, Africa and some European countries.

UMA/GAN Beyond Dual-mode

While UMA is nearly always associated with dual-mode GSM/Wi-Fi services, it is actually a 'generic' access network technology that provides a generic method for extending the services and applications in an operator's mobile core (voice, data, IMS) over IP and the public Internet.

GAN defines a secure, managed connection from the mobile core (GANC) to different devices/access points over IP.

Femtocells - The GAN standard is currently used to provide a secure, managed, standardized interface from a femtocell to the mobile core network. Recently[when?] Kineto, NEC and Motorola issued a joint proposal to the 3GPP work group studying femtocells (also known as 'Home Node B's or HNB) to propose GAN as the basis for that standard.

Analog Terminal Adaptor – T-Mobile US once offered a fixed-line VoIP service called @Home. Similar to Vonage, consumers can port their fixed phone number to T-Mobile. Then T-Mobile associates that number with an ATA (analog telephone adapter). The consumer plugs the ATA into a home broadband network and begins receiving calls to the fixed number over the IP access network. The service was discontinued in 2010, however earlier subscribers were "grandfathered" in.

Mobile VoIP Client - Consumers have started to use telephony interfaces on their PCs. Applications offer a low cost, convenient way to access telephony services while traveling. Now mobile operators can offer a similar service with a UMA-enabled mobile VoIP client. Developed by Vitendo, the client provides a mirror interface to a subscriber's existing mobile service. For the mobile operator, services can now be extended to a PC/laptop, and they can give consumers another way to use their mobile service.

Similar Technologies

GAN/UMA is not the first system to allow the use of unlicensed spectrum to connect handsets to a GSM network. The GIP/IWP standard for DECT provides similar functionality, but requires a more direct connection to the GSM network from the base station. While dual-mode DECT/GSM phones have appeared, these have generally been functionally cordless phones with a GSM handset built-in (or vice versa, depending on your point of view), rather than phones implementing DECT/GIP, due to the lack of suitable infrastructure to hook DECT base-stations supporting GIP to GSM networks on an ad-hoc basis.

GAN/UMA's ability to use the Internet to provide the "last mile" connection to the GSM network solves the major issue that DECT/GIP has faced. Had GIP emerged as a practical standard, the low power usage of DECT technology when idle would have been an advantage compared to GAN.

There is nothing preventing an operator from deploying micro- and pico-cells that use towers that connect with the home network over the Internet. Several companies have developed so-called Femtocell systems that do precisely that, broadcasting a "real" GSM or UMTS signal, bypassing the need for special handsets that require 802.11 technology. In theory, such systems are more universal, and again require lower power than 802.11, but their legality will vary depending on the jurisdiction, and will require the cooperation of the operator. Further, users may be charged at higher cell phone rates, even though they are paying for the DSL or other network that ultimately carries their traffic; in contrast, GAN/UMA providers charge reduced rates when making calls off the providers cellular phone network.

References

- Guowang Miao; Jens Zander; Ki Won Sung; Ben Slimane (2016). Fundamentals of Mobile Data Networks. Cambridge University Press. ISBN 1107143217.
- Guowang Miao, Jens Zander, Ki Won Sung, and Ben Slimane, Fundamentals of Mobile Data Networks, Cambridge University Press, ISBN 1107143217, 2016
- Guowang Miao, Jens Zander, Ki Won Sung, and Ben Slimane, Fundamentals of Mobile Data Networks, Cambridge University Press, ISBN 1107143217, 2016.
- "GSM Bands information by country". WorldTimeZone.com. 2016-01-16. Retrieved 2016-02-06.
- "WARC-92: World Administrative Radio Conference for Dealing with Frequency Allocations in Certain Parts of the Spectrum (Málaga-Torremolinos, 1992)". 1992-03-03. Retrieved 2016-03-14.
- "T-Mobile shifting 1700 MHz HSPA+ users to 1900 MHz band". TeleGeography. 2015-06-24. Retrieved 2016-02-16.

- Phil Goldstein (September 17, 2014). "Verizon Wireless plans to launch Wi-Fi calling in mid-2015". FierceWireless. Retrieved September 19, 2014.

- Greg Kumparak (2014-09-12). "AT&T To Get iPhone 6-Friendly Wi-Fi Calling In 2015". TechCrunch. Retrieved 2014-09-14.

- Kushnick, Bruce (7 January 2013). "What Are the Public Switched Telephone Networks, 'PSTN' and Why You Should Care?". Huffington Post Blog. Retrieved 11 April 2014.

- TS 36.211 rel.11, LTE, Evolved Universal Terrestrial Radio Access, Physical channels and modulation - chapters 5.2.3 and 6.2.3: Resource blocks etsi.org, January 2014.

- Bernhard Walke: „The Roots of GPRS: The First System for Mobile Packet-Based Global Internet Access", IEEE Wireless Communications, Oct. 2013, 12-23.

5

Channel Access Method: An Overview

Channel access method, which spreads the area covered, permits several terminals to be connected to the same multi-point transmission medium. Some of the examples of channel access methods are wireless networks, bus networks, ring networks and point-to-point links. This chapter also focuses on topics like time division multiple access, code division multiple access, space-division multiple access and ALOHAnet.

Channel Access Method

In telecommunications and computer networks, a channel access method or multiple access method allows several terminals connected to the same multi-point transmission medium to transmit over it and to share its capacity. Examples of shared physical media are wireless networks, bus networks, ring networks and point-to-point links operating in half-duplex mode.

A channel-access scheme is based on a multiplexing method, that allows several data streams or signals to share the same communication channel or physical medium. In this context. multiplexing is provided by the physical layer.

A channel-access scheme is also based on a multiple access protocol and control mechanism, also known as media access control (MAC). Media access control deals with issues such as addressing, assigning multiplex channels to different users, and avoiding collisions. Media access control is a sub-layer in Layer 2 (data link layer) of the OSI model and a component of the link layer of the TCP/IP model.

Fundamental Types of Channel Access Schemes

These numerous channel access schemes which generally fall into the following categories:

Frequency-division Multiple Access (FDMA)

The frequency-division multiple access (FDMA) channel-access scheme is based on the frequency-division multiplexing (FDM) scheme, which provides different frequency bands to different data-streams. In the FDMA case, the data streams are allocated to different nodes or devices. An example of FDMA systems were the first-generation (1G) cell-phone systems, where each phone call was assigned to a specific uplink frequency channel, and another downlink frequency channel. Each message signal (each phone call) is modulated on a specific carrier frequency.

A related technique is wavelength division multiple access (WDMA), based on wavelength-division multiplexing (WDM), where different datastreams get different colors in fiber-optical communications. In the WDMA case, different network nodes in a bus or hub network get a different color.

An advanced form of FDMA is the orthogonal frequency-division multiple access (OFDMA) scheme, for example used in 4G cellular communication systems. In OFDMA, each node may use several sub-carriers, making it possible to provide different quality of service (different data rates) to different users. The assignment of sub-carriers to users may be changed dynamically, based on the current radio channel conditions and traffic load.

Time Division Multiple Access (TDMA)

The time division multiple access (TDMA) channel access scheme is based on the time-division multiplexing (TDM) scheme, which provides different time-slots to different data-streams (in the TDMA case to different transmitters) in a cyclically repetitive frame structure. For example, node 1 may use time slot 1, node 2 time slot 2, etc. until the last transmitter. Then it starts all over again, in a repetitive pattern, until a connection is ended and that slot becomes free or assigned to another node. An advanced form is Dynamic TDMA (DTDMA), where a scheduling may give different time sometimes but some times node 1 may use time slot 1 in first frame and use another time slot in next frame.

As an example, 2G cellular systems are based on a combination of TDMA and FDMA. Each frequency channel is divided into eight timeslots, of which seven are used for seven phone calls, and one for signalling data.

Statistical time division multiplexing multiple-access is typically also based on time-domain multiplexing, but not in a cyclically repetitive frame structure. Due to its random character it can be categorised as statistical multiplexing methods, making it possible to provide dynamic bandwidth allocation. This requires a media access control (MAC) protocol, i.e. a principle for the nodes to take turns on the channel and to avoid collisions. Common examples are CSMA/CD, used in Ethernet bus networks and hub networks, and CSMA/CA, used in wireless networks such as IEEE 802.11.

Code Division Multiple Access (CDMA)/Spread Spectrum Multiple Access (SSMA)

The code division multiple access (CDMA) scheme is based on spread spectrum, meaning that a wider radio spectrum in Hertz is used than the data rate of each of the transferred bit streams, and several message signals are transferred simultaneously over the same carrier frequency, utilizing different spreading codes. The wide bandwidth makes it possible to send with a very poor signal-to-noise ratio of much less than 1 (less than 0 dB) according to the Shannon-Heartly formula, meaning that the transmission power can be reduced to a level below the level of the noise and co-channel interference (cross talk) from other message signals sharing the same frequency.

One form is direct sequence spread spectrum (DS-CDMA), used for example in 3G cell phone systems. Each information bit (or each symbol) is represented by a long code sequence of several pulses, called chips. The sequence is the spreading code, and each message signal (for example each phone call) uses a different spreading code.

Another form is frequency-hopping (FH-CDMA), where the channel frequency is changing very rapidly according to a sequence that constitutes the spreading code. As an example, the Bluetooth communication system is based on a combination of frequency-hopping and either CSMA/ CA statistical time division multiplexing communication (for data communication applications) or TDMA (for audio transmission). All nodes belonging to the same user (to the same virtual private area network or piconet) use the same frequency hopping sequency synchronously, meaning that they send on the same frequency channel, but CDMA/CA or TDMA is used to avoid collisions within the VPAN. Frequency-hopping is used to reduce the cross-talk and collision probability between nodes in different VPANs.

Subdivisions of FH-CDMA are "fast hopping" where the frequency of hopping is much higher than the message frequency content and "slow hopping" where the hopping frequency is comparable to message frequency content. The subdivision is necessary as they are considerably different.

Space Division Multiple Access (SDMA)

Space-division multiple access (SDMA) transmits different information in different physical areas. Examples include simple cellular radio systems and more advanced cellular systems which use directional antennas and power modulation to refine spatial transmission patterns.

Power Division Multiple Access (PDMA)

Power-division multiple access (PDMA) scheme is based on using variable transmission power between users in order to share the available power on the channel. Examples include multiple SCPC modems on a satellite transponder, where users get on demand a larger share of the power budget to transmit at higher data rates.

List of Channel Access Methods

Circuit Mode and Channelization Methods

The following are common circuit mode and channelization channel access methods:

- *Frequency-division multiple access (FDMA),* based on frequency-division multiplexing (FDM)
    - o Wavelength division multiple access (WDMA)
    - o Orthogonal frequency-division multiple access (OFDMA), based on Orthogonal frequency-division multiplexing (OFDM)
    - o Single-carrier FDMA (SC-FDMA), a.k.a. linearly-precoded OFDMA (LP-OFDMA), based on single-carrier frequency-domain-equalization (SC-FDE).
- *Time-division multiple access (TDMA),* based on time-division multiplexing (TDM)
    - o Multi-Frequency Time Division Multiple Access (MF-TDMA)
- *Code division multiple access (CDMA),* a.k.a. Spread spectrum multiple access (SSMA)
    - o Direct-sequence CDMA (DS-CDMA), based on Direct-sequence spread spectrum (DSSS)

- o Frequency-hopping CDMA (FH-CDMA), based on Frequency-hopping spread spectrum (FHSS)

- o Orthogonal frequency-hopping multiple access (OFHMA)

- o Multi-carrier code division multiple access (MC-CDMA)

- *Space-division multiple access (SDMA)*

- Power-division multiple access (PDMA)

Packet Mode Methods

The following are examples of packet mode channel access methods:

- *Contention based random multiple access methods*

  - o Aloha

  - o Slotted Aloha

  - o Multiple Access with Collision Avoidance (MACA)

  - o Multiple Access with Collision Avoidance for Wireless (MACAW)

  - o Carrier sense multiple access (CSMA)

  - o Carrier sense multiple access with collision detection (CSMA/CD) - suitable for wired networks

  - o Carrier sense multiple access with collision avoidance (CSMA/CA) - suitable for wireless networks

    - ▪ Distributed Coordination Function (DCF)

  - o Carrier sense multiple access with collision avoidance and Resolution using Priorities (CSMA/CARP)

  - o Carrier Sense Multiple Access/Bitwise Arbitration (CSMA/BA) Based on constructive interference (CAN-bus)

- *Token passing:*

  - o Token ring

  - o Token bus

- *Polling*

- *Resource reservation (scheduled) packet-mode protocols*

  - o Dynamic Time Division Multiple Access (Dynamic TDMA)

  - o Packet reservation multiple access (PRMA)

  - o Reservation ALOHA (R-ALOHA)

Duplexing Methods

Where these methods are used for dividing forward and reverse communication channels, they are known as duplexing methods, such as:

- Time division duplex (TDD)

- Frequency division duplex (FDD)

Hybrid Channel Access Scheme Application Examples

Note that hybrids of these techniques can be - and frequently are - used. Some examples:

- The GSM cellular system combines the use of frequency division duplex (FDD) to prevent interference between outward and return signals, with FDMA and TDMA to allow multiple handsets to work in a single cell.

- GSM with the GPRS packet switched service combines FDD and FDMA with slotted Aloha for reservation inquiries, and a Dynamic TDMA scheme for transferring the actual data.

- Bluetooth packet mode communication combines frequency hopping (for shared channel access among several private area networks in the same room) with CSMA/CA (for shared channel access inside a medium).

- IEEE 802.11b wireless local area networks (WLANs) are based on FDMA and DS-CDMA for avoiding interference among adjacent WLAN cells or access points. This is combined with CSMA/CA for multiple access within the cell.

- HIPERLAN/2 wireless networks combine FDMA with dynamic TDMA, meaning that resource reservation is achieved by packet scheduling.

- G.hn, an ITU-T standard for high-speed networking over home wiring (power lines, phone lines and coaxial cables) employs a combination of TDMA, Token passing and CSMA/CARP to allow multiple devices to share the medium.

Definition within Certain Application Areas

Local and Metropolitan Area Networks

In local area networks (LANs) and metropolitan area networks (MANs), multiple access methods enable bus networks, ring networks, hubbed networks, wireless networks and half duplex point-to-point communication, but are not required in full duplex point-to-point serial lines between network switches and routers, or in switched networks (logical star topology). The most common multiple access method is CSMA/CD, which is used in Ethernet. Although today's Ethernet installations typically are switched, CSMA/CD is utilized anyway to achieve compatibility with hubs.

Satellite Communications

In satellite communications, multiple access is the capability of a communications satellite to function as a portion of a communications link between more than one pair of satellite terminals concurrently. Three types of multiple access presently used with communications satellites are code-division, frequency-division, and time-division multiple access.

Switching Centers

In telecommunication switching centers, multiple access is the connection of a user to two or more switching centers by separate access lines using a single message routing indicator or telephone number.

Classifications in the Literature

Several ways of categorizing multiple-access schemes and protocols have been used in the literature. For example, Daniel Minoli (2009) identifies five principal types of multiple-access schemes: FDMA, TDMA, CDMA, SDMA, and Random access. R. Rom and M. Sidi (1990) categorize the protocols into *Conflict-free access protocols*, *Aloha protocols*, and *Carrier Sensing protocols*.

The Telecommunications Handbook (Terplan and Morreale, 2000) identifies the following MAC categories:

- Fixed assigned: TDMA, FDMA+WDMA, CDMA, SDMA

- Demand assigned (DA)

    ○ Reservation: DA/TDMA, DA/FDMA+DA/WDMA, DA/CDMA, DA/SDMA

    ○ Polling: Generalized polling, Distributed polling, Token Passing, Implicit polling, Slotted access

- Random access (RA): Pure RA (ALOHA, GRA), Adaptive RA (TRA), CSMA, CSMA/CD, CSMA/CA

Frequency-division Multiple Access

Frequency division multiple access or FDMA is a channel access method used in multiple-access protocols as a channelization protocol. FDMA gives users an individual allocation of one or several frequency bands, or channels. It is particularly commonplace in satellite communication. FDMA, like other multiple access systems, coordinates access between multiple users. Alternatives include TDMA, CDMA, or SDMA. These protocols are utilized differently, at different levels of the theoretical OSI model.

Disadvantage: Crosstalk may cause interference among

- In FDMA, all users share the satellite transponder or frequency channel simultaneously but each user transmits at single frequency.

- FDMA can be used with both analog and digital signal.

- FDMA requires high-performing filters in the radio hardware, in contrast to TDMA and CDMA.

- FDMA is not vulnerable to the timing problems that TDMA has. Since a predetermined frequency band is available for the entire period of communication, stream data (a continuous flow of data that may not be packetized) can easily be used with FDMA.

- Due to the frequency filtering, FDMA is not sensitive to near-far problem which is pronounced for CDMA.

- Each user transmits and receives at different frequencies as each user gets a unique frequency slots.

FDMA is distinct from frequency division duplexing (FDD). While FDMA allows multiple users simultaneous access to a transmission system, FDD refers to how the radio channel is shared between the uplink and downlink (for instance, the traffic going back and forth between a mobile-phone and a mobile phone base station). Frequency-division multiplexing (FDM) is also distinct from FDMA. FDM is a physical layer technique that combines and transmits low-bandwidth channels through a high-bandwidth channel. FDMA, on the other hand, is an access method in the data link layer.

FDMA also supports demand assignment in addition to fixed assignment. *Demand assignment* allows all users apparently continuous access of the radio spectrum by assigning carrier frequencies on a temporary basis using a statistical assignment process. The first FDMA *demand-assignment* system for satellite was developed by COMSAT for use on the *Intelsat* series *IVA* and *V* satellites.

There are two main techniques:

- Multi-channel per-carrier (MCPC)

- Single-channel per-carrier (SCPC)

Time Division Multiple Access

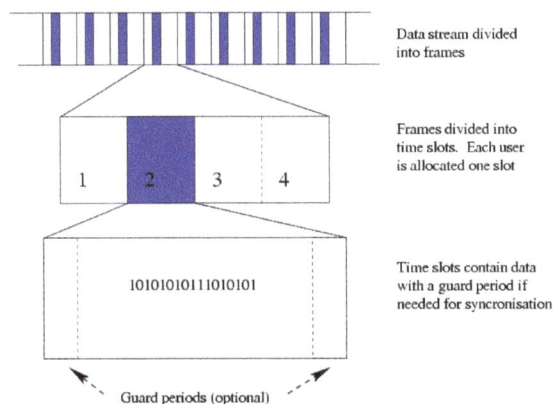

Data stream divided into frames

Frames divided into time slots. Each user is allocated one slot

Time slots contain data with a guard period if needed for syncronisation

1010101011010101

Guard periods (optional)

TDMA frame structure showing a data stream divided into frames and those frames divided into time slots.

Time division multiple access (TDMA) is a channel access method for shared medium networks. It allows several users to share the same frequency channel by dividing the signal into different time slots. The users transmit in rapid succession, one after the other, each using its own time slot. This allows multiple stations to share the same transmission medium (e.g. radio frequency channel) while using only a part of its channel capacity. TDMA is used in the digital 2G cellular systems such as Global System for Mobile Communications (GSM), IS-136, Personal Digital Cellular (PDC) and iDEN, and in the Digital Enhanced Cordless Telecommunications (DECT) standard for portable phones. It is also used extensively in satellite systems, combat-net radio systems, and PON networks for upstream traffic from premises to the operator. For usage of Dynamic TDMA packet mode communication.

TDMA is a type of time-division multiplexing, with the special point that instead of having one transmitter connected to one receiver, there are multiple transmitters. In the case of the *uplink* from a mobile phone to a base station this becomes particularly difficult because the mobile phone can move around and vary the *timing advance* required to make its transmission match the gap in transmission from its peers.

TDMA Characteristics

- Shares single carrier frequency with multiple users

- Non-continuous transmission makes handoff simpler

- Slots can be assigned on demand in dynamic TDMA

- Less stringent power control than CDMA due to reduced intra cell interference

- Higher synchronization overhead than CDMA

- Advanced equalization may be necessary for high data rates if the channel is "frequency selective" and creates Intersymbol interference

- Cell breathing (borrowing resources from adjacent cells) is more complicated than in CDMA

- Frequency/slot allocation complexity

- Pulsating power envelope: interference with other devices

TDMA in Mobile Phone Systems

2G Systems

Most 2G cellular systems, with the notable exception of IS-95, are based on TDMA. GSM, D-AMPS, PDC, iDEN, and PHS are examples of TDMA cellular systems. GSM combines TDMA with Frequency Hopping and wideband transmission to minimize common types of interference.

In the GSM system, the synchronization of the mobile phones is achieved by sending timing advance commands from the base station which instructs the mobile phone to transmit earlier and by how much. This compensates for the propagation delay resulting from the light speed velocity of radio waves. The mobile phone is not allowed to transmit for its entire time slot, but there is a guard interval at the end of each time slot. As the transmission moves into the guard period, the mobile network adjusts the timing advance to synchronize the transmission.

Initial synchronization of a phone requires even more care. Before a mobile transmits there is no way to actually know the offset required. For this reason, an entire time slot has to be dedicated to mobiles attempting to contact the network; this is known as the random-access channel (RACH) in GSM. The mobile attempts to broadcast at the beginning of the time slot, as received from the network. If the mobile is located next to the base station, there will be no time delay and this will succeed. If, however, the mobile phone is at just less than 35 km from the base station, the time delay will mean the mobile's broadcast arrives at the very end of the time slot. In that case, the mobile will be instructed to broadcast its messages starting nearly a whole time slot earlier than would be expected otherwise. Finally, if the mobile is beyond the 35 km cell range in GSM, then the RACH will arrive in a neighbouring time slot and be ignored. It is this feature, rather than limitations of power, that limits the range of a GSM cell to 35 km when no special extension techniques are used. By changing the synchronization between the uplink and downlink at the base station, however, this limitation can be overcome.

3G Systems

Although most major 3G systems are primarily based upon CDMA, time division duplexing (TDD), packet scheduling (dynamic TDMA) and packet oriented multiple access schemes are available in 3G form, combined with CDMA to take advantage of the benefits of both technologies.

While the most popular form of the UMTS 3G system uses CDMA and frequency division duplexing (FDD) instead of TDMA, TDMA is combined with CDMA and Time Division Duplexing in two standard UMTS UTRA

TDMA in Wired Networks

The ITU-T G.hn standard, which provides high-speed local area networking over existing home wiring (power lines, phone lines and coaxial cables) is based on a TDMA scheme. In G.hn, a "master" device allocates "Contention-Free Transmission Opportunities" (CFTXOP) to other "slave" devices in the network. Only one device can use a CFTXOP at a time, thus avoiding collisions. FlexRay protocol which is also a wired network used for safety-critical communication in modern cars, uses the TDMA method for data transmission control.

Comparison with other Multiple-access Schemes

In radio systems, TDMA is usually used alongside Frequency-division multiple access (FDMA) and Frequency division duplex (FDD); the combination is referred to as FDMA/TDMA/FDD. This is the case in both GSM and IS-136 for example. Exceptions to this include the DECT and PHS micro-cellular systems, UMTS-TDD UMTS variant, and China's TD-SCDMA, which use Time Division duplexing, where different time slots are allocated for the base station and handsets on the same frequency.

A major advantage of TDMA is that the radio part of the mobile only needs to listen and broadcast for its own time slot. For the rest of the time, the mobile can carry out measurements on the network, detecting surrounding transmitters on different frequencies. This allows safe inter frequency handovers, something which is difficult in CDMA systems, not supported at all in IS-95 and supported through complex system additions in Universal Mobile Telecommunications System (UMTS). This in turn allows for co-existence of microcell layers with macrocell layers.

CDMA, by comparison, supports "soft hand-off" which allows a mobile phone to be in communication with up to 6 base stations simultaneously, a type of "same-frequency handover". The incoming packets are compared for quality, and the best one is selected. CDMA's "cell breathing" characteristic, where a terminal on the boundary of two congested cells will be unable to receive a clear signal, can often negate this advantage during peak periods.

A disadvantage of TDMA systems is that they create interference at a frequency which is directly connected to the time slot length. This is the buzz which can sometimes be heard if a TDMA phone is left next to a radio or speakers. Another disadvantage is that the "dead time" between time slots limits the potential bandwidth of a TDMA channel. These are implemented in part because of the difficulty in ensuring that different terminals transmit at exactly the times required. Handsets that are moving will need to constantly adjust their timings to ensure their transmission is received at precisely the right time, because as they move further from the base station, their signal will take longer to arrive. This also means that the major TDMA systems have hard limits on cell sizes in terms of range, though in practice the power levels required to receive and transmit over distances greater than the supported range would be mostly impractical anyway.

Dynamic TDMA

In dynamic time division multiple access, a scheduling algorithm dynamically reserves a variable number of time slots in each frame to variable bit-rate data streams, based on the traffic demand of each data stream. Dynamic TDMA is used in

- HIPERLAN/2 broadband radio access network.

- IEEE 802.16a WiMax

- Bluetooth

- Military Radios / Tactical Data Link

- TD-SCDMA

- ITU-T G.hn

- Simulation of TDMA / DTMA links

Code Division Multiple Access

Code division multiple access (CDMA) is a channel access method used by various radio communication technologies.

CDMA is an example of multiple access, where several transmitters can send information simultaneously over a single communication channel. This allows several users to share a band of frequencies. To permit this without undue interference between the users, CDMA employs spread-spectrum technology and a special coding scheme (where each transmitter is assigned a code).

CDMA is used as the access method in many mobile phone standards. IS-95, also called "cdma-One", and its 3G evolution CDMA2000, are often simply referred to as "CDMA"', but UMTS, the 3G standard used by GSM carriers, also uses "wideband CDMA", or W-CDMA, as well as TD-CDMA and TD-SCDMA, as its radio technologies.

History

The technology of code division multiple access channels has long been known. In the Soviet Union (USSR), the first work devoted to this subject was published in 1935 by professor Dmitriy V. Ageev. It was shown that through the use of linear methods, there are three types of signal separation: frequency, time and compensatory. The technology of CDMA was used in 1957, when the young military radio engineer Leonid Kupriyanovich in Moscow, made an experimental model of a wearable automatic mobile phone, called LK-1 by him, with a base station. LK-1 has a weight of 3 kg, 20–30 km operating distance, and 20–30 hours of battery life. The base station, as described by the author, could serve several customers. In 1958, Kupriyanovich made the new experimental "pocket" model of mobile phone. This phone weighed 0.5 kg. To serve more customers, Kupriyanovich proposed the device, named by him as correllator. In 1958, the USSR also started the development of the "Altai" national civil mobile phone service for cars, based on the Soviet MRT-1327 standard. The phone system weighed 11 kg (24 lb). It was placed in the trunk of the vehicles of high-ranking officials and used a standard handset in the passenger compartment. The main developers of the Altai system were VNIIS (Voronezh Science Research Institute of Communications) and GSPI (State Specialized Project Institute). In 1963 this service started in Moscow and in 1970 Altai service was used in 30 USSR cities.

Uses

A CDMA2000 mobile phone

- One of the early applications for code division multiplexing is in the Global Positioning System (GPS). This predates and is distinct from its use in mobile phones.

- The Qualcomm standard IS-95, marketed as cdmaOne.

- The Qualcomm standard IS-2000, known as CDMA2000, is used by several mobile phone companies, including the Globalstar satellite phone network.

- The UMTS 3G mobile phone standard, which uses W-CDMA.

- CDMA has been used in the OmniTRACS satellite system for transportation logistics.

Steps in CDMA Modulation

CDMA is a spread-spectrum multiple access technique. A spread spectrum technique spreads the bandwidth of the data uniformly for the same transmitted power. A spreading code is a pseudo-random code that has a narrow ambiguity function, unlike other narrow pulse codes. In CDMA a locally generated code runs at a much higher rate than the data to be transmitted. Data for transmission is combined via bitwise XOR (exclusive OR) with the faster code. The figure shows how a spread spectrum signal is generated. The data signal with pulse duration of T_b (symbol period) is XOR'ed with the code signal with pulse duration of T_c (chip period). (Note: bandwidth is proportional to $1/T$, where T = bit time.) Therefore, the bandwidth of the data signal is $1/T_b$ and the bandwidth of the spread spectrum signal is $1/T_c$. Since T_c is much smaller than T_b, the bandwidth of the spread spectrum signal is much larger than the bandwidth of the original signal. The ratio T_b/T_c is called the spreading factor or processing gain and determines to a certain extent the upper limit of the total number of users supported simultaneously by a base station.

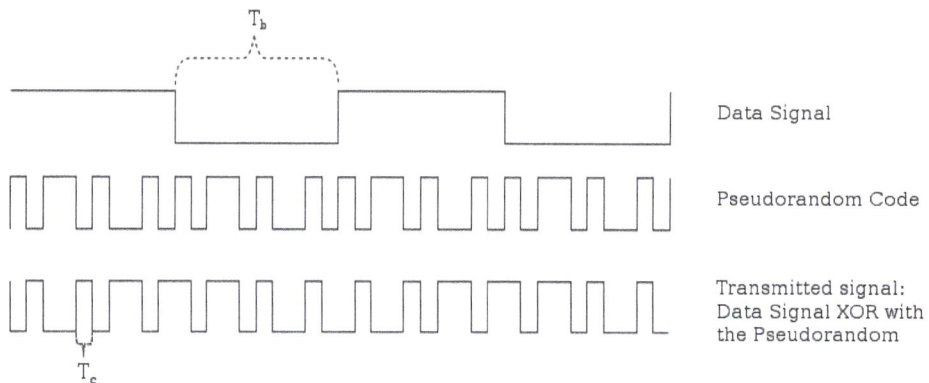

Generation of a CDMA signal

Each user in a CDMA system uses a different code to modulate their signal. Choosing the codes used to modulate the signal is very important in the performance of CDMA systems. The best performance will occur when there is good separation between the signal of a desired user and the signals of other users. The separation of the signals is made by correlating the received signal with the locally generated code of the desired user. If the signal matches the desired user's code then the correlation function will be high and the system can extract that signal. If the desired user's code has nothing in common with the signal the correlation should be as close to zero as possible (thus eliminating the signal); this is referred to as cross-correlation. If the code is correlated with the signal at any time offset other than zero, the correlation should be as close to zero as possible. This is referred to as auto-correlation and is used to reject multi-path interference.

An analogy to the problem of multiple access is a room (channel) in which people wish to talk to each other simultaneously. To avoid confusion, people could take turns speaking (time division), speak at different pitches (frequency division), or speak in different languages (code division).

CDMA is analogous to the last example where people speaking the same language can understand each other, but other languages are perceived as noise and rejected. Similarly, in radio CDMA, each group of users is given a shared code. Many codes occupy the same channel, but only users associated with a particular code can communicate.

In general, CDMA belongs to two basic categories: synchronous (orthogonal codes) and asynchronous (pseudorandom codes).

Code Division Multiplexing (Synchronous Cdma)

The digital modulation method is analogous to those used in simple radio transceivers. In the analog case, a low frequency data signal is time multiplied with a high frequency pure sine wave carrier, and transmitted. This is effectively a frequency convolution (Wiener–Khinchin theorem) of the two signals, resulting in a carrier with narrow sidebands. In the digital case, the sinusoidal carrier is replaced by Walsh functions. These are binary square waves that form a complete orthonormal set. The data signal is also binary and the time multiplication is achieved with a simple XOR function. This is usually a Gilbert cell mixer in the circuitry.

Synchronous CDMA exploits mathematical properties of orthogonality between vectors representing the data strings. For example, binary string *1011* is represented by the vector (1, 0, 1, 1). Vectors can be multiplied by taking their dot product, by summing the products of their respective components (for example, if u = (a, b) and v = (c, d), then their dot product u·v = ac + bd). If the dot product is zero, the two vectors are said to be *orthogonal* to each other. Some properties of the dot product aid understanding of how W-CDMA works. If vectors a and b are orthogonal, then $\mathbf{a}\cdot\mathbf{b}=0$ and:

$$\mathbf{a}\cdot(\mathbf{a}+\mathbf{b}) = \|\mathbf{a}\|^2 \quad \text{since} \quad \mathbf{a}\cdot\mathbf{a}+\mathbf{a}\cdot\mathbf{b} = \|\mathbf{a}\|^2 + 0$$

$$\mathbf{a}\cdot(-\mathbf{a}+\mathbf{b}) = -\|\mathbf{a}\|^2 \quad \text{since} \quad -\mathbf{a}\cdot\mathbf{a}+\mathbf{a}\cdot\mathbf{b} = -\|\mathbf{a}\|^2 + 0$$

$$\mathbf{b}\cdot(\mathbf{a}+\mathbf{b}) = \|\mathbf{b}\|^2 \quad \text{since} \quad \mathbf{b}\cdot\mathbf{a}+\mathbf{b}\cdot\mathbf{b} = 0 + \|\mathbf{b}\|^2$$

$$\mathbf{b}\cdot(\mathbf{a}-\mathbf{b}) = -\|\mathbf{b}\|^2 \quad \text{since} \quad \mathbf{b}\cdot\mathbf{a}-\mathbf{b}\cdot\mathbf{b} = 0 - \|\mathbf{b}\|^2$$

Each user in synchronous CDMA uses a code orthogonal to the others' codes to modulate their signal. An example of four mutually orthogonal digital signals is shown in the figure. Orthogonal codes have a cross-correlation equal to zero; in other words, they do not interfere with each other. In the case of IS-95 64 bit Walsh codes are used to encode the signal to separate different users. Since each of the 64 Walsh codes are orthogonal to one another, the signals are channelized into 64 orthogonal signals. The following example demonstrates how each user's signal can be encoded and decoded.

Example

Start with a set of vectors that are mutually orthogonal. (Although mutual orthogonality is the only condition, these vectors are usually constructed for ease of decoding, for example columns or rows from Walsh matrices.) An example of orthogonal functions is shown in the picture on the right. These vectors will be assigned to individual users and are called the *code*, *chip code*, or *chipping code*. In the interest of brevity, the rest of this example uses codes, **v**, with only two bits.

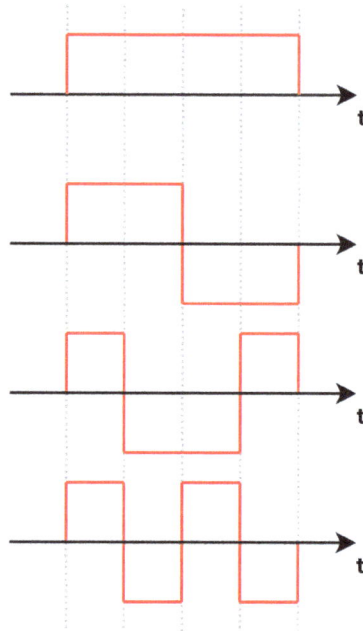

An example of four mutually orthogonal digital signals.

Each user is associated with a different code, say **v**. A 1 bit is represented by transmitting a positive code, **v**, and a 0 bit is represented by a negative code, −**v**. For example, if $\mathbf{v} = (v_0, v_1) = (1, -1)$ and the data that the user wishes to transmit is $(1, 0, 1, 1)$, then the transmitted symbols would be $(\mathbf{v}, -\mathbf{v}, \mathbf{v}, \mathbf{v}) = (v_0, v_1, -v_0, -v_1, v_0, v_1, v_0, v_1) = (1, -1, -1, 1, 1, -1, 1, -1)$. For the purposes of this article, we call this constructed vector the *transmitted vector*.

Each sender has a different, unique vector **v** chosen from that set, but the construction method of the transmitted vector is identical.

Now, due to physical properties of interference, if two signals at a point are in phase, they add to give twice the amplitude of each signal, but if they are out of phase, they subtract and give a signal that is the difference of the amplitudes. Digitally, this behaviour can be modelled by the addition of the transmission vectors, component by component.

If sender0 has code $(1, -1)$ and data $(1, 0, 1, 1)$, and sender1 has code $(1, 1)$ and data $(0, 0, 1, 1)$, and both senders transmit simultaneously, then this table describes the coding steps:

Step	Encode sender0	Encode sender1
0	code0 = $(1, -1)$, data0 = $(1, 0, 1, 1)$	code1 = $(1, 1)$, data1 = $(0, 0, 1, 1)$
1	encode0 = $2(1, 0, 1, 1) - (1, 1, 1, 1) = (1, -1, 1, 1)$	encode1 = $2(0, 0, 1, 1) - (1, 1, 1, 1) = (-1, -1, 1, 1)$
2	signal0 = encode0 $\otimes$ code0 = $(1, -1, 1, 1) \otimes (1, -1)$ = $(1, -1, -1, 1, 1, -1, 1, -1)$	signal1 = encode1 $\otimes$ code1 = $(-1, -1, 1, 1) \otimes (1, 1)$ = $(-1, -1, -1, -1, 1, 1, 1, 1)$

Because signal0 and signal1 are transmitted at the same time into the air, they add to produce the raw signal:

$$(1, -1, -1, 1, 1, -1, 1, -1) + (-1, -1, -1, -1, 1, 1, 1, 1) = (0, -2, -2, 0, 2, 0, 2, 0)$$

This raw signal is called an interference pattern. The receiver then extracts an intelligible signal for any known sender by combining the sender's code with the interference pattern. The following table explains how this works, and shows that the signals do not interfere with one another:

Step	Decode sender0	Decode sender1
0	code0 = (1, −1), signal = (0, −2, −2, 0, 2, 0, 2, 0)	code1 = (1, 1), signal = (0, −2, −2, 0, 2, 0, 2, 0)
1	decode0 = pattern.vector0	decode1 = pattern.vector1
2	decode0 = ((0, −2), (−2, 0), (2, 0), (2, 0)).(1, −1)	decode1 = ((0, −2), (−2, 0), (2, 0), (2, 0)).(1, 1)
3	decode0 = ((0 + 2), (−2 + 0), (2 + 0), (2 + 0))	decode1 = ((0 − 2), (−2 + 0), (2 + 0), (2 + 0))
4	data0=(2, −2, 2, 2), meaning (1, 0, 1, 1)	data1=(−2, −2, 2, 2), meaning (0, 0, 1, 1)

Further, after decoding, all values greater than 0 are interpreted as 1 while all values less than zero are interpreted as 0. For example, after decoding, data0 is (2, −2, 2, 2), but the receiver interprets this as (1, 0, 1, 1). Values of exactly 0 means that the sender did not transmit any data, as in the following example:

Assume signal0 = (1, −1, −1, 1, 1, −1, 1, −1) is transmitted alone. The following table shows the decode at the receiver:

Step	Decode sender0	Decode sender1
0	code0 = (1, −1), signal = (1, −1, −1, 1, 1, −1, 1, −1)	code1 = (1, 1), signal = (1, −1, −1, 1, 1, −1, 1, −1)
1	decode0 = pattern.vector0	decode1 = pattern.vector1
2	decode0 = ((1, −1), (−1, 1), (1, −1), (1, −1)).(1, −1)	decode1 = ((1, −1), (−1, 1), (1, −1), (1, −1)).(1, 1)
3	decode0 = ((1 + 1), (−1 − 1), (1 + 1), (1 + 1))	decode1 = ((1 − 1), (−1 + 1), (1 − 1), (1 − 1))
4	data0 = (2, −2, 2, 2), meaning (1, 0, 1, 1)	data1 = (0, 0, 0, 0), meaning no data

When the receiver attempts to decode the signal using sender1's code, the data is all zeros, therefore the cross correlation is equal to zero and it is clear that sender1 did not transmit any data.

Asynchronous CDMA

When mobile-to-base links cannot be precisely coordinated, particularly due to the mobility of the handsets, a different approach is required. Since it is not mathematically possible to create signature sequences that are both orthogonal for arbitrarily random starting points and which make full use of the code space, unique "pseudo-random" or "pseudo-noise" (PN) sequences are used in *asynchronous* CDMA systems. A PN code is a binary sequence that appears random but can be reproduced in a deterministic manner by intended receivers. These PN codes are used to encode and decode a user's signal in Asynchronous CDMA in the same manner as the orthogonal codes in synchronous CDMA. These PN sequences are statistically uncorrelated, and the sum of a large number of PN sequences results in *multiple access interference* (MAI) that is approximated by a Gaussian noise process (following the central limit theorem in statistics). Gold codes are an example of a PN suitable for this purpose, as there is low correlation between the codes. If all of the users are received with the same power level, then the variance (e.g., the noise power) of the MAI increases in direct proportion to the number of users. In other words, unlike synchronous CDMA, the signals of other users will appear as noise to the signal of interest and interfere slightly with the desired signal in proportion to number of users.

All forms of CDMA use spread spectrum process gain to allow receivers to partially discriminate against unwanted signals. Signals encoded with the specified PN sequence (code) are received, while signals with different codes (or the same code but a different timing offset) appear as wideband noise reduced by the process gain.

Since each user generates MAI, controlling the signal strength is an important issue with CDMA transmitters. A CDM (synchronous CDMA), TDMA, or FDMA receiver can in theory completely reject arbitrarily strong signals using different codes, time slots or frequency channels due to the orthogonality of these systems. This is not true for Asynchronous CDMA; rejection of unwanted signals is only partial. If any or all of the unwanted signals are much stronger than the desired signal, they will overwhelm it. This leads to a general requirement in any asynchronous CDMA system to approximately match the various signal power levels as seen at the receiver. In CDMA cellular, the base station uses a fast closed-loop power control scheme to tightly control each mobile's transmit power.

Advantages of Asynchronous CDMA Over Other Techniques

Efficient Practical Utilization of the Fixed Frequency Spectrum

In theory CDMA, TDMA and FDMA have exactly the same spectral efficiency but practically, each has its own challenges – power control in the case of CDMA, timing in the case of TDMA, and frequency generation/filtering in the case of FDMA.

TDMA systems must carefully synchronize the transmission times of all the users to ensure that they are received in the correct time slot and do not cause interference. Since this cannot be perfectly controlled in a mobile environment, each time slot must have a guard-time, which reduces the probability that users will interfere, but decreases the spectral efficiency. Similarly, FDMA systems must use a guard-band between adjacent channels, due to the unpredictable doppler shift of the signal spectrum because of user mobility. The guard-bands will reduce the probability that adjacent channels will interfere, but decrease the utilization of the spectrum.

Flexible Allocation of Resources

Asynchronous CDMA offers a key advantage in the flexible allocation of resources i.e. allocation of a PN codes to active users. In the case of CDM (synchronous CDMA), TDMA, and FDMA the number of simultaneous orthogonal codes, time slots and frequency slots respectively are fixed hence the capacity in terms of number of simultaneous users is limited. There are a fixed number of orthogonal codes, time slots or frequency bands that can be allocated for CDM, TDMA, and FDMA systems, which remain underutilized due to the bursty nature of telephony and packetized data transmissions. There is no strict limit to the number of users that can be supported in an asynchronous CDMA system, only a practical limit governed by the desired bit error probability, since the SIR (Signal to Interference Ratio) varies inversely with the number of users. In a bursty traffic environment like mobile telephony, the advantage afforded by asynchronous CDMA is that the performance (bit error rate) is allowed to fluctuate randomly, with an average value determined by the number of users times the percentage of utilization. Suppose there are 2N users that only talk half of the time, then 2N users can be accommodated with the same *average* bit error probability as N users that talk all of the time. The key difference here is that the bit error probability for N users talking all of the time is constant, whereas it is a *random* quantity (with the same mean) for 2N users talking half of the time.

In other words, asynchronous CDMA is ideally suited to a mobile network where large numbers of transmitters each generate a relatively small amount of traffic at irregular intervals. CDM (synchronous CDMA), TDMA, and FDMA systems cannot recover the underutilized resources inherent to bursty traffic due to the fixed number of orthogonal codes, time slots or frequency channels that can be assigned to individual transmitters. For instance, if there are N time slots in a TDMA system and 2N users that talk half of the time, then half of the time there will be more than N users needing to use more than N time slots. Furthermore, it would require significant overhead to continually allocate and deallocate the orthogonal code, time slot or frequency channel resources. By comparison, asynchronous CDMA transmitters simply send when they have something to say, and go off the air when they don't, keeping the same PN signature sequence as long as they are connected to the system.

Spread-spectrum Characteristics of CDMA

Most modulation schemes try to minimize the bandwidth of this signal since bandwidth is a limited resource. However, spread spectrum techniques use a transmission bandwidth that is several orders of magnitude greater than the minimum required signal bandwidth. One of the initial reasons for doing this was military applications including guidance and communication systems. These systems were designed using spread spectrum because of its security and resistance to jamming. Asynchronous CDMA has some level of privacy built in because the signal is spread using a pseudo-random code; this code makes the spread spectrum signals appear random or have noise-like properties. A receiver cannot demodulate this transmission without knowledge of the pseudo-random sequence used to encode the data. CDMA is also resistant to jamming. A jamming signal only has a finite amount of power available to jam the signal. The jammer can either spread its energy over the entire bandwidth of the signal or jam only part of the entire signal.

CDMA can also effectively reject narrow band interference. Since narrow band interference affects only a small portion of the spread spectrum signal, it can easily be removed through notch filtering without much loss of information. Convolution encoding and interleaving can be used to assist in recovering this lost data. CDMA signals are also resistant to multipath fading. Since the spread spectrum signal occupies a large bandwidth only a small portion of this will undergo fading due to multipath at any given time. Like the narrow band interference this will result in only a small loss of data and can be overcome.

Another reason CDMA is resistant to multipath interference is because the delayed versions of the transmitted pseudo-random codes will have poor correlation with the original pseudo-random code, and will thus appear as another user, which is ignored at the receiver. In other words, as long as the multipath channel induces at least one chip of delay, the multipath signals will arrive at the receiver such that they are shifted in time by at least one chip from the intended signal. The correlation properties of the pseudo-random codes are such that this slight delay causes the multipath to appear uncorrelated with the intended signal, and it is thus ignored.

Some CDMA devices use a rake receiver, which exploits multipath delay components to improve the performance of the system. A rake receiver combines the information from several correlators, each one tuned to a different path delay, producing a stronger version of the signal than a simple receiver with a single correlation tuned to the path delay of the strongest signal.

Frequency reuse is the ability to reuse the same radio channel frequency at other cell sites within a cellular system. In the FDMA and TDMA systems frequency planning is an important consideration. The frequencies used in different cells must be planned carefully to ensure signals from different cells do not interfere with each other. In a CDMA system, the same frequency can be used in every cell, because channelization is done using the pseudo-random codes. Reusing the same frequency in every cell eliminates the need for frequency planning in a CDMA system; however, planning of the different pseudo-random sequences must be done to ensure that the received signal from one cell does not correlate with the signal from a nearby cell.

Since adjacent cells use the same frequencies, CDMA systems have the ability to perform soft hand offs. Soft hand offs allow the mobile telephone to communicate simultaneously with two or more cells. The best signal quality is selected until the hand off is complete. This is different from hard hand offs utilized in other cellular systems. In a hard hand off situation, as the mobile telephone approaches a hand off, signal strength may vary abruptly. In contrast, CDMA systems use the soft hand off, which is undetectable and provides a more reliable and higher quality signal.

Collaborative CDMA

In a recent study, a novel collaborative multi-user transmission and detection scheme called Collaborative CDMA has been investigated for the uplink that exploits the differences between users' fading channel signatures to increase the user capacity well beyond the spreading length in multiple access interference (MAI) limited environment. The authors show that it is possible to achieve this increase at a low complexity and high bit error rate performance in flat fading channels, which is a major research challenge for overloaded CDMA systems. In this approach, instead of using one sequence per user as in conventional CDMA, the authors group a small number of users to share the same spreading sequence and enable group spreading and despreading operations. The new collaborative multi-user receiver consists of two stages: group multi-user detection (MUD) stage to suppress the MAI between the groups and a low complexity maximum-likelihood detection stage to recover jointly the co-spread users' data using minimum Euclidean distance measure and users' channel gain coefficients. In CDMA, signal security is high.

Space-division Multiple Access

Space-division multiple access (SDMA) is a channel access method based on creating parallel spatial pipes next to higher capacity pipes through spatial multiplexing and/or diversity, by which it is able to offer superior performance in radio multiple access communication systems. In traditional mobile cellular network systems, the base station has no information on the position of the mobile units within the cell and radiates the signal in all directions within the cell in order to provide radio coverage. This results in wasting power on transmissions when there are no mobile units to reach, in addition to causing interference for adjacent cells using the same frequency, so called co-channel cells. Likewise, in reception, the antenna receives signals coming from all directions including noise and interference signals. By using smart antenna technology and differing spatial locations of mobile units within the cell, space-division multiple access techniques offer attractive performance enhancements. The radiation pattern of the base station, both in transmission and reception, is adapted to each user to obtain highest gain in the direction of that user. This is often done using phased array techniques.

In GSM cellular networks, the base station is aware of the distance (but not direction) of a mobile phone by use of a technique called "timing advance" (TA). The base transceiver station (BTS) can determine how distant the mobile station (MS) is by interpreting the reported TA. This information, along with other parameters, can then be used to power down the BTS or MS, if a power control feature is implemented in the network. The power control in either BTS or MS is implemented in most modern networks, especially on the MS, as this ensures a better battery life for the MS. This is also why having a BTS close to the user results in less exposure to electromagnetic radiation.

This is why one may actually be safer to have a BTS close to them as their MS will be powered down as much as possible. For example, there is more power being transmitted from the MS than what one would receive from the BTS even if they were 6 meters away from a BTS mast. However, this estimation might not consider all the Mobile stations that a particular BTS is supporting with EM radiation at any given time.

In the same manner, 5th generation mobile networks will be focused in utilizing the given position of the MS in relation to BTS in order to focus all MS Radio frequency power to the BTS direction and vice versa, thus enabling power savings for the Mobile Operator, reducing MS SAR index, reducing the EM field around base stations since beam forming will concentrate rf power when it will be actually used rather than spread uniformly around the BTS, reducing health and safety concerns, enhancing spectral efficiency, and decreased MS battery consumption.

ALOHAnet

ALOHAnet, also known as the ALOHA System, or simply ALOHA, was a pioneering computer networking system developed at the University of Hawaii. ALOHAnet became operational in June, 1971, providing the first public demonstration of a wireless packet data network. ALOHA originally stood for Additive Links On-line Hawaii Area.

The ALOHAnet used a new method of medium access (ALOHA random access) and experimental ultra high frequency (UHF) for its operation, since frequency assignments for communications to and from a computer were not available for commercial applications in the 1970s. But even before such frequencies were assigned there were two other media available for the application of an ALOHA channel – cables and satellites. In the 1970s ALOHA random access was employed in the widely used Ethernet cable based network and then in the Marisat (now Inmarsat) satellite network.

In the early 1980s frequencies for mobile networks became available, and in 1985 frequencies suitable for what became known as Wi-Fi were allocated in the US. These regulatory developments made it possible to use the ALOHA random-access techniques in both Wi-Fi and in mobile telephone networks.

ALOHA channels were used in a limited way in the 1980s in 1G mobile phones for signaling and control purposes. In the late 1980s, the European standardisation group GSM who worked on the Pan-European Digital mobile communication system GSM greatly expanded the use of ALOHA channels for access to radio channels in mobile telephony. In addition SMS message texting was implemented in 2G mobile phones. In the early 2000s additional ALOHA channels were added to 2.5G and 3G mobile phones with the widespread introduction of GPRS, using a slotted-ALOHA random-access channel combined with a version of the Reservation ALOHA scheme first analyzed by a group at BBN.

Overview

One of the early computer networking designs, development of the ALOHA network was begun in September 1968 at the University of Hawaii under the leadership of Norman Abramson along with Thomas Gaarder, Franklin Kuo, Shu Lin, Wesley Peterson and Edward Wheldon. The goal was to use low-cost commercial radio equipment to connect users on Oahu and the other Hawaiian islands with a central time-sharing computer on the main Oahu campus. The first packet broadcasting unit went into operation in June 1971. Terminals were connected to a special purpose "terminal connection unit" using RS-232 at 9600 bit/s.

The original version of ALOHA used two distinct frequencies in a hub/star configuration, with the hub machine broadcasting packets to everyone on the "outbound" channel, and the various client machines sending data packets to the hub on the "inbound" channel. If data was received correctly at the hub, a short acknowledgment packet was sent to the client; if an acknowledgment was not received by a client machine after a short wait time, it would automatically retransmit the data packet after waiting a randomly selected time interval. This acknowledgment mechanism was used to detect and correct for "collisions" created when two client machines both attempted to send a packet at the same time.

ALOHAnet's primary importance was its use of a shared medium for client transmissions. Unlike the ARPANET where each node could only talk directly to a node at the other end of a wire or satellite circuit, in ALOHAnet all client nodes communicated with the hub on the same frequency. This meant that some sort of mechanism was needed to control who could talk at what time. The ALOHAnet solution was to allow each client to send its data without controlling when it was sent, with an acknowledgment/retransmission scheme used to deal with collisions. This approach radically reduced the complexity of the protocol and the networking hardware, since nodes do not need negotiate "who" is allowed to speak.

This solution became known as a pure ALOHA, or random-access channel, and was the basis for subsequent Ethernet development and later Wi-Fi networks. Various versions of the ALOHA protocol (such as Slotted ALOHA) also appeared later in satellite communications, and were used in wireless data networks such as ARDIS, Mobitex, CDPD, and GSM.

Also important was ALOHAnet's use of the outgoing hub channel to broadcast packets directly to all clients on a second shared frequency, using an address in each packet to allow selective receipt at each client node. Two frequencies were used so that a device could both receive acknowledgments regardless of transmissions. The Aloha network introduced the mechanism of randomized multiple access, which resolved device transmission collisions by transmitting a package immediately if no acknowledgement is present, and if no acknowledgment was received, the transmission was repeated after a random waiting time.

ALOHA Protocol

Pure ALOHA

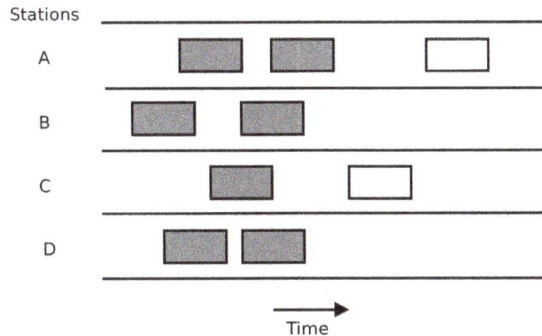

Pure ALOHA protocol. Boxes indicate frames. Shaded boxes indicate frames which have collided.

The first version of the protocol (now called "Pure ALOHA", and the one implemented in ALOHAnet) was quite simple:

- If you have data to send, send the data

- If, while you are transmitting data, you receive any data from another station, there has been a message collision. All transmitting stations will need to try resending "later".

Note that the first step implies that Pure ALOHA does not check whether the channel is busy before transmitting. Since collisions can occur and data may have to be sent again, ALOHA cannot use 100% of the capacity of the communications channel. How long a station waits until it transmits, and the likelihood a collision occurs are interrelated, and both affect how efficiently the channel can be used. This means that the concept of "transmit later" is a critical aspect: the quality of the backoff scheme chosen significantly influences the efficiency of the protocol, the ultimate channel capacity, and the predictability of its behavior.

To assess Pure ALOHA, there is a need to predict its throughput, the rate of (successful) transmission of frames. (This discussion of Pure ALOHA's performance follows Tanenbaum.) First, let's make a few simplifying assumptions:

- All frames have the same length.

- Stations cannot generate a frame while transmitting or trying to transmit. (That is, if a station keeps trying to send a frame, it cannot be allowed to generate more frames to send.)

- The population of stations attempts to transmit (both new frames and old frames that collided) according to a Poisson distribution.

Let "T" refer to the time needed to transmit one frame on the channel, and let's define "frame-time" as a unit of time equal to T. Let "G" refer to the mean used in the Poisson distribution over transmission-attempt amounts: that is, on average, there are G transmission-attempts per frame-time.

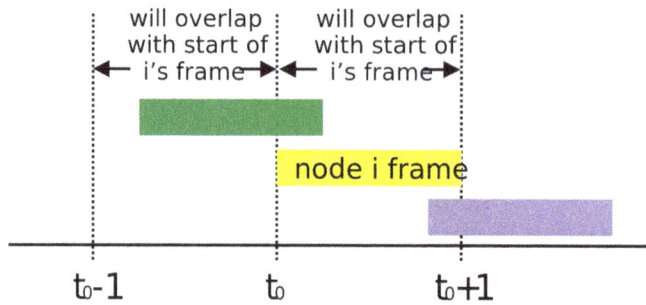

Overlapping frames in the pure ALOHA protocol. Frame-time is equal to 1 for all frames.

Consider what needs to happen for a frame to be transmitted successfully. Let "t" refer to the time at which it is intended to send a frame. It is preferable to use the channel for one frame-time beginning at t, and all other stations to refrain from transmitting during this time.

For any frame-time, the probability of there being k transmission-attempts during that frame-time is:

$$\frac{G^k e^{-G}}{k!}$$

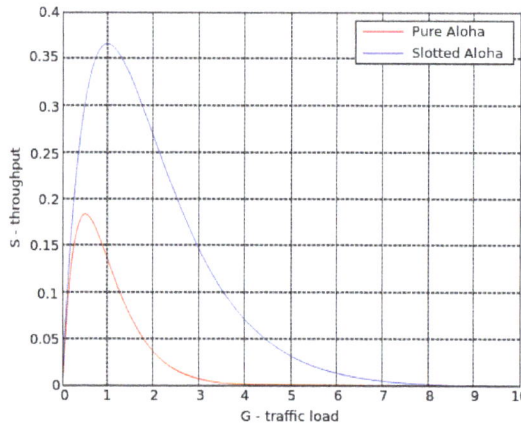

Comparison of Pure Aloha and Slotted Aloha shown on Throughput vs. Traffic Load plot.

The average amount of transmission-attempts for 2 consecutive frame-times is $2G$. Hence, for any pair of consecutive frame-times, the probability of there being k transmission-attempts during those two frame-times is:

$$\frac{(2G)^k e^{-2G}}{k!}$$

Therefore, the probability ($Prob_{pure}$) of there being zero transmission-attempts between $t-T$ and $t+T$ (and thus of a successful transmission for us) is:

$$Prob_{pure} = e^{-2G}$$

The throughput can be calculated as the rate of transmission-attempts multiplied by the probability of success, and it can be concluded that the throughput (S_{pure}) is:

$$S_{pure} = Ge^{-2G} \text{ Vulnerable time=2*T.}$$

The maximum throughput is *0.5/e* frames per frame-time (reached when $G = 0.5$), which is approximately 0.184 frames per frame-time. This means that, in Pure ALOHA, only about 18.4% of the time is used for successful transmissions.

Another simple and mathematical way to establish the equation for throughput in Pure ALOHA (and in Slotted ALOHA) is as follows:

Consider what needs to happen for frames to be transmitted successfully. Let T represents the frame time. For simplicity, it is assumed that the contention begins at t=0. Then if exactly one node sends during interval t=0 to t=T and no node tries between t=T to t=2T, then the frame will be transmitted successfully. Similarly during all next time intervals t=2nT to t=(2n+1)T, exactly one node sends and during t=(2n+1)T to t=(2n+2)T no node tries to send where n=1,2,3, …, then the frames are successfully transmitted. But in pure ALOHA, the nodes begin transmission whenever they want to do so without checking that what other nodes are doing at that time. Thus sending frames are independent events, that is, transmission by any particular node neither affects nor is affected by the time of start of transmission by other nodes. Let G be the average number of nodes that begin transmission within period T (the frame time). If a large number of nodes is trying to transmit, then by using Poisson distribution, the probability that exactly x nodes begin transmission during period T is

$$P[X = x] = \frac{G^x e^{-G}}{x!}$$

Therefore, the probability that during any particular period from t=2nT to t=(2n+1)T, (that is for any particular non-zero integral value of n) exactly one node will begin transmission is

$$P[X = 1] = \frac{G^1 e^{-G}}{1!} = Ge^{-G}$$

And the probability that during any particular period t=(2n+1)T to t=(2n+2)T, no node will begin transmission is

$$P[X = 0] = \frac{G^0 e^{-G}}{0!} = e^{-G}$$

But for successful transmission of a frame, both the events should occur simultaneously. That is during period t=2nT to t=(2n+1)T, exactly one node begins transmission and during t=(2n+1)T to t=(2n+2)T no node begins transmission. Hence the probability that both the independent events will occur simultaneously is

$$P = P(0) \times P(1) = Ge^{-G} \times e^{-G} = Ge^{-2G}$$

This is the throughput. Throughput is intended to mean the probability of successful transmission during minimum possible period. Therefore, the throughput in pure ALOHA,

$$S_{pure} = Ge^{-2G}$$

Similarly for slotted ALOHA, a frame will be successfully transmitted, if exactly one node will begin transmission at the beginning of any particular time slot (equal to frame time T). But the probability that one node will begin during any particular time slot is

$$P[X = 1] = \frac{G^1 e^{-G}}{1!} = Ge^{-G}$$

This is the throughput in slotted ALOHA. Thus,

$$S_{slotted} = Ge^{-G}$$

Disadvantages of Pure ALOHA:

1) Time is wasted

2) Data is lost

Slotted ALOHA

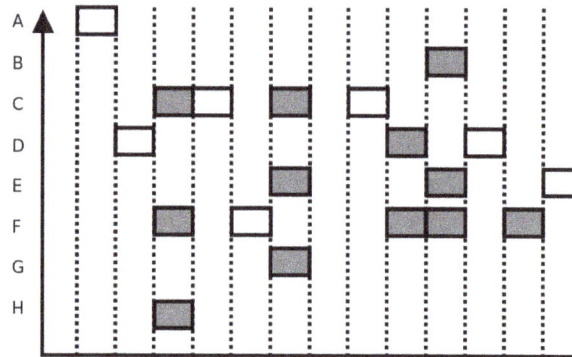

Slotted ALOHA protocol (shaded slots indicate collision)

Slotted ALOHA protocol. Boxes indicate frames. Shaded boxes indicate frames which are in the same slots.

An improvement to the original ALOHA protocol was "Slotted ALOHA", which introduced discrete timeslots and increased the maximum throughput. A station can send only at the beginning of a timeslot, and thus collisions are reduced. In this case, only transmission-attempts within 1 frame-time and not 2 consecutive frame-times need to be considered, since collisions can only occur during each timeslot. Thus, the probability of there being zero transmission-attempts in a single timeslot is:

$$Prob_{slotted} = e^{-G}$$

the probability of k packets is:

$$Prob_{slotted}k = e^{-G}(1-e^{-G})^{k-1}$$

The throughput is:

$$S_{slotted} = Ge^{-G}$$

The maximum throughput is *1/e* frames per frame-time (reached when $G = 1$), which is approximately 0.368 frames per frame-time, or 36.8%.

Slotted ALOHA is used in low-data-rate tactical satellite communications networks by military forces, in subscriber-based satellite communications networks, mobile telephony call setup, set-top box communications and in the contactless RFID technologies.

Other Protocol

The use of a random-access channel in ALOHAnet led to the development of carrier sense multiple access (CSMA), a "listen before send" random-access protocol that can be used when all nodes send and receive on the same channel. The first implementation of CSMA was Ethernet. CSMA in radio channels was extensively modeled. The AX.25 packet radio protocol is based on the CSMA approach with collision recovery, based on the experience gained from ALOHAnet.

ALOHA and the other random-access protocols have an inherent variability in their throughput and delay performance characteristics. For this reason, applications which need highly deterministic load behavior sometimes used polling or token-passing schemes (such as token ring) instead of contention systems. For instance ARCNET was popular in embedded data applications in the 1980 network.

Design

Network Architecture

Two fundamental choices which dictated much of the ALOHAnet design were the two-channel star configuration of the network and the use of random accessing for user transmissions.

The two-channel configuration was primarily chosen to allow for efficient transmission of the relatively dense total traffic stream being returned to users by the central time-sharing computer. An additional reason for the star configuration was the desire to centralize as many communication functions as possible at the central network node (the Menehune), minimizing the cost of the original all-hardware terminal control unit (TCU) at each user node.

The random-access channel for communication between users and the Menehune was designed specifically for the traffic characteristics of interactive computing. In a conventional communication system a user might be assigned a portion of the channel on either a frequency-division multiple access (FDMA) or time-division multiple access (TDMA) basis. Since it was well known that in time-sharing systems [circa 1970], computer and user data are bursty, such fixed assignments are generally wasteful of bandwidth because of the high peak-to-average data rates that characterize the traffic.

To achieve a more efficient use of bandwidth for bursty traffic, ALOHAnet developed the random-access packet switching method that has come to be known as a *pure ALOHA* channel. This approach effectively dynamically allocates bandwidth immediately to a user who has data to send, using the acknowledgment/retransmission mechanism described earlier to deal with occasional access collisions. While the average channel loading must be kept below about 10% to maintain a low collision rate, this still results in better bandwidth efficiency than when fixed allocations are used in a bursty traffic context.

Two 100 kHz channels in the experimental UHF band were used in the implemented system, one for the user-to-computer random-access channel and one for the computer-to-user broadcast channel. The system was configured as a star network, allowing only the central node to receive transmissions in the random-access channel. All user TCUs received each transmission made by the central node in the broadcast channel. All transmissions were made in bursts at 9600 bit/s, with data and control information encapsulated in packets.

Each packet consisted of a 32-bit header and a 16-bit header parity check word, followed by up to 80 bytes of data and a 16-bit parity check word for the data. The header contained address information identifying a particular user so that when the Menehune broadcast a packet, only the intended user's node would accept it.

Menehune

The central node communications processor was an HP 2100 minicomputer called the Menehune, which is the Hawaiian language word for "imp", or dwarf people, and was named for its similar role to the original ARPANET Interface Message Processor (IMP) which was being deployed at about the same time. In the original system, the Menehune forwarded correctly received user data to the UH central computer, an IBM System 360/65 time-sharing system. Outgoing messages from the 360 were converted into packets by the Menehune, which were queued and broadcast to the remote users at a data rate of 9600 bit/s. Unlike the half-duplex radios at the user TCUs, the Menehune was interfaced to the radio channels with full-duplex radio equipment.

Remote Units

The original user interface developed for the system was an all-hardware unit called an ALOHAnet Terminal Control Unit (TCU), and was the sole piece of equipment necessary to connect a terminal into the ALOHA channel. The TCU was composed of a UHF antenna, transceiver, modem, buffer and control unit. The buffer was designed for a full line length of 80 characters, which allowed handling of both the 40- and 80-character fixed-length packets defined for the system. The typical user terminal in the original system consisted of a Teletype Model 33 or a dumb CRT user terminal connected to the TCU using a standard RS-232C interface. Shortly after the original ALOHA network went into operation, the TCU was redesigned with one of the first Intel microprocessors, and the resulting upgrade was called a PCU (Programmable Control Unit).

Additional basic functions performed by the TCU's and PCU's were generation of a cyclic-parity-check code vector and decoding of received packets for packet error-detection purposes, and generation of packet retransmissions using a simple random interval generator. If an acknowledgment was not received from the Menehune after the prescribed number of automatic retransmissions, a flashing light was used as an indicator to the human user. Also, since the TCU's and PCU's did not send acknowledgments to the Menehune, a steady warning light was displayed to the human user when an error was detected in a received packet. Thus it can be seen that considerable simplification was incorporated into the initial design of the TCU as well as the PCU, making use of the fact that it was interfacing a human user into the network.

Later Developments

In later versions of the system, simple radio relays were placed in operation to connect the main network on the island of Oahu to other islands in Hawaii, and Menehune routing capabilities were expanded to allow user nodes to exchange packets with other user nodes, the ARPANET, and an experimental satellite network.

References

- Guowang Miao; Jens Zander; Ki Won Sung; Ben Slimane (2016). Fundamentals of Mobile Data Networks. Cambridge University Press. ISBN 1107143217.

- Halit Eren (Nov 16, 2005). Wireless Sensors and Instruments: Networks, Design, and Applications. CRC Press. p. 112. ISBN 9781420037401.

- Daniel Minoli (3 February 2009). Satellite Systems Engineering in an IPv6 Environment. CRC Press. pp. 136–. ISBN 978-1-4200-7868-8. Retrieved 1 June 2012.

- Kornel Terplan (2000). The Telecommunications Handbook. CRC Press. pp. 266–. ISBN 978-0-8493-3137-4. Retrieved 1 June 2012.

- Walrand, Jean; Parekh, Shyam (2010). Communication Networks: A Concise Introduction. University of California, Berkeley: Morgan & Claypool Publishers series. pp. 28–29. ISBN 9781608450947.

- Abramson, Norman (Mar 1985). "Development of the ALOHANET". IEEE Transactions on Information Theory. 31 (2): 119–123. doi:10.1109/TIT.1985.1057021. Retrieved August 2, 2015.

- Kamins, Robert M.; Potter, Robert E. (1998). Måalamalama: A History of the University of Hawai'i. University of Hawaii Press. p. 159. Retrieved August 2, 2015.

- "AX.25 Link Access Protocol for Amateur Packet Radio" (PDF). Tucson Amateur Packet Radio. 1997. p. 39. Retrieved 2014-01-06.

- "Fundamentals of Communications Access Technologies: FDMA, TDMA, CDMA, OFDMA, AND SDMA". Electronic Design. 2013-01-22. Retrieved 2014-08-28.

6

Mobile Security and Concerns

Mobile security is very important in contemporary times. The concern of security arises with the concern for the security of personal and business information that is stored in the smartphones. The following section helps the reader in understanding the importance of security in mobile phones.

Mobile Security

Mobile security or mobile phone security has become increasingly important in mobile computing. Of particular concern is the security of personal and business information now stored on smartphones.

More and more users and businesses employ smartphones as communication tools, but also as a means of planning and organizing their work and private life. Within companies, these technologies are causing profound changes in the organization of information systems and therefore they have become the source of new risks. Indeed, smartphones collect and compile an increasing amount of sensitive information to which access must be controlled to protect the privacy of the user and the intellectual property of the company.

All smartphones, as computers, are preferred targets of attacks. These attacks exploit weaknesses related to smartphones that can come from means of communication like Short Message Service (SMS, aka text messaging), Multimedia Messaging Service (MMS), Wi-Fi networks, Bluetooth and GSM, the de facto global standard for mobile communications. There are also attacks that exploit software vulnerabilities from both the web browser and operating system. Finally, there are forms of malicious software that rely on the weak knowledge of average users.

Different security counter-measures are being developed and applied to smartphones, from security in different layers of software to the dissemination of information to end users. There are good practices to be observed at all levels, from design to use, through the development of operating systems, software layers, and downloadable apps.

Challenges of Mobile Security

Threats

A smartphone user is exposed to various threats when they use their phone. In just the last two quarters of 2012, the number of unique mobile threats grew by 261%, according to ABI Research. These threats can disrupt the operation of the smartphone, and transmit or modify user data. For these reasons, the applications deployed there must guarantee privacy and integrity of the information they handle. In addition, since some apps could themselves be malware, their functionality and activities should be limited (for example, restricting the apps from accessing location information via GPS, blocking access to the user's address book, preventing the transmission of data on the network, sending SMS messages that are billed to the user, etc.).

There are three prime targets for attackers:

- Data: smartphones are devices for data management, therefore they may contain sensitive data like credit card numbers, authentication information, private information, activity logs (calendar, call logs);

- Identity: smartphones are highly customizable, so the device or its contents are associated with a specific person. For example, every mobile device can transmit information related to the owner of the mobile phone contract, and an attacker may want to steal the identity of the owner of a smartphone to commit other offenses;

- Availability: by attacking a smartphone one can limit access to it and deprive the owner of the service.

The source of these attacks are the same actors found in the non-mobile computing space:

- Professionals, whether commercial or military, who focus on the three targets mentioned above. They steal sensitive data from the general public, as well as undertake industrial espionage. They will also use the identity of those attacked to achieve other attacks;

- Thieves who want to gain income through data or identities they have stolen. The thieves will attack many people to increase their potential income;

- Black hat hackers who specifically attack availability. Their goal is to develop viruses, and cause damage to the device. In some cases, hackers have an interest in stealing data on devices.

- Grey hat hackers who reveal vulnerabilities. Their goal is to expose vulnerabilities of the device. Grey hat hackers do not intend on damaging the device or stealing data.

Consequences

When a smartphone is infected by an attacker, the attacker can attempt several things:

- The attacker can manipulate the smartphone as a zombie machine, that is to say, a machine with which the attacker can communicate and send commands which will be used to send unsolicited messages (spam) via sms or email;

- The attacker can easily force the smartphone to make phone calls. For example, one can use the API (library that contains the basic functions not present in the smartphone) Phone-MakeCall by Microsoft, which collects telephone numbers from any source such as yellow pages, and then call them. But the attacker can also use this method to call paid services, resulting in a charge to the owner of the smartphone. It is also very dangerous because the smartphone could call emergency services and thus disrupt those services;

- A compromised smartphone can record conversations between the user and others and send them to a third party. This can cause user privacy and industrial security problems;

- An attacker can also steal a user's identity, usurp their identity (with a copy of the user's sim card or even the telephone itself), and thus impersonate the owner. This raises security concerns in countries where smartphones can be used to place orders, view bank accounts or are used as an identity card;

- The attacker can reduce the utility of the smartphone, by discharging the battery. For example, they can launch an application that will run continuously on the smartphone processor, requiring a lot of energy and draining the battery. One factor that distinguishes mobile computing from traditional desktop PCs is their limited performance. Frank Stajano and Ross Anderson first described this form of attack, calling it an attack of "battery exhaustion" or "sleep deprivation torture";

- The attacker can prevent the operation and/or starting of the smartphone by making it unusable. This attack can either delete the boot scripts, resulting in a phone without a functioning OS, or modify certain files to make it unusable (e.g. a script that launches at startup that forces the smartphone to restart) or even embed a startup application that would empty the battery;

- The attacker can remove the personal (photos, music, videos, etc.) or professional data of the user.

Attacks based on Communication

Attack based on SMS and MMS

Some attacks derive from flaws in the management of SMS and MMS.

Some mobile phone models have problems in managing binary SMS messages. It is possible, by sending an ill-formed block, to cause the phone to restart, leading to denial of service attacks. If a user with a Siemens S55 received a text message containing a Chinese character, it would lead to a denial of service. In another case, while the standard requires that the maximum size of a Nokia Mail address is 32 characters, some Nokia phones did not verify this standard, so if a user enters an email address over 32 characters, that leads to complete dysfunction of the e-mail handler and puts it out of commission. This attack is called "curse of silence". A study on the safety of the SMS infrastructure revealed that SMS messages sent from the Internet can be used to perform a distributed denial of service (DDoS) attack against the mobile telecommunications infrastructure of a big city. The attack exploits the delays in the delivery of messages to overload the network.

Another potential attack could begin with a phone that sends an MMS to other phones, with an attachment. This attachment is infected with a virus. Upon receipt of the MMS, the user can choose to open the attachment. If it is opened, the phone is infected, and the virus sends an MMS with an infected attachment to all the contacts in the address book. There is a real-world example of this attack: the virus Commwarrior uses the address book and sends MMS messages including an infected file to recipients. A user installs the software, as received via MMS message. Then, the virus began to send messages to recipients taken from the address book.

Attacks based on Communication Networks

Attacks based on the GSM networks

The attacker may try to break the encryption of the mobile network. The GSM network encryption algorithms belong to the family of algorithms called A5. Due to the policy of security through obscurity it has not been possible to openly test the robustness of these algorithms. There were originally two variants of the algorithm: A5/1 and A5/2 (stream ciphers), where the former was designed to be relatively strong, and the latter was designed to be weak on purpose to allow easy cryptanalysis and eavesdropping. ETSI forced some countries (typically outside Europe) to use A5/2. Since the encryption algorithm was made public, it was proved it was possible to break the encryption: A5/2 could be broken on the fly, and A5/1 in about 6 hours . In July 2007, the 3GPP approved a change request to prohibit the implementation of A5/2 in any new mobile phones, which means that is has been decommissioned and is no longer implemented in mobile phones. Stronger public algorithms have been added to the GSM standard, the A5/3 and A5/4 (Block ciphers), otherwise known as KASUMI or UEA1 published by the ETSI. If the network does not support A5/1, or any other A5 algorithm implemented by the phone, then the base station can specify A5/0 which is the null-algorithm, whereby the radio traffic is sent unencrypted. Even in case mobile phones are able to use 3G or 4G which have much stronger encryption than 2G GSM, the base station can downgrade the radio communication to 2G GSM and specify A5/0 (no encryption) . This is the basis for eavesdropping attacks on mobile radio networks using a fake base station commonly called an IMSI catcher.

In addition, tracing of mobile terminals is difficult since each time the mobile terminal is accessing or being accessed by the network, a new temporary identity (TMSI) is allocated to the mobile terminal. The TSMI is used as identity of the mobile terminal the next time it accesses the network. The TMSI is sent to the mobile terminal in encrypted messages.

Once the encryption algorithm of GSM is broken, the attacker can intercept all unencrypted communications made by the victim's smartphone.

Attacks based on Wi-Fi

An attacker can try to eavesdrop on Wi-Fi communications to derive information (e.g. username, password). This type of attack is not unique to smartphones, but they are very vulnerable to these attacks because very often the Wi-Fi is the only means of communication they have to access the internet. The security of wireless networks (WLAN) is thus an important subject. Initially wireless networks were secured by WEP keys. The weakness of WEP is a short encryption key which is the same for all connected clients. In addition, several reductions in the search space of the keys have been found by researchers. Now, most wireless networks are protected by the WPA security protocol. WPA is based on the "Temporal Key Integrity Protocol (TKIP)" which was designed to allow migration from WEP to WPA on the equipment already deployed. The major improvements in security are the dynamic encryption keys.

For small networks, the WPA is a "pre-shared key" which is based on a shared key. Encryption can be vulnerable if the length of the shared key is short. With limited opportunities for input (i.e. only the numeric keypad) mobile phone users might define short encryption keys that contain only numbers.

This increases the likelihood that an attacker succeeds with a brute-force attack. The successor to WPA, called WPA2, is supposed to be safe enough to withstand a brute force attack.

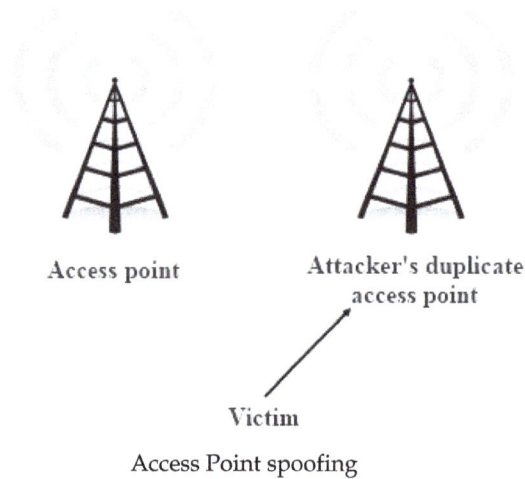

Access point   Attacker's duplicate
                              access point

Victim

Access Point spoofing

As with GSM, if the attacker succeeds in breaking the identification key, it will be possible to attack not only the phone but also the entire network it is connected to.

Many smartphones for wireless LANs remember they are already connected, and this mechanism prevents the user from having to re-identify with each connection. However, an attacker could create a WIFI access point twin with the same parameters and characteristics as the real network. Using the fact that some smartphones remember the networks, they could confuse the two networks and connect to the network of the attacker who can intercept data if it does not transmit its data in encrypted form.

Lasco is a worm that initially infects a remote device using the SIS file format. SIS file format (Software Installation Script) is a script file that can be executed by the system without user interaction. The smartphone thus believes the file to come from a trusted source and downloads it, infecting the machine.

Principle of Bluetooth-based Attacks

Security issues related to Bluetooth on mobile devices have been studied and have shown numerous problems on different phones. One easy to exploit vulnerability: unregistered services do not require authentication, and vulnerable applications have a virtual serial port used to control the phone. An attacker only needed to connect to the port to take full control of the device. Another example: a phone must be within reach and Bluetooth in discovery mode. The attacker sends a file via Bluetooth. If the recipient accepts, a virus is transmitted. For example: Cabir is a worm that spreads via Bluetooth connection. The worm searches for nearby phones with Bluetooth in discoverable mode and sends itself to the target device. The user must accept the incoming file and install the program. After installing, the worm infects the machine.

Attacks based on Vulnerabilities in Software Applications

Other attacks are based on flaws in the OS or applications on the phone.

Web Browser

The mobile web browser is an emerging attack vector for mobile devices. Just as common Web browsers, mobile web browsers are extended from pure web navigation with widgets and plug-ins, or are completely native mobile browsers.

Jailbreaking the iPhone with firmware 1.1.1 was based entirely on vulnerabilities on the web browser. As a result, the exploitation of the vulnerability described here underlines the importance of the Web browser as an attack vector for mobile devices. In this case, there was a vulnerability based on a stack-based buffer overflow in a library used by the web browser (Libtiff).

A vulnerability in the web browser for Android was discovered in October 2008. As the iPhone vulnerability above, it was due to an obsolete and vulnerable library. A significant difference with the iPhone vulnerability was Android's sandboxing architecture which limited the effects of this vulnerability to the Web browser process.

Smartphones are also victims of classic piracy related to the web: phishing, malicious websites, etc. The big difference is that smartphones do not yet have strong antivirus software available.

Operating System

Sometimes it is possible to overcome the security safeguards by modifying the operating system itself. As real-world examples, this section covers the manipulation of firmware and malicious signature certificates. These attacks are difficult.

In 2004, vulnerabilities in virtual machines running on certain devices were revealed. It was possible to bypass the bytecode verifier and access the native underlying operating system. The results of this research were not published in detail. The firmware security of Nokia's Symbian Platform Security Architecture (PSA) is based on a central configuration file called SWIPolicy. In 2008 it was possible to manipulate the Nokia firmware before it is installed, and in fact in some downloadable versions of it, this file was human readable, so it was possible to modify and change the image of the firmware. This vulnerability has been solved by an update from Nokia.

In theory smartphones have an advantage over hard drives since the OS files are in ROM, and cannot be changed by malware. However, in some systems it was possible to circumvent this: in the Symbian OS it was possible to overwrite a file with a file of the same name. On the Windows OS, it was possible to change a pointer from a general configuration file to an editable file.

When an application is installed, the signing of this application is verified by a series of certificates. One can create a valid signature without using a valid certificate and add it to the list. In the Symbian OS all certificates are in the directory. With firmware changes explained above it is very easy to insert a seemingly valid but malicious certificate.

Attacks based on Hardware Vulnerabilities

Electromagnetic Waveforms

In 2015, researchers at the French government agency ANSSI demonstrated the capability to trigger the voice interface of certain smartphones remotely by using "specific electromagnetic waveforms". The exploit took advantage of antenna-properties of headphone wires while plugged into the audio-output jacks of the vulnerable smartphones and effectively spoofed audio input to inject commands via the audio interface.

Juice Jacking

Juice Jacking is a method of physical or a hardware vulnerability specific to mobile platforms. Utilizing the dual purpose of the USB charge port, many devices have been susceptible to having data ex-filtrated from, or malware installed on to a mobile device by utilizing malicious charging kiosks set up in public places, or hidden in normal charge adapters.

Password Cracking

In 2010, researcher from the University of Pennsylvania investigated the possibility of cracking a device's password through a smudge attack (literally imaging the finger smudges on the screen to discern the user's password). The researchers were able to discern the device password up to 68% of the time under certain conditions. Outsiders may perform over-the-shoulder on victims, such as watching specific keystrokes or pattern gestures, to unlock device password or passcode.

Malicious Software (Malware)

As smartphones are a permanent point of access to the internet (mostly on), they can be compromised as easily as computers with malware. A malware is a computer program that aims to harm the system in which it resides. Trojans, worms and viruses are all considered malware. A Trojan is a program that is on the smartphone and allows external users to connect discreetly. A worm is a program that reproduces on multiple computers across a network. A virus is malicious software designed to spread to other computers by inserting itself into legitimate programs and running programs in parallel. However, it must be said that the malware are far less numerous and important to smartphones as they are to computers.

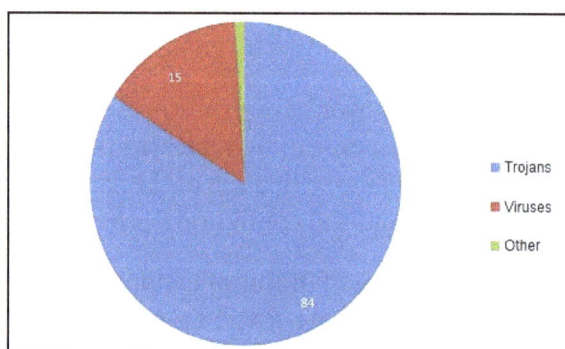

Nonetheless, recent studies show that the evolution of malware in smartphones have rocketed in the last few years posing a threat to analysis and detection.

The Three Phases of Malware Attacks

Typically an attack on a smartphone made by malware takes place in 3 phases: the infection of a host, the accomplishment of its goal, and the spread of the malware to other systems. Malware often use the resources offered by the infected smartphones. It will use the output devices such as Bluetooth or infrared, but it may also use the address book or email address of the person to infect the user's acquaintances. The malware exploits the trust that is given to data sent by an acquaintance.

Infection

Infection is the means used by the malware to get into the smartphone, it can either use one of the faults previously presented or may use the gullibility of the user. Infections are classified into four classes according to their degree of user interaction:

Explicit permission

> the most benign interaction is to ask the user if it is allowed to infect the machine, clearly indicating its potential malicious behavior. This is typical behavior of a proof of concept malware.

Implied permission

> this infection is based on the fact that the user has a habit of installing software. Most trojans try to seduce the user into installing attractive applications (games, useful applications etc.) that actually contain malware.

Common interaction

> this infection is related to a common behavior, such as opening an MMS or email.

No interaction

> the last class of infection is the most dangerous. Indeed, a worm that could infect a smartphone and could infect other smartphones without any interaction would be catastrophic.

Accomplishment of its Goal

Once the malware has infected a phone it will also seek to accomplish its goal, which is usually one of the following: monetary damage, damage data and/or device, and concealed damage:

Monetary damages

> the attacker can steal user data and either sell them to the same user, or sell to a third party.

Damage

> malware can partially damage the device, or delete or modify data on the device.

Concealed damage

> the two aforementioned types of damage are detectable, but the malware can also leave a backdoor for future attacks or even conduct wiretaps.

Spread to other Systems

Once the malware has infected a smartphone, it always aims to spread one way or another:

- It can spread through proximate devices using Wi-Fi, Bluetooth and infrared;

- It can also spread using remote networks such as telephone calls or SMS or emails.

Examples of Malware

Here are various malware that exist in the world of smartphones with a short description of each.

Viruses and Trojans

- Cabir (also known as Caribe, SybmOS/Cabir, Symbian/Cabir and EPOC.cabir) is the name of a computer worm developed in 2004 that is designed to infect mobile phones running Symbian OS. It is believed to be the first computer worm that can infect mobile phones

- Commwarrior, found March 7, 2005, is the first worm that can infect many machines from MMS. It is sent in the form of an archive file COMMWARRIOR.ZIP that contains a file COMMWARRIOR.SIS. When this file is executed, Commwarrior attempts to connect to nearby devices by Bluetooth or infrared under a random name. It then attempts to send MMS message to the contacts in the smartphone with different header messages for each person, who receive the MMS and often open them without further verification.

- Phage is the first Palm OS virus that was discovered. It transfers to the Palm from a PC via synchronization. It infects all applications that are in the smartphone and it embeds its own code to function without the user and the system detecting it. All that the system will detect is that its usual applications are functioning.

- RedBrowser is a Trojan which is based on java. The Trojan masquerades as a program called "RedBrowser" which allows the user to visit WAP sites without a WAP connection. During application installation, the user sees a request on their phone that the application needs permission to send messages. Therefore, if the user accepts, RedBrowser can send sms to paid call centers. This program uses the smartphone's connection to social networks (Facebook, Twitter, etc.) to get the contact information for the user's acquaintances (provided the required permissions have been given) and will send them messages.

- WinCE.PmCryptic.A is a malicious software on Windows Mobile which aims to earn money for its authors. It uses the infestation of memory cards that are inserted in the smartphone to spread more effectively.

- CardTrap is a virus that is available on different types of smartphone, which aims to deactivate the system and third party applications. It works by replacing the files used to start the smartphone and applications to prevent them from executing. There are different variants of this virus such as Cardtrap.A for SymbOS devices. It also infects the memory card with malware capable of infecting Windows.

- Ghost Push is a malicious software on Android OS which automatically root the android device and installs malicious applications directly to system partition then unroots the device to prevent users from removing the threat by master reset (The threat can be removed only by reflashing). It cripples the system resources, executes quickly, and harder to detect.

Ransomware

Mobile ransomware is a type of malware that locks users out of their mobile devices in a pay-to-unlock-your-device ploy, it has grown by leaps and bounds as a threat category since 2014. Specific to mobile computing platforms, users are often less security-conscious, particularly as it pertains to scrutinizing applications and web links trusting the native protection capability of the mobile device operating system. Mobile ransomware poses a significant threat to businesses reliant on instant access and availability of their proprietary information and contacts. The likelihood of a traveling businessman paying a ransom to unlock their device is significantly higher since they are at a disadvantage given inconveniences such as timeliness and less likely direct access to IT staff.

Spyware

- Flexispy is an application that can be considered as a trojan, based on Symbian. The program sends all information received and sent from the smartphone to a Flexispy server. It was originally created to protect children and spy on adulterous spouses.

Number of Malware

Below is a diagram which loads the different behaviors of smartphone malware in terms of their effects on smartphones:

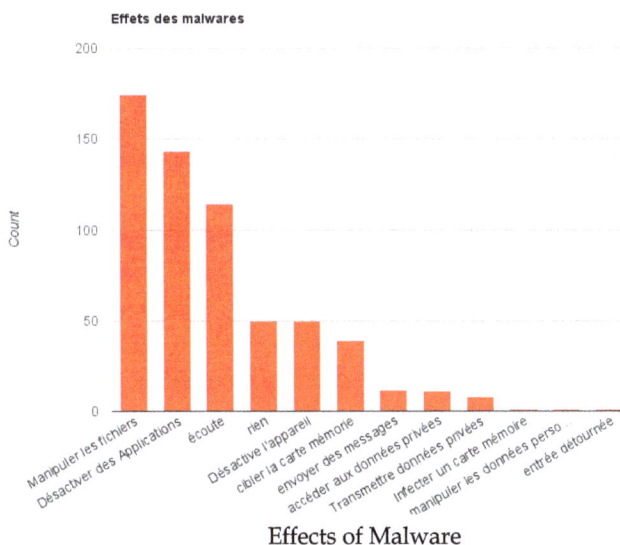

Effects of Malware

We can see from the graph that at least 50 malwares exhibit no negative behavior, except their ability to spread.

Portability of Malware across Platforms

There is a multitude of malware. This is partly due to the variety of operating systems on smartphones. However attackers can also choose to make their malware target multiple platforms, and malware can be found which attacks an OS but is able to spread to different systems.

To begin with, malware can use runtime environments like Java virtual machine or the .NET

Framework. They can also use other libraries present in many operating systems. Other malware carry several executable files in order to run in multiple environments and they utilize these during the propagation process. In practice, this type of malware requires a connection between the two operating systems to use as an attack vector. Memory cards can be used for this purpose, or synchronization software can be used to propagate the virus.

Countermeasures

The security mechanisms in place to counter the threats described above are presented in this section. They are divided into different categories, as all do not act at the same level, and they range from the management of security by the operating system to the behavioral education of the user. The threats prevented by the various measures are not the same depending on the case. Considering the two cases mentioned above, in the first case one would protect the system from corruption by an application, and in the second case the installation of a suspicious software would be prevented.

Security in Operating Systems

The first layer of security within a smartphone is at the level of the operating system (OS). Beyond the usual roles of an operating system (e.g. resource management, scheduling processes) on a smartphone, it must also establish the protocols for introducing external applications and data without introducing risk.

A central idea found in the mobile operating systems is the idea of a sandbox. Since smartphones are currently being designed to accommodate many applications, they must put in place mechanisms to ensure these facilities are safe for themselves, for other applications and data on the system, and the user. If a malicious program manages to reach a device, it is necessary that the vulnerable area presented by the system be as small as possible. Sandboxing extends this idea to compartmentalize different processes, preventing them from interacting and damaging each other. Based on the history of operating systems, sandboxing has different implementations. For example, where iOS will focus on limiting access to its public API for applications from the App Store by default, Managed Open In allows you to restrict which apps can access which types of data. Android bases its sandboxing on its legacy of Linux and TrustedBSD.

The following points highlight mechanisms implemented in operating systems, especially Android.

Rootkit Detectors

> The intrusion of a rootkit in the system is a great danger in the same way as on a computer. It is important to prevent such intrusions, and to be able to detect them as often as possible. Indeed, there is concern that with this type of malicious program, the result could be a partial or complete bypass of the device security, and the acquisition of administrator rights by the attacker. If this happens, then nothing prevents the attacker from studying or disabling the safety features that were circumvented, deploying the applications they want, or disseminating a method of intrusion by a rootkit to a wider audience. We can cite, as a defense mechanism, the Chain of trust in iOS. This mechanism relies on the signature of the different applications required to start the operating system, and a certificate signed by

Apple. In the event that the signature checks are inconclusive, the device detects this and stops the boot-up. If the Operating System is compromised due to Jailbreaking, root kit detection may not work if it is disabled by the Jailbreak method or software is loaded after Jailbreak disables Rootkit Detection.

Process isolation

Android uses mechanisms of user process isolation inherited from Linux. Each application has a user associated with it, and a tuple (UID, GID). This approach serves as a sandbox: while applications can be malicious, they can not get out of the sandbox reserved for them by their identifiers, and thus cannot interfere with the proper functioning of the system. For example, since it is impossible for a process to end the process of another user, an application can thus not stop the execution of another.

File permissions

From the legacy of Linux, there are also filesystem permissions mechanisms. They help with sandboxing: a process can not edit any files it wants. It is therefore not possible to freely corrupt files necessary for the operation of another application or system. Furthermore, in Android there is the method of locking memory permissions. It is not possible to change the permissions of files installed on the SD card from the phone, and consequently it is impossible to install applications.

Memory Protection

In the same way as on a computer, memory protection prevents privilege escalation. Indeed, if a process managed to reach the area allocated to other processes, it could write in the memory of a process with rights superior to their own, with root in the worst case, and perform actions which are beyond its permissions on the system. It would suffice to insert function calls are authorized by the privileges of the malicious application.

Development through runtime environments

Software is often developed in high-level languages, which can control what is being done by a running program. For example, Java Virtual Machines continuously monitor the actions of the execution threads they manage, monitor and assign resources, and prevent malicious actions. Buffer overflows can be prevented by these controls.

Security Software

Above the operating system security, there is a layer of security software. This layer is composed of individual components to strengthen various vulnerabilities: prevent malware, intrusions, the identification of a user as a human, and user authentication. It contains software components that have learned from their experience with computer security; however, on smartphones, this software must deal with greater constraints.

Antivirus and firewall

An antivirus software can be deployed on a device to verify that it is not infected by a known threat, usually by signature detection software that detects malicious executable files. A firewall, meanwhile, can watch over the existing traffic on the network and ensure that a malicious application does not seek to communicate through it. It may equally verify that an installed application does not seek to establish suspicious communication, which may prevent an intrusion attempt.

Visual Notifications

In order to make the user aware of any abnormal actions, such as a call they did not initiate, one can link some functions to a visual notification that is impossible to circumvent. For example, when a call is triggered, the called number should always be displayed. Thus, if a call is triggered by a malicious application, the user can see, and take appropriate action.

Turing test

In the same vein as above, it is important to confirm certain actions by a user decision. The Turing test is used to distinguish between a human and a virtual user, and it often comes as a captcha.

Biometric identification

Another method to use is biometrics. Biometrics is a technique of identifying a person by means of their morphology(by recognition of the eye or face, for example) or their behavior (their signature or way of writing for example). One advantage of using biometric security is that users can avoid having to remember a password or other secret combination to authenticate and prevent malicious users from accessing their device. In a system with strong biometric security, only the primary user can access the smartphone.

Resource Monitoring in the Smartphone

When an application passes the various security barriers, it can take the actions for which it was designed. When such actions are triggered, the activity of a malicious application can be sometimes detected if one monitors the various resources used on the phone. Depending on the goals of the malware, the consequences of infection are not always the same; all malicious applications are not intended to harm the devices on which they are deployed. The following sections describe different ways to detect suspicious activity.

Battery

Some malware is aimed at exhausting the energy resources of the phone. Monitoring the energy consumption of the phone can be a way to detect certain malware applications.

Memory usage

Memory usage is inherent in any application. However, if one finds that a substantial proportion of memory is used by an application, it may be flagged as suspicious.

Network traffic

On a smartphone, many applications are bound to connect via the network, as part of their normal operation. However, an application using a lot of bandwidth can be strongly suspected of attempting to communicate a lot of information, and disseminate data to many other devices. This observation only allows a suspicion, because some legitimate applications can be very resource-intensive in terms of network communications, the best example being streaming video.

Services

One can monitor the activity of various services of a smartphone. During certain moments, some services should not be active, and if one is detected, the application should be suspected. For example, the sending of an SMS when the user is filming video: this communication does not make sense and is suspicious; malware may attempt to send SMS while its activity is masked.

The various points mentioned above are only indications and do not provide certainty about the legitimacy of the activity of an application. However, these criteria can help target suspicious applications, especially if several criteria are combined.

Network Surveillance

Network traffic exchanged by phones can be monitored. One can place safeguards in network routing points in order to detect abnormal behavior. As the mobile's use of network protocols is much more constrained than that of a computer, expected network data streams can be predicted (e.g. the protocol for sending an SMS), which permits detection of anomalies in mobile networks.

Spam filters

As is the case with email exchanges, we can detect a spam campaign through means of mobile communications (SMS, MMS). It is therefore possible to detect and minimize this kind of attempt by filters deployed on network infrastructure that is relaying these messages.

Encryption of stored or transmitted information

Because it is always possible that data exchanged can be intercepted, communications, or even information storage, can rely on encryption to prevent a malicious entity from using any data obtained during communications. However, this poses the problem of key exchange for encryption algorithms, which requires a secure channel.

Telecom network monitoring

The networks for SMS and MMS exhibit predictable behavior, and there is not as much liberty compared with what one can do with protocols such as TCP or UDP. This implies that one cannot predict the use made of the common protocols of the web; one might generate very little traffic by consulting simple pages, rarely, or generate heavy traffic by using video streaming. On the other hand, messages exchanged via mobile phone have a framework and a specific model, and the user does not, in a normal case, have the freedom to intervene in the details of these communications. Therefore, if an abnormality is found in the flux of network data in the mobile networks, the potential threat can be quickly detected.

Manufacturer Surveillance

In the production and distribution chain for mobile devices, it is the responsibility of manufacturers to ensure that devices are delivered in a basic configuration without vulnerabilities. Most users are not experts and many of them are not aware of the existence of security vulnerabilities, so the device configuration as provided by manufacturers will be retained by many users. Below are listed several points which manufacturers should consider.

Remove debug mode

> Phones are sometimes set in a debug mode during manufacturing, but this mode must be disabled before the phone is sold. This mode allows access to different features, not intended for routine use by a user. Due to the speed of development and production, distractions occur and some devices are sold in debug mode. This kind of deployment exposes mobile devices to exploits that utilize this oversight.

Default settings

> When a smartphone is sold, its default settings must be correct, and not leave security gaps. The default configuration is not always changed, so a good initial setup is essential for users. There are, for example, default configurations that are vulnerable to denial of service attacks.

Security audit of apps

> Along with smart phones, appstores have emerged. A user finds themselves facing a huge range of applications. This is especially true for providers who manage appstores because they are tasked with examining the apps provided, from different points of view (e.g. security, content). The security audit should be particularly cautious, because if a fault is not detected, the application can spread very quickly within a few days, and infect a significant number of devices.

Detect suspicious applications demanding rights

> When installing applications, it is good to warn the user against sets of permissions that, grouped together, seem potentially dangerous, or at least suspicious. Frameworks like such as Kirin, on Android, attempt to detect and prohibit certain sets of permissions.

Revocation procedures

> Along with appstores appeared a new feature for mobile apps: remote revocation. First developed by Android, this procedure can remotely and globally uninstall an application, on any device that has it. This means the spread of a malicious application that managed to evade security checks can be immediately stopped when the threat is discovered.

Avoid heavily customized systems

> Manufacturers are tempted to overlay custom layers on existing operating systems, with the dual purpose of offering customized options and disabling or charging for certain features. This has the dual effect of risking the introduction of new bugs in the system, coupled

with an incentive for users to modify the systems to circumvent the manufacturer's restrictions. These systems are rarely as stable and reliable as the original, and may suffer from phishing attempts or other exploits.

Improve software patch processes

New versions of various software components of a smartphone, including operating systems, are regularly published. They correct many flaws over time. Nevertheless, manufacturers often do not deploy these updates to their devices in a timely fashion, and sometimes not at all. Thus, vulnerabilities persist when they could be corrected, and if they are not, since they are known, they are easily exploitable.

User awareness

Much malicious behavior is allowed by the carelessness of the user. From simply not leaving the device without a password, to precise control of permissions granted to applications added to the smartphone, the user has a large responsibility in the cycle of security: to not be the vector of intrusion. This precaution is especially important if the user is an employee of a company that stores business data on the device. Detailed below are some precautions that a user can take to manage security on a smartphone.

A recent survey by internet security experts BullGuard showed a lack of insight into the rising number of malicious threats affecting mobile phones, with 53% of users claiming that they are unaware of security software for Smartphones. A further 21% argued that such protection was unnecessary, and 42% admitted it hadn't crossed their mind ("Using APA," 2011). These statistics show consumers are not concerned about security risks because they believe it is not a serious problem. The key here is to always remember smartphones are effectively handheld computers and are just as vulnerable.

Being skeptical

A user should not believe everything that may be presented, as some information may be phishing or attempting to distribute a malicious application. It is therefore advisable to check the reputation of the application that they want to buy before actually installing it.

Permissions given to applications

The mass distribution of applications is accompanied by the establishment of different permissions mechanisms for each operating system. It is necessary to clarify these permissions mechanisms to users, as they differ from one system to another, and are not always easy to understand. In addition, it is rarely possible to modify a set of permissions requested by an application if the number of permissions is too great. But this last point is a source of risk because a user can grant rights to an application, far beyond the rights it needs. For example, a note taking application does not require access to the geolocation service. The user must ensure the privileges required by an application during installation and should not accept the installation if requested rights are inconsistent.

Be careful

Protection of a user's phone through simple gestures and precautions, such as locking the smartphone when it is not in use, not leaving their device unattended, not trusting applications, not storing sensitive data, or encrypting sensitive data that cannot be separated from the device.

Ensure data

Smartphones have a significant memory and can carry several gigabytes of data. The user must be careful about what data it carries and whether they should be protected. While it is usually not dramatic if a song is copied, a file containing bank information or business data can be more risky. The user must have the prudence to avoid the transmission of sensitive data on a smartphone, which can be easily stolen. Furthermore, when a user gets rid of a device, they must be sure to remove all personal data first.

These precautions are measures that leave no easy solution to the intrusion of people or malicious applications in a smartphone. If users are careful, many attacks can be defeated, especially phishing and applications seeking only to obtain rights on a device.

Centralized Storage of text messages

One form of mobile protection allows companies to control the delivery and storage of text messages, by hosting the messages on a company server, rather than on the sender or receiver's phone. When certain conditions are met, such as an expiration date, the messages are deleted.

Limitations of Certain Security Measures

The security mechanisms mentioned in this article are to a large extent inherited from knowledge and experience with computer security. The elements composing the two device types are similar, and there are common measures that can be used, such as antivirus and firewall. However, the implementation of these solutions is not necessarily possible or at least highly constrained within a mobile device. The reason for this difference is the technical resources offered by computers and mobile devices: even though the computing power of smartphones is becoming faster, they have other limitations than their computing power.

- Single-task system: Some operating systems, including some still commonly used, are single-tasking. Only the foreground task is executed. It is difficult to introduce applications such as antivirus and firewall on such systems, because they could not perform their monitoring while the user is operating the device, when there would be most need of such monitoring.

- Energy autonomy: A critical one for the use of a smartphone is energy autonomy. It is important that the security mechanisms not consume battery resources, without which the autonomy of devices will be affected dramatically, undermining the effective use of the smartphone.

- Network Directly related to battery life, network utilization should not be too high. It is indeed one of the most expensive resources, from the point of view of energy consumption. Nonetheless, some calculations may need to be relocated to remote servers in order to preserve the battery. This balance can make implementation of certain intensive computation mechanisms a delicate proposition.

Furthermore, it should be noted that it is common to find that updates exist, or can be developed or deployed, but this is not always done. One can, for example, find a user who does not know that there is a newer version of the operating system compatible with the smartphone, or a user may discover known vulnerabilities that are not corrected until the end of a long development cycle, which allows time to exploit the loopholes.

Next Generation of Mobile Security

There is expected to be four mobile environments that will make up the security framework:

Rich operating system

> In this category will fall traditional Mobile OS like Android, iOS, Symbian OS or Windows Phone. They will provide the traditional functionaity and security of an OS to the applications.

Secure Operating System (Secure OS)

> A secure kernel which will run in parallel with a fully featured Rich OS, on the same processor core. It will include drivers for the Rich OS ("normal world") to communicate with the secure kernel ("secure world"). The trusted infrastructure could include interfaces like the display or keypad to regions of PCI-E address space and memories.

Trusted Execution Environment (TEE)

> Made up of hardware and software. It helps in the control of access rights and houses sensitive applications, which need to be isolated from the Rich OS. It effectively acts as a firewall between the "normal world" and "secure world".

Secure Element (SE)

> The SE consists of tamper resistant hardware and associated software. It can provide high levels of security and work in tandem with the TEE. The SE will be mandatory for hosting proximity payment applications or official electronic signatures.

Malware

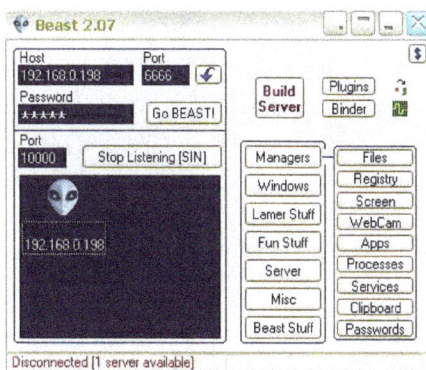

Beast, a Windows-based backdoor Trojan horse.

Malware, short for malicious software, is any software used to disrupt computer operations, gather sensitive information, gain access to private computer systems, or display unwanted advertising. Before the term malware was coined by Yisrael Radai in 1990, malicious software was referred to as computer viruses. The first category of malware propagation concerns parasitic software fragments that attach themselves to some existing executable content. The fragment may be machine code that infects some existing application, utility, or system program, or even the code used to boot a computer system. Malware is defined by its malicious intent, acting against the requirements of the computer user, and does not include software that causes unintentional harm due to some deficiency.

Malware may be stealthy, intended to steal information or spy on computer users for an extended period without their knowledge, as for example Regin, or it may be designed to cause harm, often as sabotage (e.g., Stuxnet), or to extort payment (CryptoLocker). 'Malware' is an umbrella term used to refer to a variety of forms of hostile or intrusive software, including computer viruses, worms, trojan horses, ransomware, spyware, adware, scareware, and other malicious programs. It can take the form of executable code, scripts, active content, and other software. Malware is often disguised as, or embedded in, non-malicious files. As of 2011 the majority of active malware threats were worms or trojans rather than viruses.

In law, malware is sometimes known as a computer contaminant, as in the legal codes of several U.S. states.

Spyware or other malware is sometimes found embedded in programs supplied officially by companies, e.g., downloadable from websites, that appear useful or attractive, but may have, for example, additional hidden tracking functionality that gathers marketing statistics. An example of such software, which was described as illegitimate, is the Sony rootkit, a Trojan embedded into CDs sold by Sony, which silently installed and concealed itself on purchasers' computers with the intention of preventing illicit copying; it also reported on users' listening habits, and unintentionally created vulnerabilities that were exploited by unrelated malware.

Software such as anti-virus, anti-malware, and firewalls are used to protect against activity identified as malicious, and to recover from attacks.

Purposes

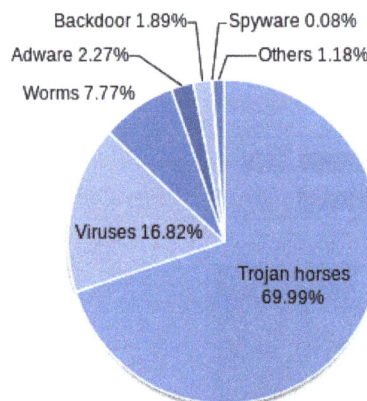

Malware by categories on 16 March 2011.

Many early infectious programs, including the first Internet Worm, were written as experiments or pranks. Today, malware is used by both black hat hackers and governments, to steal personal, financial, or business information.

Malware is sometimes used broadly against government or corporate websites to gather guarded information, or to disrupt their operation in general. However, malware is often used against individuals to gain information such as personal identification numbers or details, bank or credit card numbers, and passwords. Left unguarded, personal and networked computers can be at considerable risk against these threats. (These are most frequently defended against by various types of firewall, anti-virus software, and network hardware).

Since the rise of widespread broadband Internet access, malicious software has more frequently been designed for profit. Since 2003, the majority of widespread viruses and worms have been designed to take control of users' computers for illicit purposes. Infected "zombie computers" are used to send email spam, to host contraband data such as child pornography, or to engage in distributed denial-of-service attacks as a form of extortion.

Programs designed to monitor users' web browsing, display unsolicited advertisements, or redirect affiliate marketing revenues are called spyware. Spyware programs do not spread like viruses; instead they are generally installed by exploiting security holes. They can also be hidden and packaged together with unrelated user-installed software.

Ransomware affects an infected computer in some way, and demands payment to reverse the damage. For example, programs such as CryptoLocker encrypt files securely, and only decrypt them on payment of a substantial sum of money.

Some malware is used to generate money by click fraud, making it appear that the computer user has clicked an advertising link on a site, generating a payment from the advertiser. It was estimated in 2012 that about 60 to 70% of all active malware used some kind of click fraud, and 22% of all ad-clicks were fraudulent.

Malware is usually used for criminal purposes, but can be used for sabotage, often without direct benefit to the perpetrators. One example of sabotage was Stuxnet, used to destroy very specific industrial equipment. There have been politically motivated attacks that have spread over and shut down large computer networks, including massive deletion of files and corruption of master boot records, described as "computer killing". Such attacks were made on Sony Pictures Entertainment (25 November 2014, using malware known as Shamoon or W32.Disttrack) and Saudi Aramco (August 2012).

Proliferation

Preliminary results from Symantec published in 2008 suggested that "the release rate of malicious code and other unwanted programs may be exceeding that of legitimate software applications." According to F-Secure, "As much malware [was] produced in 2007 as in the previous 20 years altogether." Malware's most common pathway from criminals to users is through the Internet: primarily by e-mail and the World Wide Web.

The prevalence of malware as a vehicle for Internet crime, along with the challenge of anti-malware software to keep up with the continuous stream of new malware, has seen the adoption of a new mindset for individuals and businesses using the Internet. With the amount of malware currently being distributed, some percentage of computers are currently assumed to be infected. For businesses, especially those that sell mainly over the Internet, this means they need to find a way to operate despite security concerns. The result is a greater emphasis on back-office protection designed to protect against advanced malware operating on customers' computers. A 2013 Webroot study shows that 64% of companies allow remote access to servers for 25% to 100% of their workforce and that companies with more than 25% of their employees accessing servers remotely have higher rates of malware threats.

On 29 March 2010, Symantec Corporation named Shaoxing, China, as the world's malware capital. A 2011 study from the University of California, Berkeley, and the Madrid Institute for Advanced Studies published an article in *Software Development Technologies*, examining how entrepreneurial hackers are helping enable the spread of malware by offering access to computers for a price. Microsoft reported in May 2011 that one in every 14 downloads from the Internet may now contain malware code. Social media, and Facebook in particular, are seeing a rise in the number of tactics used to spread malware to computers.

A 2014 study found that malware is being increasingly aimed at mobile devices such as smartphones as they increase in popularity.

Infectious Malware: Viruses and Worms

The best-known types of malware, viruses and worms, are known for the manner in which they spread, rather than any specific types of behavior. The term *computer virus* is used for a program that embeds itself in some other executable software (including the operating system itself) on the target system without the user's consent and when that is run causes the virus to spread to other executables. On the other hand, a *worm* is a stand-alone malware program that *actively* transmits itself over a network to infect other computers. These definitions lead to the observation that a virus requires the user to run an infected program or operating system for the virus to spread, whereas a worm spreads itself.

Concealment: Viruses, Trojan Horses, Rootkits, Backdoors and Evasion

These categories are not mutually exclusive, so malware may use multiple techniques. This section only applies to malware designed to operate undetected, not sabotage and ransomware.

Viruses

A computer program usually hidden within another seemingly innocuous program that produces copies of itself and inserts them into other programs or files, and that usually performs a malicious action (such as destroying data).

Trojan Horses

In computing, Trojan horse, or Trojan, is any malicious computer program which misrepresents itself to appear useful, routine, or interesting in order to persuade a victim to install it. The term is derived from the Ancient Greek story of the wooden horse that was used to help Greek troops invade the city of Troy by stealth.

Trojans are generally spread by some form of social engineering, for example where a user is duped into executing an e-mail attachment disguised to be unsuspicious, (e.g., a routine form to be filled in), or by drive-by download. Although their payload can be anything, many modern forms act as a backdoor, contacting a controller which can then have unauthorized access to the affected computer. While Trojans and backdoors are not easily detectable by themselves, computers may appear to run slower due to heavy processor or network usage.

Unlike computer viruses and worms, Trojans generally do not attempt to inject themselves into other files or otherwise propagate themselves.

Rootkits

Once a malicious program is installed on a system, it is essential that it stays concealed, to avoid detection. Software packages known as *rootkits* allow this concealment, by modifying the host's operating system so that the malware is hidden from the user. Rootkits can prevent a malicious process from being visible in the system's list of processes, or keep its files from being read.

Some malicious programs contain routines to defend against removal, not merely to hide themselves. An early example of this behavior is recorded in the Jargon File tale of a pair of programs infesting a Xerox CP-V time sharing system:

> Each ghost-job would detect the fact that the other had been killed, and would start a new copy of the recently stopped program within a few milliseconds. The only way to kill both ghosts was to kill them simultaneously (very difficult) or to deliberately crash the system.

Backdoors

A backdoor is a method of bypassing normal authentication procedures, usually over a connection to a network such as the Internet. Once a system has been compromised, one or more backdoors may be installed in order to allow access in the future, invisibly to the user.

The idea has often been suggested that computer manufacturers preinstall backdoors on their systems to provide technical support for customers, but this has never been reliably verified. It was reported in 2014 that US government agencies had been diverting computers purchased by those considered "targets" to secret workshops where software or hardware permitting remote access by the agency was installed, considered to be among the most productive operations to obtain access to networks around the world. Backdoors may be installed by Trojan horses, worms, implants, or other methods.

Evasion

Since the beginning of 2015, a sizable portion of malware utilizes a combination of many techniques designed to avoid detection and analysis.

- The most common evasion technique is when the malware evades analysis and detection by fingerprinting the environment when executed.

- The second most common evasion technique is confusing automated tools' detection methods. This allows malware to avoid detection by technologies such as signature-based antivirus software by changing the server used by the malware.

- The third most common evasion technique is timing-based evasion. This is when malware runs at certain times or following certain actions taken by the user, so it executes during certain vulnerable periods, such as during the boot process, while remaining dormant the rest of the time.

- The fourth most common evasion technique is done by obfuscating internal data so that automated tools do not detect the malware.

- An increasingly common technique is adware that uses stolen certificates to disable anti-malware and virus protection; technical remedies are available to deal with the adware.

Nowadays, one of the most sophisticated and stealthy ways of evasion is to use information hiding techniques, namely stegomalware.

Vulnerability to Malware

- In this context, and throughout, what is called the "system" under attack may be anything from a single application, through a complete computer and operating system, to a large network.

- Various factors make a system more vulnerable to malware:

Security Defects in Software

Malware exploits security defects (security bugs or vulnerabilities) in the design of the operating system, in applications (such as browsers, e.g. older versions of Microsoft Internet Explorer supported by Windows XP), or in vulnerable versions of browser plugins such as Adobe Flash Player, Adobe Acrobat or Reader, or Java SE. Sometimes even installing new versions of such plugins does not automatically uninstall old versions. Security advisories from plug-in providers announce security-related updates. Common vulnerabilities are assigned CVE IDs and listed in the US National Vulnerability Database. Secunia PSI is an example of software, free for personal use, that will check a PC for vulnerable out-of-date software, and attempt to update it.

Malware authors target bugs, or loopholes, to exploit. A common method is exploitation of a buffer overrun vulnerability, where software designed to store data in a specified region of memory does not prevent more data than the buffer can accommodate being supplied. Malware may provide data that overflows the buffer, with malicious executable code or data after the end; when this payload is accessed it does what the attacker, not the legitimate software, determines.

Insecure Ign or User Error

Early PCs had to be booted from floppy disks. When built-in hard drives became common, the operating system was normally started from them, but it was possible to boot from another boot device if available, such as a floppy disk, CD-ROM, DVD-ROM, USB flash drive or network. It was common to configure the computer to boot from one of these devices when available. Normally none would be available; the user would intentionally insert, say, a CD into the optical drive to boot the computer in some special way, for example, to install an operating system. Even without booting, computers can be configured to execute software on some media as soon as they become available, e.g. to autorun a CD or USB device when inserted.

Malicious software distributors would trick the user into booting or running from an infected device or medium. For example, a virus could make an infected computer add autorunnable code to any USB stick plugged into it. Anyone who then attached the stick to another computer set to autorun from USB would in turn become infected, and also pass on the infection in the same way. More generally, any device that plugs into a USB port - even lights, fans, speakers, toys, or peripherals such as a digital microscope - can be used to spread malware. Devices can be infected during manufacturing or supply if quality control is inadequate.

This form of infection can largely be avoided by setting up computers by default to boot from the internal hard drive, if available, and not to autorun from devices. Intentional booting from another device is always possible by pressing certain keys during boot.

Older email software would automatically open HTML email containing potentially malicious JavaScript code. Users may also execute disguised malicious email attachments and infected executable files supplied in other ways.

Over-privileged users and over-privileged Code

In computing, privilege refers to how much a user or program is allowed to modify a system. In poorly designed computer systems, both users and programs can be assigned more privileges than they should be, and malware can take advantage of this. The two ways that malware does this is through overprivileged users and overprivileged code.

Some systems allow all users to modify their internal structures, and such users today would be considered over-privileged users. This was the standard operating procedure for early microcomputer and home computer systems, where there was no distinction between an *administrator* or *root*, and a regular user of the system. In some systems, non-administrator users are over-privileged by design, in the sense that they are allowed to modify internal structures of the system. In some environments, users are over-privileged because they have been inappropriately granted administrator or equivalent status.

Some systems allow code executed by a user to access all rights of that user, which is known as over-privileged code. This was also standard operating procedure for early microcomputer and home computer systems. Malware, running as over-privileged code, can use this privilege to subvert the system. Almost all currently popular operating systems, and also many scripting applications allow code too many privileges, usually in the sense that when a user executes code, the system allows that code all rights of that user. This makes users vulnerable to malware in the form of e-mail attachments, which may or may not be disguised.

Use of the Same Operating System

- Homogeneity can be a vulnerability. For example, when all computers in a network run the same operating system, upon exploiting one, one worm can exploit them all: In particular,

Microsoft Windows or Mac OS X have such a large share of the market that an exploited vulnerability concentrating on either operating system could subvert a large number of systems. Introducing diversity purely for the sake of robustness, such as adding Linux computers, could increase short-term costs for training and maintenance. However, as long as all the nodes are not part of the same directory service for authentication, having a few diverse nodes could deter total shutdown of the network and allow those nodes to help with recovery of the infected nodes. Such separate, functional redundancy could avoid the cost of a total shutdown, at the cost of increased complexity and reduced usability in terms of single sign-on authentication.

Anti-malware Strategies

As malware attacks become more frequent, attention has begun to shift from viruses and spyware protection, to malware protection, and programs that have been specifically developed to combat malware.

Anti-virus and Anti-malware Software

A specific component of anti-virus and anti-malware software, commonly referred to as an on-access or real-time scanner, hooks deep into the operating system's core or kernel and functions in a manner similar to how certain malware itself would attempt to operate, though with the user's informed permission for protecting the system. Any time the operating system accesses a file, the on-access scanner checks if the file is a 'legitimate' file or not. If the file is identified as malware by the scanner, the access operation will be stopped, the file will be dealt with by the scanner in a pre-defined way (how the anti-virus program was configured during/post installation), and the user will be notified. This may have a considerable performance impact on the operating system, though the degree of impact is dependent on how well the scanner was programmed. The goal is to stop any operations the malware may attempt on the system before they occur, including activities which might exploit bugs or trigger unexpected operating system behavior.

Anti-malware programs can combat malware in two ways:

1. They can provide real time protection against the installation of malware software on a computer. This type of malware protection works the same way as that of antivirus protection in that the anti-malware software scans all incoming network data for malware and blocks any threats it comes across.

2. Anti-malware software programs can be used solely for detection and removal of malware software that has already been installed onto a computer. This type of anti-malware software scans the contents of the Windows registry, operating system files, and installed programs on a computer and will provide a list of any threats found, allowing the user to choose which files to delete or keep, or to compare this list to a list of known malware components, removing files that match.

Real-time protection from malware works identically to real-time antivirus protection: the software scans disk files at download time, and blocks the activity of components known to represent malware. In some cases, it may also intercept attempts to install start-up items or to modify browser settings. Because many malware components are installed as a result of browser exploits or user error, using security software (some of which are anti-malware, though many are not) to "sandbox" browsers (essentially isolate the browser from the computer and hence any malware induced change) can also be effective in helping to restrict any damage done.

Examples of Microsoft Windows antivirus and anti-malware software include the optional Microsoft Security Essentials (for Windows XP, Vista, and Windows 7) for real-time protection, the Windows Malicious Software Removal Tool (now included with Windows (Security) Updates on "Patch Tuesday", the second Tuesday of each month), and Windows Defender (an optional download in the case of Windows XP, incorporating MSE functionality in the case of Windows 8 and later). Additionally, several capable antivirus software programs are available for free download from the Internet (usually restricted to non-commercial use). Tests found some free programs to be competitive with commercial ones. Microsoft's System File Checker can be used to check for and repair corrupted system files.

Some viruses disable System Restore and other important Windows tools such as Task Manager and Command Prompt. Many such viruses can be removed by rebooting the computer, entering Windows safe mode with networking, and then using system tools or Microsoft Safety Scanner.

Hardware implants can be of any type, so there can be no general way to detect them.

Website Security Scans

As malware also harms the compromised websites (by breaking reputation, blacklisting in search engines, etc.), some websites offer vulnerability scanning. Such scans check the website, detect malware, may note outdated software, and may report known security issues.

"Air Gap" Isolation or "Parallel Network"

As a last resort, computers can be protected from malware, and infected computers can be prevented from disseminating trusted information, by imposing an "air gap" (i.e. completely disconnecting them from all other networks). However, malware can still cross the air gap in some situations. For example, removable media can carry malware across the gap. In December 2013 researchers in Germany showed one way that an apparent air gap can be defeated.

"AirHopper", "BitWhisper", "GSMem" and "Fansmitter" are four techniques introduced by researchers that can leak data from air-gapped computers using electromagnetic, thermal and acoustic emissions.

Grayware

Grayware is a term applied to unwanted applications or files that are not classified as malware, but can worsen the performance of computers and may cause security risks.

It describes applications that behave in an annoying or undesirable manner, and yet are less serious or troublesome than malware. Grayware encompasses spyware, adware, fraudulent dialers, joke programs, remote access tools and other unwanted programs that harm the performance of computers or cause inconvenience. The term came into use around 2004.

Another term, PUP, which stands for *Potentially Unwanted Program* (or PUA *Potentially Unwanted Application*), refers to applications that would be considered unwanted despite often having been downloaded by the user, possibly after failing to read a download agreement. PUPs include spyware, adware, and fraudulent dialers. Many security products classify unauthorised key generators as grayware, although they frequently carry true malware in addition to their ostensible purpose.

Software maker Malwarebytes lists several criteria for classifying a program as a PUP. Some adware (using stolen certificates) disables anti-malware and virus protection; technical remedies are available.

History of Viruses and Worms

Before Internet access became widespread, viruses spread on personal computers by infecting the executable boot sectors of floppy disks. By inserting a copy of itself into the machine code instructions in these executables, a virus causes itself to be run whenever a program is run or the disk is booted. Early computer viruses were written for the Apple II and Macintosh, but they became more widespread with the dominance of the IBM PC and MS-DOS system. Executable-infecting viruses are dependent on users exchanging software or boot-able floppies and thumb drives so they spread rapidly in computer hobbyist circles.

The first worms, network-borne infectious programs, originated not on personal computers, but on multitasking Unix systems. The first well-known worm was the Internet Worm of 1988, which infected SunOS and VAX BSD systems. Unlike a virus, this worm did not insert itself into other programs. Instead, it exploited security holes (vulnerabilities) in network server programs and started itself running as a separate process. This same behavior is used by today's worms as well.

With the rise of the Microsoft Windows platform in the 1990s, and the flexible macros of its applications, it became possible to write infectious code in the macro language of Microsoft Word and similar programs. These *macro viruses* infect documents and templates rather than applications (executables), but rely on the fact that macros in a Word document are a form of executable code.

Today, worms are most commonly written for the Windows OS, although a few like Mare-D and the L10n worm are also written for Linux and Unix systems. Worms today work in the same basic way as 1988's Internet Worm: they scan the network and use vulnerable computers to replicate. Because they need no human intervention, worms can spread with incredible speed. The SQL Slammer infected thousands of computers in a few minutes in 2003.

Academic Research

The notion of a self-reproducing computer program can be traced back to initial theories about the operation of complex automata. John von Neumann showed that in theory a program could reproduce itself. This constituted a plausibility result in computability theory. Fred Cohen experimented with computer viruses and confirmed Neumann's postulate and investigated other properties of malware such as detectability and self-obfuscation using rudimentary encryption. His doctoral dissertation was on the subject of computer viruses. The combination of cryptographic technology as part of the payload of the virus, exploiting it for attack purposes was initialized and investigated from the mid 1990s, and includes initial ransomware and evasion ideas.

Mobile Malware

Mobile malware is malicious software that targets mobile phones or wireless-enabled Personal digital assistants (PDA), by causing the collapse of the system and loss or leakage of confidential information. As wireless phones and PDA networks have become more and more common and have grown in complexity, it has become increasingly difficult to ensure their safety and security against electronic attacks in the form of viruses or other malware.

History

Cell phone malware were initially demonstrated by Brazilian software engineer Marcos Velasco. He created a virus that could be used by anyone in order to educate the public of the threat.

The first known mobile virus, "Timofonica", originated in Spain and was identified by antivirus labs in Russia and Finland in June 2000. "Timofonica" sent SMS messages to GSM mobile phones that read (in Spanish) "Information for you: Telefónica is fooling you." These messages were sent through the Internet SMS gate of the MoviStar mobile operator.

In June 2004, it was discovered that a company called Ojam had engineered an anti-piracy Trojan virus in older versions of its mobile phone game, *Mosquito*. This virus sent SMS text messages to the company without the user's knowledge. Although this malware was removed from the game's more recent versions, it still exists in older, unlicensed versions, and these may still be distributed on file-sharing networks and free software download web sites.

In July 2004, computer hobbyists released a proof-of-concept mobile virus *Cabir*, that replicates and spreads itself on Bluetooth wireless networks and infects mobile phones running the Symbian OS.

In March 2005, it was reported that a computer worm called Commwarrior-A had been infecting Symbian series 60 mobile phones. This specific worm replicated itself through the phone's Multimedia Messaging Service (MMS), sending copies of itself to other phone owners listed in the phone user's address book. Although the worm is not considered harmful, experts agree that it heralded a new age of electronic attacks on mobile phones.

In August 2010, Kaspersky Lab reported a trojan designated Trojan-SMS.AndroidOS.FakePlayer.a. This was the first malicious program classified as a Trojan SMS that affects smartphones running on Google's Android operating system, and which had already infected a number of mobile devices, sending SMS messages to premium rate numbers without the owner's knowledge or consent, and accumulating huge bills.

Currently, various antivirus software companies like Trend Micro, AVG, avast!, Comodo, Kaspersky Lab, PSafe, and Softwin are working to adapt their programs to the mobile operating systems that are most at risk. Meanwhile, operating system developers try to curb the spread of infections with quality control checks on software and content offered through their digital application distribution platforms, such as Google Play or Apple's App Store. Recent studies however show that mobile antivirus programs are ineffective due to the rapid evolution of mobile malware.

Taxonomy

Four types of the most common malicious programs are known to affect mobile devices:

- Expander: Expanders target mobile meters for additional phone billing and profit

- Worm: The main objective of this stand-alone type of malware is to endlessly reproduce itself and spread to other devices. Worms may also contain harmful and misleading instructions. Mobile worms may be transmitted via text messages SMS or MMS and typically do not require user interaction for execution.

- Trojan: Unlike worms, a Trojan horse always requires user interaction to be activated. This kind of virus is usually inserted into seemingly attractive and non-malicious executable files or applications that are downloaded to the device and executed by the user. Once activated, the malware can cause serious damage by infecting and deactivating other applications or the phone itself, rendering it paralyzed after a certain period of time or a certain number of operations. Usurpation data (spyware) synchronizes with calendars, email accounts, notes, and any other source of information before it is sent to a remote server.

- Spyware: This malware poses a threat to mobile devices by collecting, using, and spreading a user's personal or sensitive information without the user's consent or knowledge. It is mostly classified into four categories: system monitors, trojans, adware, and tracking cookies.

- Ghost Push: This is a kind of malware which infects the Android OS by automatically gaining root access, downloading malicious software, converting to a system app and then losing root access which makes it virtually impossible to remove the infection with a factory reset unless the firmware is reflashed. The malware hogs all system resources making it unresponsive and drains the battery. The advertisements appear anytime either in full screen, as part of a display, or in the status bar. The unnecessary apps automatically activate and sometimes download more malicious software when connected to the internet. It is hard to detect and remove. It steals the personal data of the user from the phone.

Notable Mobile Malicious Programs

- Cabir: This malware infects mobile phones running on Symbian OS and was first identified in June 2004. When a phone is infected, the message 'Caribe' is displayed on the phone's screen and is displayed every time the phone is turned on. The worm then attempts to spread to other phones in the area using wireless Bluetooth signals, although the recipient has to confirm this manually.

- Duts: This parasitic file infector virus is the first known virus for the Pocket PC platform. It attempts to infect all EXE files that are larger than 4096 bytes in the current directory.

- Skulls: A trojan horse piece of code that targets mainly Symbian OS. Once downloaded, the virus replaces all phone desktop icons with images of a skull. It also renders all phone applications useless. This malware also tends to mass text messages containing malicious links to all contacts accessible through the device in order to spread the damage. This mass texting can also give rise to high expenses.

- Commwarrior: This malware was identified in 2005. It was the first worm to use MMS messages and can spread through Bluetooth as well. It infects devices running under OS Symbian Series 60. The executable worm file, once launched, hunts for accessible Bluetooth devices and sends the infected files under a random name to various devices.

- Gingermaster: A trojan developed for an Android platform that propagates by installing applications that incorporate a hidden malware for installation in the background. It exploits the frailty in the version Gingerbread (2.3) of the operating system to use super-user permissions by privileged escalation. It then creates a service that steals information from infected terminals (user ID, number SIM, phone number, IMEI, IMSI, screen resolution and local time) by sending it to a remote server through petitions HTTP.

- DroidKungFu: A trojan content in Android applications, which when executed, obtains root privileges and installs the file com.google. ssearch.apk, which contains a back door that allows files to be removed, open home pages to be supplied, and 'open web and download and install' application packages. This virus collects and sends to a remote server all available data on the terminal.

- Ikee: The first worm known for iOS platforms. It only works on terminals that were previously made a process of jailbreak, and spreads by trying to access other devices using the SSH protocol, first through the subnet that is connected to the device. Then, it repeats the process generating a random range and finally uses some preset ranges corresponding to the IP address of certain telephone companies. Once the computer is infected, the wallpaper is replaced by a photograph of the singer Rick Astley, a reference to the Rickroll phenomenon.

- Gunpoder : This worm file infector virus is the first known virus that officially infected the Google Play Store in few countries, including Brazil.

- Shedun: adware serving malware able to root Android devices.

- HummingBad - has infected over 10 million Android operating systems. User details are sold and adverts are tapped on without the user's knowledge thereby generating fraudulent advertising revenue.

Dendroid (Malware)

Dendroid is malware that affects Android OS and targets the mobile platform.

It was first discovered in early of 2014 by Symantec and appeared in the underground for sale mi- for $300. Some things were noted in Dendroid, such as being able to hide from emulators at the time. When first discovered in 2014 it was one of the most sophisticated Android remote administration tools known at that time. It was one of the first Trojan applications to get past Google's Bouncer and caused researchers to warn about it being easier to create Android malware due to it. It also seems to have follow in the footsteps of Zeus and SpyEye by having simple-to-use command and control panels. The code appeared to be leaked somewhere around 2014. It was noted that an apk binder was included in the leak, which provided a simple way to bind Dendroid to legitimate applications.

It is capable of:

- Deleting call logs

- Opening web pages

- Dialing any number

- Recording calls

- SMS intercepting

- Uploading images and video

- Opening an application

- Performing denial-of-service attacks

- Changing the command and control server

Spyware

Spyware is software that aims to gather information about a person or organization without their knowledge and that may send such information to another entity without the consumer's consent, or that asserts control over a computer without the consumer's knowledge.

"Spyware" is mostly classified into four types: system monitors, trojans, adware, and tracking cookies. Spyware is mostly used for the purposes of tracking and storing Internet users' movements on the Web and serving up pop-up ads to Internet users.

Whenever spyware is used for malicious purposes, its presence is typically hidden from the user and can be difficult to detect. Some spyware, such as keyloggers, may be installed by the owner of a shared, corporate, or public computer intentionally in order to monitor users.

While the term *spyware* suggests software that monitors a user's computing, the functions of spyware can extend beyond simple monitoring. Spyware can collect almost any type of data, including personal information like internet surfing habits, user logins, and bank or credit account information. Spyware can also interfere with user control of a computer by installing additional software or redirecting web browsers. Some spyware can change computer settings, which can result in slow Internet connection speeds, un-authorized changes in browser settings, or changes to software settings.

Sometimes, spyware is included along with genuine software, and may come from a malicious website or may have been added to the intentional functionality of genuine software. In response to the emergence of spyware, a small industry has sprung up dealing in anti-spyware software. Running anti-spyware software has become a widely recognized element of computer security practices, especially for computers running Microsoft Windows. A number of jurisdictions have passed anti-spyware laws, which usually target any software that is surreptitiously installed to control a user's computer.

In German-speaking countries, spyware used or made by the government is called *govware* by computer experts (in common parlance: *Regierungstrojaner*, literally 'Government Trojan'). Govware is typically a trojan horse software used to intercept communications from the target computer. Some countries like Switzerland and Germany have a legal framework governing the use of such software. In the US, the term policeware has been used for similar purposes.

Use of the term "spyware" has eventually declined as the practice of tracking users has been pushed ever further into the mainstream by major websites and data mining companies; these generally break no known laws and compel users to be tracked, not by fraudulent practices per se, but by the default settings created for users and the language of terms-of-service agreements. As one documented example, on March 7, 2011, CBS/Cnet News reported on a Wall Street Journal analysis revealing the practice of Facebook and other websites of tracking users' browsing activity, linked to their identity, far beyond users' visit and activity within the Facebook site itself. The report stated "Here's how it works. You go to Facebook, you log in, you spend some time there, and then ... you move on without logging out. Let's say the next site you go to is New York Times. Those buttons, without you clicking on them, have just reported back to Facebook and Twitter that you went there and also your identity within those accounts. Let's say you moved on to something like a site about depression. This one also has a tweet button, a Google widget, and those, too, can report back who you are and that you went there." The WSJ analysis was researched by Brian Kennish, founder of Disconnect, Inc.

Routes of Infection

Spyware does not necessarily spread in the same way as a virus or worm because infected systems generally do not attempt to transmit or copy the software to other computers. Instead, spyware installs itself on a system by deceiving the user or by exploiting software vulnerabilities.

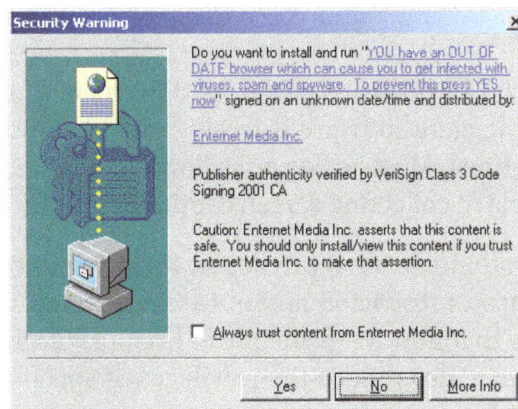

Malicious websites attempt to install spyware on readers' computers.

Most spyware is installed without knowledge, or by using deceptive tactics. Spyware may try to deceive users by bundling itself with desirable software. Other common tactics are using a Trojan horse, spy gadgets that look like normal devices but turn out to be something else, such as a USB Keylogger. These devices actually are connected to the device as memory units but are capable of recording each stroke made on the keyboard. Some spyware authors infect a system through security holes in the Web browser or in other software. When the user navigates to a Web page controlled by the spyware author, the page contains code which attacks the browser and forces the download and installation of spyware.

The installation of spyware frequently involves Internet Explorer. Its popularity and history of security issues have made it a frequent target. Its deep integration with the Windows environment make it susceptible to attack into the Windows operating system. Internet Explorer also serves as a point of attachment for spyware in the form of Browser Helper Objects, which modify the browser's behavior to add toolbars or to redirect traffic.

Effects and Behaviors

A spyware program is rarely alone on a computer: an affected machine usually has multiple infections. Users frequently notice unwanted behavior and degradation of system performance. A spyware infestation can create significant unwanted CPU activity, disk usage, and network traffic. Stability issues, such as applications freezing, failure to boot, and system-wide crashes are also common. Spyware, which interferes with networking software, commonly causes difficulty connecting to the Internet.

In some infections, the spyware is not even evident. Users assume in those situations that the performance issues relate to faulty hardware, Windows installation problems, or another infection. Some owners of badly infected systems resort to contacting technical support experts, or even buying a new computer because the existing system "has become too slow". Badly infected systems may require a clean reinstallation of all their software in order to return to full functionality.

Moreover, some types of spyware disable software firewalls and anti-virus software, and/or reduce browser security settings, which further open the system to further opportunistic infections. Some spyware disables or even removes competing spyware programs, on the grounds that more spyware-related annoyances make it even more likely that users will take action to remove the programs.

Keyloggers are sometimes part of malware packages downloaded onto computers without the owners' knowledge. Some keyloggers software is freely available on the internet while others are commercial or private applications. Most keyloggers allow not only keyboard keystrokes to be captured but also are often capable of collecting screen captures from the computer.

A typical Windows user has administrative privileges, mostly for convenience. Because of this, any program the user runs has unrestricted access to the system. As with other operating systems, Windows users are able to follow the principle of least privilege and use non-administrator accounts. Alternatively, they can also reduce the privileges of specific vulnerable Internet-facing processes such as Internet Explorer.

Since Windows Vista, by default, a computer administrator runs everything under limited user privileges. When a program requires administrative privileges, a User Account Control pop-up will prompt the user to allow or deny the action. This improves on the design used by previous versions of Windows.

Remedies and Prevention

As the spyware threat has worsened, a number of techniques have emerged to counteract it. These include programs designed to remove or block spyware, as well as various user practices which reduce the chance of getting spyware on a system.

Nonetheless, spyware remains a costly problem. When a large number of pieces of spyware have infected a Windows computer, the only remedy may involve backing up user data, and fully re-installing the operating system. For instance, some spyware cannot be completely removed by Symantec, Microsoft, PC Tools.

Anti-spyware Programs

Many programmers and some commercial firms have released products dedicated to remove or block spyware. Programs such as PC Tools' Spyware Doctor, Lavasoft's *Ad-Aware SE* and Patrick Kolla's *Spybot - Search & Destroy* rapidly gained popularity as tools to remove, and in some cases intercept, spyware programs. On December 16, 2004, Microsoft acquired the *GIANT AntiSpyware* software, rebranding it as *Windows AntiSpyware beta* and releasing it as a free download for Genuine Windows XP and Windows 2003 users. (In 2006 it was renamed Windows Defender).

Major anti-virus firms such as Symantec, PC Tools, McAfee and Sophos have also added anti-spyware features to their existing anti-virus products. Early on, anti-virus firms expressed reluctance to add anti-spyware functions, citing lawsuits brought by spyware authors against the authors of web sites and programs which described their products as "spyware". However, recent versions of these major firms' home and business anti-virus products do include anti-spyware functions, albeit treated differently from viruses. Symantec Anti-Virus, for instance, categorizes spyware programs as "extended threats" and now offers real-time protection against these threats.

How anti-spyware Software Works

Anti-spyware programs can combat spyware in two ways:

1. They can provide real-time protection in a manner similar to that of anti-virus protection: they scan all incoming network data for spyware and blocks any threats it detects.

2. Anti-spyware software programs can be used solely for detection and removal of spyware software that has already been installed into the computer. This kind of anti-spyware can often be set to scan on a regular schedule.

Such programs inspect the contents of the Windows registry, operating system files, and installed programs, and remove files and entries which match a list of known spyware. Real-time protection from spyware works identically to real-time anti-virus protection: the software scans disk files at download time, and blocks the activity of components known to represent spyware. In some cases, it may also intercept attempts to install start-up items or to modify browser settings. Earlier versions of anti-spyware programs focused chiefly on detection and removal. Javacool Software's SpywareBlaster, one of the first to offer real-time protection, blocked the installation of ActiveX-based spyware.

Like most anti-virus software, many anti-spyware/adware tools require a frequently updated database of threats. As new spyware programs are released, anti-spyware developers discover and evaluate them, adding to the list of known spyware, which allows the software to detect and remove new spyware. As a result, anti-spyware software is of limited usefulness without regular updates. Updates may be installed automatically or manually.

A popular generic spyware removal tool used by those that requires a certain degree of expertise is HijackThis, which scans certain areas of the Windows OS where spyware often resides and presents a list with items to delete manually. As most of the items are legitimate windows files/registry entries it is advised for those who are less knowledgeable on this subject to post a HijackThis log on the numerous antispyware sites and let the experts decide what to delete.

If a spyware program is not blocked and manages to get itself installed, it may resist attempts to terminate or uninstall it. Some programs work in pairs: when an anti-spyware scanner (or the user) terminates one running process, the other one respawns the killed program. Likewise, some spyware will detect attempts to remove registry keys and immediately add them again. Usually, booting the infected computer in safe mode allows an anti-spyware program a better chance of removing persistent spyware. Killing the process tree may also work.

Security Practices

To detect spyware, computer users have found several practices useful in addition to installing anti-spyware programs. Many users have installed a web browser other than Internet Explorer, such as Mozilla Firefox or Google Chrome. Though no browser is completely safe, Internet Explorer was once at a greater risk for spyware infection due to its large user base as well as vulnerabilities such as ActiveX but these three major browsers are now close to equivalent when it comes to security.

Some ISPs—particularly colleges and universities—have taken a different approach to blocking spyware: they use their network firewalls and web proxies to block access to Web sites known to install spyware. On March 31, 2005, Cornell University's Information Technology department released a report detailing the behavior of one particular piece of proxy-based spyware, *Marketscore*, and the steps the university took to intercept it. Many other educational institutions have taken similar steps.

Individual users can also install firewalls from a variety of companies. These monitor the flow of information going to and from a networked computer and provide protection against spyware and malware. Some users install a large hosts file which prevents the user's computer from connecting to known spyware-related web addresses. Spyware may get installed via certain shareware programs offered for download. Downloading programs only from reputable sources can provide some protection from this source of attack.

Applications

"Stealware" and Affiliate Fraud

A few spyware vendors, notably 180 Solutions, have written what the *New York Times* has dubbed "stealware", and what spyware researcher Ben Edelman terms *affiliate fraud*, a form of click fraud. Stealware diverts the payment of affiliate marketing revenues from the legitimate affiliate to the spyware vendor.

Spyware which attacks affiliate networks places the spyware operator's affiliate tag on the user's activity – replacing any other tag, if there is one. The spyware operator is the only party that gains from this. The user has their choices thwarted, a legitimate affiliate loses revenue, networks' reputations are injured, and vendors are harmed by having to pay out affiliate revenues to an "affiliate" who is not party to a contract. Affiliate fraud is a violation of the terms of service of most affiliate marketing networks. As a result, spyware operators such as 180 Solutions have been terminated from affiliate networks including LinkShare and ShareSale. Mobile devices can also be vulnerable to chargeware, which manipulates users into illegitimate mobile charges.

Identity Theft and Fraud

In one case, spyware has been closely associated with identity theft. In August 2005, researchers from security software firm Sunbelt Software suspected the creators of the common CoolWebSearch spyware had used it to transmit "chat sessions, user names, passwords, bank information, etc."; however it turned out that "it actually (was) its own sophisticated criminal little trojan that's independent of CWS." This case is currently under investigation by the FBI.

The Federal Trade Commission estimates that 27.3 million Americans have been victims of identity theft, and that financial losses from identity theft totaled nearly $48 billion for businesses and financial institutions and at least $5 billion in out-of-pocket expenses for individuals.

Digital Rights Management

Some copy-protection technologies have borrowed from spyware. In 2005, Sony BMG Music Entertainment was found to be using rootkits in its XCP digital rights management technology Like spyware, not only was it difficult to detect and uninstall, it was so poorly written that most efforts to remove it could have rendered computers unable to function. Texas Attorney General Greg Abbott filed suit, and three separate class-action suits were filed. Sony BMG later provided a workaround on its website to help users remove it.

Beginning on April 25, 2006, Microsoft's Windows Genuine Advantage Notifications application was installed on most Windows PCs as a "critical security update". While the main purpose of this deliberately uninstallable application is to ensure the copy of Windows on the machine was lawfully purchased and installed, it also installs software that has been accused of "phoning home" on a daily basis, like spyware. It can be removed with the RemoveWGA tool.

Personal Relationships

Spyware has been used to monitor electronic activities of partners in intimate relationships. At least one software package, Loverspy, was specifically marketed for this purpose. Depending on local laws regarding communal/marital property, observing a partner's online activity without their consent may be illegal; the author of Loverspy and several users of the product were indicted in California in 2005 on charges of wiretapping and various computer crimes.

Browser Cookies

Anti-spyware programs often report Web advertisers' HTTP cookies, the small text files that track browsing activity, as spyware. While they are not always inherently malicious, many users object to third parties using space on their personal computers for their business purposes, and many anti-spyware programs offer to remove them.

Examples

These common spyware programs illustrate the diversity of behaviors found in these attacks. Note that as with computer viruses, researchers give names to spyware programs which may not be used by their creators. Programs may be grouped into "families" based not on shared program code, but on common behaviors, or by "following the money" of apparent financial or business connections. For instance, a number of the spyware programs distributed by Claria are collectively known as "Gator". Likewise, programs that are frequently installed together may be described as parts of the same spyware package, even if they function separately.

- CoolWebSearch, a group of programs, takes advantage of Internet Explorer vulnerabilities. The package directs traffic to advertisements on Web sites including coolwebsearch.com. It displays pop-up ads, rewrites search engine results, and alters the infected computer's hosts file to direct DNS lookups to these sites.

- FinFisher, sometimes called FinSpy is a high-end surveillance suite sold to law enforcement and intelligence agencies. Support services such as training and technology updates are part of the package.

- HuntBar, aka WinTools or Adware.Websearch, was installed by an ActiveX drive-by download at affiliate Web sites, or by advertisements displayed by other spyware programs—an example of how spyware can install more spyware. These programs add toolbars to IE, track aggregate browsing behavior, redirect affiliate references, and display advertisements.

- Internet Optimizer, also known as DyFuCa, redirects Internet Explorer error pages to advertising. When users follow a broken link or enter an erroneous URL, they see a page of advertisements. However, because password-protected Web sites (HTTP Basic authentication) use the same mechanism as HTTP errors, Internet Optimizer makes it impossible for the user to access password-protected sites.

- Spyware such as Look2Me hides inside system-critical processes and start up even in safe mode. With no process to terminate they are harder to detect and remove, which is a combination of both spyware and a rootkit. Rootkit technology is also seeing increasing use, as newer spyware programs also have specific countermeasures against well known anti-malware products and may prevent them from running or being installed, or even uninstall them.

- Movieland, also known as Moviepass.tv and Popcorn.net, is a movie download service that has been the subject of thousands of complaints to the Federal Trade Commission (FTC), the Washington State Attorney General's Office, the Better Business Bureau, and other agencies. Consumers complained they were held hostage by a cycle of oversized pop-up windows demanding payment of at least $29.95, claiming that they had signed up for a three-day free trial but had not cancelled before the trial period was over, and were thus obligated to pay. The FTC filed a complaint, since settled, against Movieland and eleven other defendants charging them with having "engaged in a nationwide scheme to use deception and coercion to extract payments from consumers."

- WeatherStudio has a plugin that displays a window-panel near the bottom of a browser window. The official website notes that it is easy to remove (uninstall) WeatherStudio from a computer, using its own uninstall-program, such as under C:\Program Files\WeatherStudio. Once WeatherStudio is removed, a browser returns to the prior display appearance, without the need to modify the browser settings.

- Zango (formerly 180 Solutions) transmits detailed information to advertisers about the Web sites which users visit. It also alters HTTP requests for affiliate advertisements linked from a Web site, so that the advertisements make unearned profit for the 180 Solutions company. It opens pop-up ads that cover over the Web sites of competing companies (as seen in their [Zango End User License Agreement]).

- Zlob trojan, or just Zlob, downloads itself to a computer via an ActiveX codec and reports information back to Control Server. Some information can be the search-history, the Web-sites visited, and even keystrokes. More recently, Zlob has been known to hijack routers set to defaults.

History and Development

The first recorded use of the term spyware occurred on October 16, 1995 in a Usenet post that poked fun at Microsoft's business model. *Spyware* at first denoted *software* meant for espionage purposes. However, in early 2000 the founder of Zone Labs, Gregor Freund, used the term in a press release for the ZoneAlarm Personal Firewall. Later in 2000, a parent using ZoneAlarm was alerted to the fact that "Reader Rabbit," educational software marketed to children by the Mattel toy company, was surreptitiously sending data back to Mattel. Since then, "spyware" has taken on its present sense.

According to a 2005 study by AOL and the National Cyber-Security Alliance, 61 percent of surveyed users' computers were infected with form of spyware. 92 percent of surveyed users with spyware reported that they did not know of its presence, and 91 percent reported that they had not given permission for the installation of the spyware. As of 2006, spyware has become one of the preeminent security threats to computer systems running Microsoft Windows operating systems. Computers on which Internet Explorer (IE) is the primary browser are particularly vulnerable to such attacks, not only because IE is the most widely used, but because its tight integration with Windows allows spyware access to crucial parts of the operating system.

Before Internet Explorer 6 SP2 was released as part of Windows XP Service Pack 2, the browser would automatically display an installation window for any ActiveX component that a website wanted to install. The combination of user ignorance about these changes, and the assumption by Internet Explorer that all ActiveX components are benign, helped to spread spyware significantly.

Many spyware components would also make use of exploits in JavaScript, Internet Explorer and Windows to install without user knowledge or permission.

The Windows Registry contains multiple sections where modification of key values allows software to be executed automatically when the operating system boots. Spyware can exploit this design to circumvent attempts at removal. The spyware typically will link itself from each location in the registry that allows execution. Once running, the spyware will periodically check if any of these links are removed. If so, they will be automatically restored. This ensures that the spyware will execute when the operating system is booted, even if some (or most) of the registry links are removed.

Programs Distributed with Spyware

- Kazaa
- Morpheus
- WeatherBug
- WildTangent

Programs Formerly Distributed with Spyware

- AOL Instant Messenger (AOL Instant Messenger still packages Viewpoint Media Player, and WildTangent)
- DivX
- FlashGet
- magicJack

Rogue Anti-spyware Programs

Malicious programmers have released a large number of rogue (fake) anti-spyware programs, and widely distributed Web banner ads can warn users that their computers have been infected with spyware, directing them to purchase programs which do not actually remove spyware—or else, may add more spyware of their own.

The recent proliferation of fake or spoofed antivirus products that bill themselves as antispyware can be troublesome. Users may receive popups prompting them to install them to protect their computer, when it will in fact add spyware. This software is called rogue software. It is recommended that users do not install any freeware claiming to be anti-spyware unless it is verified to be legitimate. Some known offenders include:

- AntiVirus 360
- Antivirus 2009
- AntiVirus Gold
- ContraVirus
- MacSweeper

- Pest Trap

- PSGuard

- Spy Wiper

- Spydawn

- Spylocked

- Spysheriff

- SpyShredder

- Spyware Quake

- SpywareStrike

- UltimateCleaner

- WinAntiVirus Pro 2006

- Windows Police Pro

- WinFixer

- WorldAntiSpy

Fake antivirus products constitute 15 percent of all malware.

On January 26, 2006, Microsoft and the Washington state attorney general filed suit against Secure Computer for its Spyware Cleaner product.

Legal Issues

Criminal Law

Unauthorized access to a computer is illegal under computer crime laws, such as the U.S. Computer Fraud and Abuse Act, the U.K.'s Computer Misuse Act, and similar laws in other countries. Since owners of computers infected with spyware generally claim that they never authorized the installation, a *prima facie* reading would suggest that the promulgation of spyware would count as a criminal act. Law enforcement has often pursued the authors of other malware, particularly viruses. However, few spyware developers have been prosecuted, and many operate openly as strictly legitimate businesses, though some have faced lawsuits.

Spyware producers argue that, contrary to the users' claims, users do in fact give consent to installations. Spyware that comes bundled with shareware applications may be described in the legalese text of an end-user license agreement (EULA). Many users habitually ignore these purported contracts, but spyware companies such as Claria say these demonstrate that users have consented.

Despite the ubiquity of EULAs agreements, under which a single click can be taken as consent to the entire text, relatively little caselaw has resulted from their use. It has been established in most common law jurisdictions that this type of agreement can be a binding contract *in certain circumstances*. This does not, however, mean that every such agreement is a contract, or that every term in one is enforceable.

Some jurisdictions, including the U.S. states of Iowa and Washington, have passed laws criminal-izing some forms of spyware. Such laws make it illegal for anyone other than the owner or oper-ator of a computer to install software that alters Web-browser settings, monitors keystrokes, or disables computer-security software.

In the United States, lawmakers introduced a bill in 2005 entitled the Internet Spyware Prevention Act, which would imprison creators of spyware.

Administrative Sanctions

US FTC Actions

The US Federal Trade Commission has sued Internet marketing organizations under the "unfair-ness doctrine" to make them stop infecting consumers' PCs with spyware. In one case, that against Seismic Entertainment Productions, the FTC accused the defendants of developing a program that seized control of PCs nationwide, infected them with spyware and other malicious software, bom-barded them with a barrage of pop-up advertising for Seismic's clients, exposed the PCs to security risks, and caused them to malfunction. Seismic then offered to sell the victims an "antispyware" program to fix the computers, and stop the popups and other problems that Seismic had caused. On November 21, 2006, a settlement was entered in federal court under which a $1.75 million judgment was imposed in one case and $1.86 million in another, but the defendants were insolvent

In a second case, brought against CyberSpy Software LLC, the FTC charged that CyberSpy mar-keted and sold "RemoteSpy" keylogger spyware to clients who would then secretly monitor un-suspecting consumers' computers. According to the FTC, Cyberspy touted RemoteSpy as a "100% undetectable" way to "Spy on Anyone. From Anywhere." The FTC has obtained a temporary order prohibiting the defendants from selling the software and disconnecting from the Internet any of their servers that collect, store, or provide access to information that this software has gathered. The case is still in its preliminary stages. A complaint filed by the Electronic Privacy Information Center (EPIC) brought the RemoteSpy software to the FTC's attention.

Netherlands OPTA

An administrative fine, the first of its kind in Europe, has been issued by the Independent Author-ity of Posts and Telecommunications (OPTA) from the Netherlands. It applied fines in total value of Euro 1,000,000 for infecting 22 million computers. The spyware concerned is called Dollar-Revenue. The law articles that have been violated are art. 4.1 of the Decision on universal service providers and on the interests of end users; the fines have been issued based on art. 15.4 taken together with art. 15.10 of the Dutch telecommunications law.

Civil Law

Former New York State Attorney General and former Governor of New York Eliot Spitzer has pursued spyware companies for fraudulent installation of software. In a suit brought in 2005 by Spitzer, the California firm Intermix Media, Inc. ended up settling, by agreeing to pay US$7.5 mil-lion and to stop distributing spyware.

The hijacking of Web advertisements has also led to litigation. In June 2002, a number of large Web publishers sued Claria for replacing advertisements, but settled out of court.

Courts have not yet had to decide whether advertisers can be held liable for spyware that displays their ads. In many cases, the companies whose advertisements appear in spyware pop-ups do not directly do business with the spyware firm. Rather, they have contracted with an advertising agency, which in turn contracts with an online subcontractor who gets paid by the number of "impressions" or appearances of the advertisement. Some major firms such as Dell Computer and Mercedes-Benz have sacked advertising agencies that have run their ads in spyware.

Libel Suits by Spyware Developers

Litigation has gone both ways. Since "spyware" has become a common pejorative, some makers have filed libel and defamation actions when their products have been so described. In 2003, Gator (now known as Claria) filed suit against the website PC Pitstop for describing its program as "spyware". PC Pitstop settled, agreeing not to use the word "spyware", but continues to describe harm caused by the Gator/Claria software. As a result, other anti-spyware and anti-virus companies have also used other terms such as "potentially unwanted programs" or greyware to denote these products.

WebcamGate

In the 2010 WebcamGate case, plaintiffs charged two suburban Philadelphia high schools secretly spied on students by surreptitiously and remotely activating webcams embedded in school-issued laptops the students were using at home, and therefore infringed on their privacy rights. The school loaded each student's computer with LANrev's remote activation tracking software. This included the now-discontinued "TheftTrack". While TheftTrack was not enabled by default on the software, the program allowed the school district to elect to activate it, and to choose which of the TheftTrack surveillance options the school wanted to enable.

TheftTrack allowed school district employees to secretly remotely activate the webcam embedded in the student's laptop, above the laptop's screen. That allowed school officials to secretly take photos through the webcam, of whatever was in front of it and in its line of sight, and send the photos to the school's server. The LANrev software disabled the webcams for all other uses (*e.g.*, students were unable to use Photo Booth or video chat), so most students mistakenly believed their webcams did not work at all. In addition to webcam surveillance, TheftTrack allowed school officials to take screenshots, and send them to the school's server. In addition, LANrev allowed school officials to take snapshots of instant messages, web browsing, music playlists, and written compositions. The schools admitted to secretly snapping over 66,000 webshots and screenshots, including webcam shots of students in their bedrooms.

Dirtbox (Cell Phone)

A dirtbox (or DRT box) is a cell site simulator, or a phone device mimicking a cell phone tower. The device is designed to create a signal strong enough within a short range that it forces dormant mobile phones to automatically switch over to it. It is used by the United States Marshals Service, mounted on aircraft all over the U.S., to detect and locate cell phones and thus collect information, and can be used to jam phones. The name stems from the company that originally developed it, Digital Receiver Technology, Inc., abbreviated DRT, owned by the Boeing company. Boeing describes the device as a hybrid of "jamming, managed access and detection". A similar device with a more limited range that has been widely used by United States federal entities, including the Federal Bureau of Investigation (FBI), is the controversial StingRay phone tracker.

History

It is not known when Digital Receiver Technology, Inc. (DRT) first manufactured the dirtbox. DRT does not publicly advertise the dirtbox, stating, "Due to the sensitive nature of our work, we are unable to publicly advertise many of our products" on its Web site. *The Wall Street Journal* wrote that the program by the U.S. Marshals Service "fully matured by 2007". Boeing bought DRT in 2008.

Similar devices from the Harris Corporation, like the Stingray phone tracker, have been sold around the same time. Since 2008, their airborne mounting kit for cell phone surveillance has been said to cost $9,000.

On June 11, 2010, the Boeing Company had asked the National Telecommunications and Information Administration to advise the United States Congress that the "... Communications Act of 1934 be modified to allow prison officials and state and local law enforcement to use appropriate cell phone management". It suggested that special weapons and tactics (SWAT) teams and other paramilitary tactical units could control wireless communications in a building during a raid.

The Chicago Police Department bought dirt boxes to eavesdrop on demonstrators during the 2012 NATO summit and used them in 2014 Black Lives Matter demonstrations, and in 2015 it became known that the Los Angeles Police Department had purchased them too.

Technology

The device was described as being 2-square-foot (0.19 m²) in size in the first mass media publication that shed light on it.

It acts as a fake cell phone tower, and utilizes IMSI-catcher technology, which stands for "international mobile subscriber identity". Phone networks use this system to identify an individual subscriber. The device emits a pilot signal made to appear stronger than that from the service provider cell tower, which forces all phones within its range to broadcast their IMSI number and electronic serial number (ESN). Encryption does not prevent this process. Furthermore, the devices can retrieve the encryption session keys for a cellphone in less than a second with success rates of 50 to 75% (in real world conditions)".

By mounting the device on a plane, it can locate a phone within 10 feet, while another source claims that by triangulating flights, a dirtbox can identify a phone accurately up to two feet.

The dirtbox is a hybrid of detection, managed access and jamming technologies: According to "people with knowledge of the program", the device can determine which phones belong to suspects and which belong to non-suspects, and that "cell phones not of interest, such as those belonging to prison personnel or commercial users in the area, are returned to their local network." The dirtbox can also selectively jam for brief intervals, that is, interrupt or disable calls on certain phones, preventing contraband use in prison. It can also retrieve data from a phone. The technology is "unobtrusive to legitimate wireless communications", as described by Boeing, and bypasses phone companies in its operations.

Agency use

Law Enforcement

As of November 2014, the U.S. Marshals Service Technical Operations Group has been using the device fixed on manned airplanes to track fugitives, but it can deploy it on "targets requested by other parts of the Justice Program". It operates from at least five U.S. airports "covering most of the U.S. population". It is unclear whether the U.S. Marshals Service is requesting court orders.

Specifically, the Marshals Service has used dirtboxes in the Mexican Drug War for tracking fugitives in coordination with Mexico's Naval Infantry Force and flights in Guatemala.

DRT boxes are used by the United States Special Operations Command, the Drug Enforcement Administration, the FBI and U.S. Customs and Border Protection. The U.S. Navy bought dirtboxes to mount on drones at Naval Air Weapons Station China Lake, its research and development facility in Southern California according to procurement documents. The Pentagon Washington Headquarters Services bought them in 2011.

Signal Intelligence

Based on references to DRTBox in NSA's Boundless informant screenshots leaked by Edward Snowden, dirtbox is probably used by the NSA. Le Monde wrote in 2013, that "thanks to DRTBOX, 62.5 million phone data were collected in France". The United States Naval Special Warfare Development Group Group One bought a Digital Receiver Technology 1301B System on April 2, 2007 costing over $25,000 per the United States government procurement web site.

U.S. Regulation

The National Telecommunications and Information Administration (NTIA) has known of dirtbox since at least 2010. In 2014, the United States Department of Justice refused to confirm or deny that the program exists, but an official said "it would be "utterly false" to conflate the law-enforcement program with the collection of bulk telephone records by the National Security Agency". The Federal Communications Commission, responsible for licensing and regulating cell-service providers, was not aware of this activity prior to The Wall Street Journal's exposé.

In January 2015 the US Senate Judiciary Committee asked Department of Justice and Department of Homeland Security which law enforcement agencies use the DRTbox,and the legal process/ what policies existed to protect the privacy interests of those whose information is collected.

Criticism

Dirtbox use has been criticized as violating the Fourth Amendment to the United States Constitution. including by U.S. Rep. Alan Grayson, D-Florida.

Brian Owsley, a law professor at Indiana Institute of Technology and former United States magistrate said "I think the government would need to obtain a search warrant based on probable cause consistent with the Fourth Amendment".

The Guardian quoted Michael German at New York University Law School, a former FBI agent, as saying: "The overriding problem is the excessive secrecy that hides the government's ever-expanding surveillance programs from public accountability."

Senator Edward Markey (Democrat, Massachusetts) and Al Franken (Democrat, Minnesota) have warned that Americans' privacy rights must be assured.

Juice Jacking

Juice jacking is a term used to describe a cyber attack wherein malware might be installed on to, or data surreptitiously copied from, a smart phone, tablet or other computer device using a charging port that doubles as a data connection, typically over USB.

Published Research

The Wall of Sheep, an event at Def_Con has set up and allowed public access to an informational juice jacking kiosk each year at DefCon since 2011. Their intent is to bring awareness of this attack to the general public. Each of the informational juice jacking kiosks set up at the Wall of Sheep village have included a hidden CPU which is used in some way to notify the user that they should not plug their devices in to public charging kiosks. The first informational juice jacking kiosk included a screen which would change from "Free charging station" to a warning message that the user "should not trust public charging stations with their devices". One of the researchers who designed the charging station for the Wall of Sheep has given public presentations which showcase more malicious acts which could be taken via the kiosk, such as data theft, device tracking and information on compromising existing charging kiosks.

Security researcher Kyle Osborn released an attack framework called P2P-ADB in 2012 which utilized USB On-The-Go to connect an attacker's phone to a target victim's device. This framework included examples and proof of concepts which would allow attackers to unlock locked phones, steal data from a phone including authentication keys granting the attacker access to the target device owner's google account.

Security researcher graduates and students from the Institute of Technology Georgia released a proof of concept malicious tool "Mactans" which utilized the USB charging port on Apple mobile devices at the 2013 Blackhat USA security briefings. They utilized inexpensive hardware components to construct a small sized malicious wall charger which could infect an iPhone with the then-current version of iOS with malicious software while it was being charged. The software could defeat any security measures built into iOS and mask itself in the same way Apple masks background processes in iOS.

Security researchers Karsten Nohl and Jakob Lell from srlabs published their research on BadUSB during the 2014 Blackhat USA security briefings. Their presentation on this attack mentions that a cellphone or tablet device charging on an infected computer would be one of the simplest methods of propagating the BadUSB vulnerability. They include example malicious firmware code that would infect Android devices with BadUSB.

Researchers at Aries Security and the Wall of Sheep later revisited the juice jacking concept in 2016. They set up a "Video Jacking" charging station which was able to record the mirrored screen from phones plugged in to their malicious charging station. Affected devices at the time included android devices supporting SlimPort or MHL protocols over USB, as well as the most recent iPhone using a lightning charge cable connector.

History

Brian Krebs was the first to report on this attack and coined the term "juice jacking". After seeing the informational cell phone charging kiosk set up in the Wall of Sheep at DefCon 19 in August 2011, he wrote the first article on his security journalism site Krebs on Security. The wall of sheep researchers including Brian Markus, Joseph Mlodzianowski and Robert Rowley designed the kiosk as an information tool to bring awareness to the potential attack vector and have discussed but not released tools publicly which perform malicious actions on the charging devices.

An episode of the hacking series Hak5 released in September 2012 showcased a number of attacks which can be utilized using an attack framework named P2P-ADB released by Kyle Osborn. The P2P-ADB attack framework discussed utilized one phone to attack another phone over a USB on the Go connection.

In late 2012 a document was released by the NSA warning government employees who travel about the threat of juice jacking and reminding the reader that during overseas travel only to use their personal power charging cables and not to charge in public kiosks or by utilizing other people's computers.

The Android Hackers Handbook released in March 2014 has dedicated sections discussing both juice jacking and the ADB-P2P framework.

Juice jacking was the central focus on an episode of *CSI: Cyber*. Season 1: Episode 9, "LoM1S" aired in April 2015

Mitigation

Apple's IOS has taken multiple security measures to reduce the attack surface over USB including no longer allowing the device to automatically mount as a hard drive when plugged in over USB, as well as release security patches for vulnerabilities such as those exploited by Mactans.

Android devices commonly prompt the user before allowing the device to be mounted as a hard drive when plugged in over USB. Since release 4.2.2, Android has implemented a whitelist verification step to disallow attackers to access the Android Debug Bridge without authorization.

Juice jacking is not possible if a device is charged via the AC adapter shipped with the device, a battery backup device, or by utilizing a USB cable that has its data cables removed. A tool originally called the USB Condom, and now renamed to SyncStop has been release with the sole purpose of disallowing data connections to be passed over a USB cable.

Trusted Execution Environment

The Trusted Execution Environment (TEE) is a secure area of the main processor of a smart phone (or any connected device including tablets, set-top boxes and televisions). It guarantees code and data loaded inside to be protected with respect to confidentiality and integrity. The TEE as an isolated execution environment is providing security features such as isolated execution, integrity of Trusted Applications along with confidentiality of their assets. In general terms, the TEE offers an execution space that provides a higher level of security than a rich mobile operating system (mobile OS) and more functionality than a 'secure element' (SE).

Industry associations like GlobalPlatform (working to standardize specifications for the TEE) and Trusted Computing Group (working to align GlobalPlatform TEE specification with its Trusted Platform Module (TPM) technology for enhanced mobile security) have undertaken work in recent years.

History

OMTP first defined the TEE in their 'Advanced Trusted Environment:OMTP TR1' standard, defining it as a "set of hardware and software components providing facilities necessary to support Applications" which had to meet the requirements of one of two defined security levels. The first security level, Profile 1, was targeted against only software attacks and whilst Profile 2, was targeted against both software and hardware attacks.

Commercial TEE solutions based on ARM TrustZone technology which conformed to the TR1 standard such as Trusted Foundations, developed by Trusted Logic, were later launched. This software would become part of the Trustonic joint venture, and the basis of future GlobalPlatform TEE solutions.

Work on the OMTP standards ended in mid 2010 when the group transitioned into the 'Wholesale Applications Community' (WAC).

The OMTP standards, including those defining a TEE, are hosted by GSMA.

In July 2010 GlobalPlatform first announced their own standardisation of the TEE, focusing first on the client API (the interface to the TEE within the mobile OS) which was expanded later to include the TEE internal API, and a compliance programme and standardised security level.

Details

The TEE is an isolated environment that runs in parallel with the mobile OS, providing security for the rich environment. It is more secure than the mobile OS and offers a higher level of functionality than the SE, using a hybrid approach that utilizes both hardware and software to protect data. It therefore offers a level of security sufficient for most applications. Trusted applications running in a TEE have access to the full power of a device's main processor and memory, while hardware isolation protects these from user installed apps running in a main operating system. Software and cryptographic isolation inside the TEE protect the trusted applications contained within from each other.

Service providers, mobile network operators (MNO), operating system (OS) and application developers, device manufacturers, platform providers and silicon vendors are all key stakeholders are all interested in, and contributing to, the standardization efforts and will benefit from the resulting specifications.

Uses

There are a number of use cases for the TEE:

Premium Content Protection

The TEE is an ideal environment for protecting premium content (for example, HD films) on connected devices such as smart phones and HD televisions. Premium content is defined by its perceived value which is in itself defined by the quality of the material (4K high definition films are one example), the file's proximity to its release date (as content has more value the closer it is to its release) and by consumer recognition. The TEE is used to protect the highest value content and so will be deployed into devices where this content is available:

- 4k HD televisions
- 4k HD set-top boxes
- tablet computers
- smartphones

The TEE is used to protect the content once it is on the device. The content is encrypted during transmission or streaming so it is protected. The TEE protects the content once it has been decrypted on the device as it is a secure environment.

Mobile Financial Services

As m-Commerce (mobile wallets, peer-to-peer payments, contactless payments and using a mobile device as a point of sale (POS) terminal) develops, stronger and more standardized mobile security is needed. In collaboration with near field communication (NFC) and SEs, the TEE needs to be deployed to ensure the device is secure and that consumers can carry out any financial transaction in a safe and trusted environment.

Sensitive mobile use cases often need some form of interaction with the end user, meaning that sensitive information needs to be 'exposed' in the mobile OS to the user for validation - 'to guarantee What You See Is What You Sign'. The TEE offers a safe and trusted user interface to enable authentication on a mobile device.

Authentication

The TEE is ideal for supporting natural ID (facial recognition, fingerprint sensor and voice authorization) as PINs and passwords can be easily hacked and stolen. The authentication process is split into three stages:

- Extracting an 'image' (scanning the fingerprint or capturing a voice sample, for example).
- A reference 'template' stored on the device for comparison with the extracted 'image'.

- A match engine to process the comparison between the 'image' and the 'template'.

The TEE is an ideal area within a mobile device to house the match engine and the associated processes required to authenticate the user. The increased security of this environment is able to protect the data and establish a buffer against the non-secure apps located in mobile OS. This additional security will help to satisfy the needs of service providers in addition to keeping the costs low for handset developers.

The FIDO Alliance is collaborating with GlobalPlatform to standardize the TEE for natural ID implementations.

Enterprise and Government

The TEE can be used by governments and enterprises to enable the secure handling of confidential information on a mobile device. The TEE offers a level of protection against software attacks generated in the mobile OS and assists in the control of access rights. It achieves this by housing sensitive, 'trusted' applications that need to be isolated and protected from the mobile OS and any malicious malware that may be present. Through utilizing the functionality and security levels offered by the TEE, governments and enterprises can be assured that employees using their own devices are doing so in a secure and trusted manner.

Implementations

The following embedded hardware technologies can be used to support TEE implementations:

- AMD:

  - Platform Security Processor (PSP)

- ARM:

  - TrustZone

- Intel x86-64 instruction set:

  - SGX Software Guard Extensions

Several TEE implementations are available from different TEE providers:

- Commercial implementations

  - Kinibi, a commercial implementation from Trustonic that has been qualified by GlobalPlatform

  - securiTEE, a commercial implementation from Solacia that has been qualified by GlobalPlatform

- Open-source implementations

  - OP-TEE, an open source implementation under BSD license, originally from STMicroelectronics, now owned and maintained by Linaro.

- o   TLK, an open-source implementation from Nvidia under BSD license

- o   T6, and open-source implementation and research topic under GPL license

- o   Open TEE, an open source implementation and research project from the University of Helsinki and sponsored by Intel. Provided under an Apache license

- Implementations with dual commercial/open-source licensing

- o   SierraTEE, an implementation from Sierraware available both under commercial and GPL-licensing

Standardization

While there are a number of proprietary systems, GlobalPlatform is working to standardize the TEE. Standardizing the TEE is crucial for mobile wallets, NFC payment implementations, premium content protection and bring your own device (BYOD) initiatives.

These following TEE specifications are currently available from the GlobalPlatform website:

- TEE Client API Specification v1.0 outlines the communication between applications running in a mobile OS and trusted applications residing in the TEE.

- TEE Systems Architecture v1.0 explains the hardware and software architectures behind the TEE.

- TEE Internal API Specification v1.0 specifies how to develop trusted applications.

- TEE Secure Element API Specification v1.0 specifies the syntax and semantics of the TEE Secure Element API. It is suitable for software developers implementing trusted applications running inside the TEE which need to expose an externally visible interface to client applications.

- Trusted User Interface API Specification v1.0 specifies how a trusted UI should facilitate information that will be securely configured by the end user and securely controlled by the TEE.

- TEE TA Debug Specification v1.0 specifies the GlobalPlatform TEE debug interfaces and protocols.

Joint venture formed by ARM, Gemalto and Giesecke & Devrient (G&D), Trustonic, was the first to qualify a GlobalPlatform-compliant TEE product in 2013.

Security

The GlobalPlatform TEE Protection Profile specifies the typical threats the hardware and software of the TEE needs to withstand. It also details the security objectives that are to be met in order to counter these threats and the security functional requirements that a TEE will have to comply with. A security assurance level of EAL2+ has been selected; the focus is on vulnerabilities that are subject to widespread, software-based exploitation.

The Common Criteria portal has officially listed the GlobalPlatform TEE Protection Profile on its website, under the Trusted Computing category. This important milestone means that industries using TEE technology to deliver services such as premium content and mobile wallets, or enterprises and governments establishing secure mobility solutions, can now formally request that TEE products are certified against this security framework.

GlobalPlatform is committed to ensuring a standardized level of security for embedded applications on secure chip technology. It has developed an open and thoroughly evaluated trusted execution environment (TEE) ecosystem with accredited laboratories and evaluated products. This certification scheme created to certify a TEE product in 3 months has been launched officially in June 2015

Mobile Phone Tracking

Mobile phone tracking is the ascertaining of the position or location of a mobile phone, whether stationary or moving. Localization may occur either via multilateration of radio signals between (several) cell towers of the network and the phone, or simply via GPS. To locate a mobile phone using multilateration of radio signals, it must emit at least the roaming signal to contact the next nearby antenna tower, but the process does not require an active call. The Global System for Mobile Communications (GSM) is based on the phone's signal strength to nearby antenna masts.

Mobile positioning may include location-based services that disclose the actual coordinates of a mobile phone, which is a technology used by telecommunication companies to approximate the location of a mobile phone, and thereby also its user.

Technology

The technology of locating is based on measuring power levels and antenna patterns and uses the concept that a powered mobile phone always communicates wirelessly with one of the closest base stations, so knowledge of the location of the base station implies the cell phone is nearby.

Advanced systems determine the sector in which the mobile phone is located and roughly estimate also the distance to the base station. Further approximation can be done by interpolating signals between adjacent antenna towers. Qualified services may achieve a precision of down to 50 meters in urban areas where mobile traffic and density of antenna towers (base stations) is sufficiently high. Rural and desolate areas may see miles between base stations and therefore determine locations less precisely.

GSM localization uses multilateration to determine the location of GSM mobile phones, or dedicated trackers, usually with the intent to locate the user.

The location of a mobile phone can be determined in a number of ways:

Network-based

The location of a mobile phone can be determined using the service provider's network infrastructure. The advantage of network-based techniques, from a service provider's point of view, is that they can be implemented non-intrusively without affecting handsets. Network-based techniques were developed many years prior to the widespread availability of GPS on handsets.

The accuracy of network-based techniques varies, with cell identification as the least accurate and triangulation as moderately accurate, and newer "advanced forward link trilateration" timing methods as the most accurate. The accuracy of network-based techniques is both dependent on the concentration of cell base stations, with urban environments achieving the highest possible accuracy because of the higher number of cell towers, and the implementation of the most current timing methods.

One of the key challenges of network-based techniques is the requirement to work closely with the service provider, as it entails the installation of hardware and software within the operator's infrastructure. Frequently the compulsion associated with a legislative framework, such as Enhanced 9-1-1, is required before a service provider will deploy a solution.

Handset-based

The location of a mobile phone can be determined using client software installed on the handset. This technique determines the location of the handset by putting its location by cell identification, signal strengths of the home and neighboring cells, which is continuously sent to the carrier. In addition, if the handset is also equipped with GPS then significantly more precise location information can be then sent from the handset to the carrier.

Another approach is to use a fingerprinting-based technique, where the "signature" of the home and neighboring cells signal strengths at different points in the area of interest is recorded by war-driving and matched in real-time to determine the handset location. This is usually performed independent from the carrier.

The key disadvantage of handset-based techniques, from service provider's point of view, is the necessity of installing software on the handset. It requires the active cooperation of the mobile subscriber as well as software that must be able to handle the different operating systems of the handsets. Typically, smartphones, such as one based on Symbian, Windows Mobile, Windows Phone, BlackBerry OS, iOS, or Android, would be able to run such software, e.g. Google Maps.

One proposed work-around is the installation of embedded hardware or software on the handset by the manufacturers, e.g., Enhanced Observed Time Difference (E-OTD). This avenue has not made significant headway, due to the difficulty of convincing different manufacturers to cooperate on a common mechanism and to address the cost issue. Another difficulty would be to address the issue of foreign handsets that are roaming in the network.

SIM-based

Using the subscriber identity module (SIM) in GSM and Universal Mobile Telecommunications System (UMTS) handsets, it is possible to obtain raw radio measurements from the handset. Available measurements include the serving Cell ID, round-trip time, and signal strength. The type of information obtained via the SIM can differ from that which is available from the handset. For example, it may not be possible to obtain any raw measurements from the handset directly, yet still obtain measurements via the SIM.

Wi-Fi

Crowdsourced Wi-Fi data can also be used to identify a handset's location. Poor performance of the GPS-based methods in indoor environment and increasing popularity of Wi-Fi have encouraged companies to design new and feasible methods to carry out Wi-Fi-based indoor positioning. Most smartphones combine Global Navigation Satellite Systems (GNSS), such as GPS and GLONASS, with Wi-Fi positioning systems.

Hybrid

Hybrid positioning systems use a combination of network-based and handset-based technologies for location determination. One example would be some modes of Assisted GPS, which can both use GPS and network information to compute the location. Both types of data are thus used by the telephone to make the location more accurate (i.e., A-GPS). Alternatively tracking with both systems can also occur by having the phone attain its GPS-location directly from the satellites, and then having the information sent via the network to the person that is trying to locate the telephone. Such systems include Google Maps, as well as, LTE's OTDOA and E-CellID.

There are also hybrid positioning systems which combine several different location approaches to position mobile devices by Wi-Fi, WiMAX, GSM, LTE, IP addresses, and network environment data.

Operational Purpose

In order to route calls to a phone, the cell towers listen for a signal sent from the phone and negotiate which tower is best able to communicate with the phone. As the phone changes location, the antenna towers monitor the signal, and the phone is "roamed" to an adjacent tower as appropriate. By comparing the relative signal strength from multiple antenna towers, a general location of a phone can be roughly determined. Other means make use of the antenna pattern, which supports angular determination and phase discrimination.

Newer phones may also allow the tracking of the phone even when turned on and not active in a telephone call. This results from the roaming procedures that perform hand-over of the phone from one base station to another.

Bearer Interest

A phone's location can be uploaded to a common website where one's friends and family can view one's last reported position. Newer phones may have built-in GPS receivers which could be used in a similar fashion, but with much higher accuracy. This is controversial, because data on a common website means people who are not "friends and family" may be able to view the information.

Privacy

Locating or positioning touches upon delicate privacy issues, since it enables someone to check where a person is without the person's consent. Strict ethics and security measures are strongly recommended for services that employ positioning. In 2012 Malte Spitz held a TED talk on the issue of mobile phone privacy in which he showcased his own stored data that he received from Deutsche Telekom after suing the company. He described the data, which consists of 35,830 lines of data collected during the span of Germany's data retention at the time, saying, "This is six months of my life [...] You can see where I am, when I sleep at night, what I'm doing." He partnered up with ZEIT Online and made his information publicly available in an interactive map which allows users to watch his entire movements during that time in fast-forward. Spitz concluded that technology consumers are the key to challenging privacy norms in today's society who "have to fight for self determination in the digital age."

China

China has proposed using this technology to track commuting patterns of Beijing city residents. Aggregate presence of mobile phone users could be tracked in a privacy-preserving fashion.

Europe

In Europe most countries have a constitutional guarantee on the secrecy of correspondence, and location data obtained from mobile phone networks is usually given the same protection as the communication itself.

United States

In the United States, there is no explicit constitutional guarantee on the privacy of telecommunications, so use of location data is limited by law. Law enforcement can obtain permission to position phones in emergencies where people, including criminals, are missing. In some instances, law enforcement may even access a mobile phone's internal microphone to eavesdrop on conversations while the phone is switched off.

A secret interpretation of The Patriot Act, confirmed to exist, has been linked to secret widespread location tracking.

Since 2005 the Electronic Frontier Foundation has been following some U.S. cases, including *USA v. Pen Register*, regarding government tracking of individuals. In *In re Application of the United States for Historical Cell Site Data*, 724 F.3d 600 (5th Cir. 2013), the United States Court of Appeals for the Fifth Circuit held that the government does not need a warrant to compel cell phone providers to disclose historical cell site information. However, in *United States v. Davis (2014)*, the United States Court of Appeals for the Eleventh Circuit ruled in a criminal case that obtaining cell phone location data "without a warrant is a Fourth Amendment violation."

In 2014, it was revealed that in order to find fugitives, the United States Marshals Service has been flying small aircraft with equipment that identifies all cell phones in the area.

Commercial Privacy of location Information in the United States

The U.S. does limit commercial use of location information under the (US) Telecommunications Act, at 47 CFR §222. The Telecommunications Act, at 47 CFR §222(f), requires consent from the subscriber, and prohibits telecommunication common carriers from accessing location information for purposes other than system operation without consent of the customer. Businesses such as Locaid, which provide a tracking service based on subscriber information, require mobile users' consent prior to tracking.

References

- Dunham, Ken; Abu Nimeh, Saeed; Becher, Michael (2008). Mobile Malware Attack and Defense. Syngress Media. ISBN 978-1-59749-298-0.

- Christopher Elisan (5 September 2012). Malware, Rootkits & Botnets A Beginner's Guide. McGraw Hill Professional. pp. 10–. ISBN 978-0-07-179205-9.

- Kirat, Dhilung; Vigna, Giovanni; Kruegel, Christopher (2014). Barecloud: bare-metal analysis-based evasive malware detection. ACM. pp. 287–301. ISBN 978-1-931971-15-7.

- Young, Adam; Yung, Moti (1997). "Deniable Password Snatching: On the Possibility of Evasive Electronic Espionage". Symp. on Security and Privacy. IEEE. pp. 224–235. ISBN 0-8186-7828-3.

- Hantula, Richard (2010). How Do Cell Phones Work?. Google Books. 132 West 31st Street, New York NY 10001, US: Infobase Publishing. p. 27. ISBN 978-1-43812-805-4. Retrieved 18 January 2014.

- Drake, Joshua; Lanier, Zach; Mulliner, Collin; Fora, Pau; Ridley, Stephen; Wicherski, Georg (March 2014). Android Hacker's Handbook. Wiley. p. 576. ISBN 978-1-118-60864-7.

- Paganini, Pierluigi (March 7, 2014). "Dendroid – A new Android RAT available on the underground". securityaffairs.co. Retrieved 23 October 2016.

- Leder, Felix (May 27, 2014). "Dendroid under the hood – A look inside an Android RAT kit". Blue Coat Labs. Retrieved 23 October 2016.

- Zorz, Zeljka (March 7, 2014). "Dendroid spying RAT malware found on Google Play". helpnetsecurity.com. Retrieved 23 October 2016.

Health Concerns of Mobile Technology

The effect that technology has on human health is one of the biggest concerns of our time. Wireless electronic devices cause radiation that are known to cause issues in humans. One of the concerns related to mobile phones is also the number of accidents that are caused by using mobile phones while driving. The topics discussed in the text are of great importance to broaden the existing knowledge on the subject matter.

Mobile Phone Radiation and Health

The effect of mobile phone radiation on human health is a subject of interest and study worldwide, as a result of the enormous increase in mobile phone usage throughout the world. As of 2016, there were 7.4 billion subscriptions worldwide, though the actual number of users is lower as many users own multiple mobile phones. Mobile phones use electromagnetic radiation in the microwave range (450–2100 MHz). Other digital wireless systems, such as data communication networks, produce similar radiation.

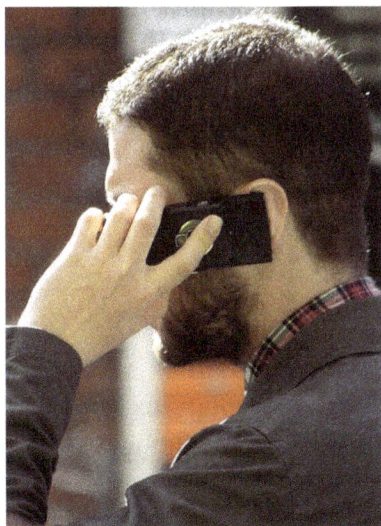

A man speaking on a mobile telephone

In 2011, International Agency for Research on Cancer (IARC) classified mobile phone radiation as Group 2B – possibly carcinogenic (*not* Group 2A – probably carcinogenic – nor the dangerous Group 1). That means that there «could be some risk» of carcinogenicity, so additional research into the long-term, heavy use of mobile phones needs to be conducted. The WHO added in June 2011 that "to date, no adverse health effects have been established as being caused by mobile phone use", a point they reiterated in October 2014. Some national radiation advisory authorities have recommended measures to minimize exposure to their citizens as a precautionary approach.

Effects

Many scientific studies have investigated possible health symptoms of mobile phone radiation. These studies are occasionally reviewed by some scientific committees to assess overall risks. A 2007 assessment published by the European Commission Scientific Committee on Emerging and Newly Identified Health Risks (SCENIHR) concludes that the three lines of evidence, *viz.* animal, *in vitro*, and epidemiological studies, indicate that "exposure to RF fields is unlikely to lead to an increase in cancer in humans".

Radiation Absorption

Part of the radio waves emitted by a mobile telephone handset are absorbed by the body. The radio waves emitted by a GSM handset are typically below a watt. The maximum power output from a mobile phone is regulated by the mobile phone standard and by the regulatory agencies in each country. In most systems the cellphone and the base station check reception quality and signal strength and the power level is increased or decreased automatically, within a certain span, to accommodate different situations, such as inside or outside of buildings and vehicles. The rate at which energy is absorbed by the human body is measured by the Specific Absorption Rate (SAR), and its maximum levels for modern handsets have been set by governmental regulating agencies in many countries. In the U.S., the Federal Communications Commission (FCC) has set a SAR limit of 1.6 W/kg, averaged over a volume of 1 gram of tissue, for the head. In Europe, the limit is 2 W/kg, averaged over a volume of 10 grams of tissue. SAR values are heavily dependent on the size of the averaging volume. Without information about the averaging volume used, comparisons between different measurements cannot be made. Thus, the European 10-gram ratings should be compared among themselves, and the American 1-gram ratings should only be compared among themselves. SAR data for specific mobile phones, along with other useful information, can be found directly on manufacturers' websites, as well as on third party web sites. It is worth noting that thermal radiation is not comparable to ionizing radiation in that it only increases the temperature in normal matter, it does not break molecular bonds or release electrons from their atoms.

Thermal Effects

One well-understood effect of microwave radiation is dielectric heating, in which any dielectric material (such as living tissue) is heated by rotations of polar molecules induced by the electromagnetic field. In the case of a person using a cell phone, most of the heating effect will occur at the surface of the head, causing its temperature to increase by a fraction of a degree. In this case, the level of temperature increase is an order of magnitude less than that obtained during the exposure of the head to direct sunlight. The brain's blood circulation is capable of disposing of excess heat by increasing local blood flow. However, the cornea of the eye does not have this temperature regulation mechanism and exposure of 2–3 hours duration has been reported to produce cataracts in rabbits' eyes at SAR values from 100–140 W/kg, which produced lenticular temperatures of 41 °C. There were no cataracts detected in the eyes of monkeys exposed under similar conditions. Premature cataracts have not been linked with cell phone use, possibly because of the lower power output of mobile phones.

Non-thermal Effects

The communications protocols used by mobile phones often result in low-frequency pulsing of the carrier signal. While the existence of effects due to the field is indisputable, whether these modulations are causing these effects or these are still of thermic nature is subject to debate.

Some researchers have argued that so-called "non-thermal effects" could be reinterpreted as a normal cellular response to an increase in temperature. The German biophysicist Roland Glaser, for example, has argued that there are several thermoreceptor molecules in cells, and that they activate a cascade of second and third messenger systems, gene expression mechanisms and production of heat shock proteins in order to defend the cell against metabolic cell stress caused by heat. The increases in temperature that cause these changes are too small to be detected by studies such as REFLEX, which base their whole argument on the apparent stability of thermal equilibrium in their cell cultures.

Other researchers believe the stress proteins are unrelated to thermal effects, since they occur for both extremely low frequencies (ELF) and radio frequencies (RF), which have very different energy levels. Another preliminary study published in 2011 by *The Journal of the American Medical Association* conducted using fluorodeoxyglucose injections and positron emission tomography concluded that exposure to radiofrequency signal waves within parts of the brain closest to the cell phone antenna resulted in increased levels of glucose metabolism, but the clinical significance of this finding is unknown.

Blood–brain Barrier Effects

Swedish researchers from Lund University (Salford, Brun, Persson, Eberhardt, and Malmgren) have studied the effects of microwave radiation on the rat brain. They found a leakage of albumin into the brain via a permeated blood–brain barrier. This confirms earlier work on the blood–brain barrier by Allan Frey, Oscar and Hawkins, and Albert and Kerns. Other groups have not confirmed these findings in vitro cell studies or whole animal studies,

Prof Leszczynski of Finland's radiation and nuclear safety authority found that, at the maximum legal limit for mobile radiation, one protein in particular, HSP 27, was affected. HSP 27 played a critical role in the integrity of the blood-brain barrier.

Cancers to Humans

In 2006, a large Danish group's study about the connection between mobile phone use and cancer incidence was published. It followed over 420,000 Danish citizens for 20 years and showed no increased risk of cancer. A 2011 follow-up confirmed these findings.

The following studies of long time exposure have been published:

- The 13 nation INTERPHONE project – the largest study of its kind ever undertaken – was published in 2010 and did not find a solid link between mobile phones and brain tumours.

The *International Journal of Epidemiology* published a combined data analysis from INTERPHONE, a multi national population-based case-control study of glioma and meningioma, the most common types of brain tumour.

The authors reported the following conclusion:

Overall, no increase in risk of glioma or meningioma was observed with use of mobile phones. There were suggestions of an increased risk of glioma at the highest exposure levels, but biases and error prevent a causal interpretation. The possible effects of long-term heavy use of mobile phones require further investigation.

In the press release accompanying the release of the paper, Dr. Christopher Wild, Director of the International Agency for Research on Cancer (IARC) said:

An increased risk of brain cancer is not established from the data from Interphone. However, observations at the highest level of cumulative call time and the changing patterns of mobile phone use since the period studied by Interphone, particularly in young people, mean that further investigation of mobile phone use and brain cancer risk is merited.

A number of independent health and government authorities have commented on this important study including The Australian Centre for Radiofrequency Bioeffects Research (ACRBR) which said in a statement that:

Until now there have been concerns that mobile phones were causing increases in brain tumours. Interphone is both large and rigorous enough to address this claim, and it has not provided any convincing scientific evidence of an association between mobile phone use and the development of glioma or meningioma. While the study demonstrates some weak evidence of an association with the highest tenth of cumulative call time (but only in those who started mobile phone use most recently), the authors conclude that biases and errors limit the strength of any conclusions in this group. It now seems clear that if there was an effect of mobile phone use on brain tumour risks in adults, this is likely to be too small to be detectable by even a large multinational study of the size of Interphone.

The Australian Radiation Protection and Nuclear Safety Agency (ARPANSA) which said in a statement that:

On the basis of current understanding of the relationship between brain cancer and use of mobile phones, including the recently published data from the INTERPHONE study, ARPANSA:

concludes that currently available data do not warrant any general recommendation to limit use of mobile phones in the adult population,

continues to inform those concerned about potential health effects that they may limit their exposure by reducing call time, by making calls where reception is good, by using hands-free devices or speaker options, or by texting; and

recommends that, due to the lack of any data relating to children and long term use of mobile phones, parents encourage their children to limit their exposure by reducing call time, by making calls where reception is good, by using hands-free devices or speaker options, or by texting.

The Cancer Council Australia said in a statement that it cautiously welcomed the results of the largest international study to date into mobile phone use, which has found no evidence that normal use of mobile phones, for a period up to 12 years, can cause brain cancer.

Chief Executive Officer, Professor Ian Olver, said findings from the Interphone study, conducted across 13 countries including Australia, were consistent with other research that had failed to find a link between mobile phones and cancer.

This supports previous research showing mobile phones don't damage cell DNA, meaning they can't cause the type of genetic mutations that develop into cancer," Professor Olver said.

However, it has been suggested that electromagnetic fields associated with mobile phones may play a role in speeding up the development of an existing cancer. The Interphone study found no evidence to support this theory.

- A Danish study (2004) that took place over 10 years found no evidence to support a link. However, this study has been criticized for collecting data from subscriptions and not necessarily from actual users. It is known that some subscribers do not use the phones themselves but provide them for family members to use. That this happens is supported by the observation that only 61% of a small sample of the subscribers reported use of mobile phones when responding to a questionnaire.

- A Swedish study (2005) that draws the conclusion that "the data do not support the hypothesis that mobile phone use is related to an increased risk of glioma or meningioma."

- A British study (2005) that draws the conclusion that "The study suggests that there is no substantial risk of acoustic neuroma in the first decade after starting mobile phone use. However, an increase in risk after longer term use or after a longer lag period could not be ruled out."

- A German study (2006) that states "In conclusion, no overall increased risk of glioma or meningioma was observed among these cellular phone users; however, for long-term cellular phone users, results need to be confirmed before firm conclusions can be drawn."

- A joint study conducted in northern Europe that draws the conclusion that "Although our results overall do not indicate an increased risk of glioma in relation to mobile phone use, the possible risk in the most heavily exposed part of the brain with long-term use needs to be explored further before firm conclusions can be drawn."

Other studies on cancer and mobile phones are:

- A Swedish scientific team at the Karolinska Institute conducted an epidemiological study (2004) that suggested that regular use of a mobile phone over a decade or more was associated with an increased risk of acoustic neuroma, a type of benign brain tumor. The increase was not noted in those who had used phones for fewer than 10 years.

- The INTERPHONE study group from Japan published the results of a study of brain tumour risk and mobile phone use. They used a new approach: determining the SAR inside a tumour by calculating the radio frequency field absorption in the exact tumour location. Cases examined included glioma, meningioma, and pituitary adenoma. They reported that the overall odds ratio (OR) was not increased and that there was no significant trend towards an increasing OR in relation to exposure, as measured by SAR.

In 2007, Dr. Lennart Hardell, from Örebro University in Sweden, reviewed published epidemiological papers (2 cohort studies and 16 case-control studies) and found that:

- Cell phone users had an increased risk of malignant gliomas.

- Cell phone use was linked to a higher rate of acoustic neuromas.

- Tumors are more likely to occur on the side of the head that the cell handset is used.

- One hour of cell phone use per day significantly increases tumor risk after ten years or more.

In a February 2008 update on the status of the INTERPHONE study IARC stated that the long-term findings '...could either be causal or artifactual, related to differential recall between cases and controls.'

A publication titled "Public health implications of wireless technologies" cites that Lennart Hardell found age is a significant factor. The report repeated the finding that the use of cell phones before age 20 increased the risk of brain tumors by 5.2, compared to 1.4 for all ages. A review by Hardell et al. concluded that current mobile phones are not safe for long-term exposure. In 2009, a meta-analysis of 23 studies on mobile phone use and tumor risk found that "there is possible evidence" that mobile phone use causes an increased risk of tumors.

In a time trends study in Europe, conducted by the Institute of Cancer Epidemiology in Copenhagen, no significant increase in brain tumors among cell phone users was found between the years of 1998 and 2003. "The lack of a trend change in incidence from 1998 to 2003 suggests that the induction period relating mobile phone use to brain tumors exceeds 5–10 years, the increased risk in this population is too small to be observed, the increased risk is restricted to subgroups of brain tumors or mobile phone users, or there is no increased risk."

On 31 May 2011, the International Agency for Research on Cancer classified radiofrequency electromagnetic fields as possibly carcinogenic to humans (Group 2B). The IARC assessed and evaluated available literature and studies about the carcinogenicity of radiofrequency electromagnetic fields (RF-EMF), and found the evidence to be "limited for carcinogenicity of RF-EMF, based on positive associations between glioma and acoustic neuroma and exposure". The conclusion of the IARC was mainly based on the INTERPHONE study, which found an increased risk for glioma in the highest category of heavy users (30 minutes per day over a 10-year period), although no increased risk was found at lower exposure and other studies could not back up the findings. The evidence for other types of cancer was found to be "inadequate". Some members of the Working Group opposed the conclusions and considered the current evidence in humans still as "inadequate", citing inconsistencies between the assessed studies.

In 2011, a review published in *Environmental Health Perspectives* found that increasing evidence suggests that mobile phone use does not cause brain tumors in adults.

In 2012, a systematic review was published which found "no statistically significant increase in risk for adult brain cancer or other head tumors from wireless phone use."

Researchers at the National Cancer Institute found that while cell phone use increased substantially over the period 1992 to 2008 (from nearly zero to almost 100 percent of the population), the U.S. trends in glioma incidence did not mirror that increase.

In 2014, a French national case-control study, CERENAT, confirmed a possible association between heavy mobile phone use and brain tumours (gliomas and meningiomas), finding an up to eight-fold increased risk of gliomas tied with cellphone use. A review published the same year concluded that its results "detract from the hypothesis that mobile phone use affects the occurrence of intracranial tumors."

In March 2015, a study on mice carried on by Prof. Alexander Lerchl of Jacobs University in Bremen and his team on behalf of the German Federal Office for Radiation Protection found out that the growth rate of liver and lung cancer generated by chemical substances raises substantially when the animals are lifelong irradiated with mobile phone-like e.m. fields. This study confirms a research carried on in 2010 at Fraunhofer Institute. Moreover, the researchers discovered a significant higher rate of lymphomas, and found out that some of the effects occur also for field intensities lower than current limits. The underlying mechanisms are unknown.

In May 2016, the National Toxicology Program released partial results from an extensive rat study finding that cell phone radiation exposure was associated with an increase in brain and heart cancer rates. Specifically, 2 to 3 percent of male rats exposed to cell phone radiation developed malignant glioma brain tumors and 5 to 7 percent of exposed rats developed schwannoma tumors in their hearts. Because of the scale of the study, scientists such as the director of the American Science Society, found the study to be based on solid grounds and serving as major advance towards resolving the important cell phone use and public health issue. Other scientists and organizations questioned the findings because of the apparent reliability problems. Only male rats indicated a significant tumor increase while there were no significant increase in female rats. Unlike the normal rat population's 2 percent malignant glioma rates, none of the control group's rats showed any malignant glioma development. What was also surprising was that the exposed male rats demonstrated longer life spans than the control group's unexposed male rats. These inconsistencies led Dr. Michael S. Lauer, deputy director for extramural research at the National Institutes of Health, to conclude that it is likely that the studies findings were based on false positive results.

Cognitive Effects

A 2009 study, examined the effects of exposure to radiofrequency radiation (RFR) emitted by standard GSM cell phones on the cognitive functions of humans. The study confirmed longer (slower) response times to a spatial working memory task when exposed to RFR from a standard GSM cellular phone placed next to the head of male subjects, and showed that longer duration of exposure to RFR may increase the effects on performance. Right-handed subjects exposed to RFR on the left side of their head on average had significantly longer response times when compared to exposure to the right side and sham-exposure.

Electromagnetic Hypersensitivity

Some users of mobile handsets have reported feeling several unspecific symptoms during and after its use; ranging from burning and tingling sensations in the skin of the head and extremities, fatigue, sleep disturbances, dizziness, loss of mental attention, reaction times and memory retentiveness, headaches, malaise, tachycardia (heart palpitations), to disturbances of the digestive system. Reports have noted that all of these symptoms can also be attributed to stress and that current research cannot separate the symptoms from nocebo effects.

Genotoxic Effects

A meta-analysis (2008) of 63 in vitro and in vivo studies from the years 1990–2005, concluded that RF radiation was genotoxic only in some conditions and that the studies reporting positive effects evidenced publication bias.

A meta-study (2009) of 101 publications on genotoxicity of RF electromagnetic fields, showed that 49 reported a genotoxic effect and 42 not. The authors found "ample evidence that RF-EMF can alter the genetic material of exposed cells in vivo and in vitro and in more than one way".

In 1995, in the journal *Bioelectromagnetics*, Henry Lai and Narenda P. Singh reported damaged DNA after two hours of microwave radiation at levels deemed safe according to U.S. government standards.

In December 2004, a pan-European study named REFLEX (Risk Evaluation of Potential Environmental Hazards from Low Energy Electromagnetic Field (EMF) Exposure Using Sensitive in vitro Methods), involving 12 collaborating laboratories in several countries showed some compelling evidence of DNA damage of cells in in-vitro cultures, when exposed between 0.3 and 2 watts/kg, whole-sample average. There were indications, but not rigorous evidence of other cell changes, including damage to chromosomes, alterations in the activity of certain genes and a boosted rate of cell division.

Research published in 2004, by a team at the University of Athens, had a reduction in reproductive capacity in fruit flies exposed to 6 minutes of 900 MHz pulsed radiation for five days.

Subsequent research, again conducted on fruit flies, was published in 2007, with the same exposure pattern but conducted at both 900 MHz and 1800 MHz, and had similar changes in reproductive capacity with no significant difference between the two frequencies.

Following additional tests published in a third article, the authors stated they thought their research suggested the changes were "...due to degeneration of large numbers of egg chambers after DNA fragmentation of their constituent cells ...".

Australian research conducted in 2009, by subjecting in vitro samples of human spermatozoa to radio-frequency radiation at 1.8 GHz and specific absorption rates (SAR) of 0.4 to 27.5 W/kg showed a correlation between increasing SAR and decreased motility and vitality in sperm, increased oxidative stress and 8-Oxo-2'-deoxyguanosine markers, stimulating DNA base adduct formation and increased DNA fragmentation.

Sleep and EEG effects

Sleep, EEG and waking rCBF have been studied in relation to RF exposure for a decade now, and the majority of papers published to date have found some form of effect. While a Finnish study failed to find any effect on sleep or other cognitive function from pulsed RF exposure, most other papers have found significant effects on sleep. Two of these papers found the effect was only present when the exposure was pulsed (amplitude modulated), and one early paper found that sleep quality (measured by the amount of participants' broken sleep) improved.

While some papers were inconclusive or inconsistent, a number of studies have now demonstrated reversible EEG and rCBF alterations from exposure to pulsed RF exposure. German research from 2006 found that statistically significant EEG changes could be consistently found, but only in a relatively low proportion of study participants (12–30%).

Behavioural Effects

A study on mice offspring suggested that cell phone use during pregnancy may cause behavioural problems that resemble the effects of ADHD.

Sperm count and Sperm Quality

A number of studies have shown relationships between mobile telephone use and reduced sperm count and sperm quality. Peer reviewed studies have shown relationships using statistical questionnaire techniques, controlled experiments on living humans, and controlled experiments on sperm outside the body.

The Environmental Working Group (EWG) has a web page entitled "Cell Phone Radiation Damages Sperm, Studies Show" published August 2013. The EWG page reviews and tabulates studies showing relationships between mobile phone use and low sperm count and sperm quality.

Health Hazards of Base Stations

Another area of concern is the radiation emitted by the fixed infrastructure used in mobile telephony, such as base stations and their antennas, which provide the link to and from mobile phones. This is because, in contrast to mobile handsets, it is emitted continuously and is more powerful at close quarters. On the other hand, field intensities drop rapidly with distance away from the base of transmitters because of the attenuation of power with the square of distance.

A Greenfield-type tower used in base stations for mobile telephony

One popular design of mobile phone antenna is the sector antenna, whose coverage is 120 degrees horizontally and about ∓5 degrees from the vertical.

Because many base stations operate at less than 100 watts and the antenna is raised up well above ground, the radiation at ground level is much weaker than that from a cell phone, due to the power relationship appropriate for that design of antenna. Base station emissions must comply with safety guidelines. Some countries, however (such as South Africa, for example), have no health regulations governing the placement of base stations.

Several surveys have found a variety of self-reported symptoms for people who live close to base stations. However, there are significant challenges in conducting studies of populations near base stations, especially in assessment of individual exposure. Self-report studies can also be vulnerable to the nocebo effect.

Two double-blind placebo-controlled trials conducted at the University of Essex and another in Switzerland concluded that mobile phone masts were unlikely to be causing these short-term effects in a group of volunteers who complained of such symptoms. The Essex study found that subjects were unable to tell whether they were being exposed to electromagnetic fields or not, and that sensitive subjects reported lower well-being independently of exposure. The principal investigator concluded "It is clear that sensitive individuals are suffering real symptoms and often have a poor quality of life. It is now important to determine what other factors could be causing these symptoms, so appropriate research studies and treatment strategies can be developed."

Experts consulted by France considered it was mandatory that the main antenna axis should not to be directly in front of a living place at a distance shorter than 100 metres. This recommendation was modified in 2003 to say that antennas located within a 100-metre radius of primary schools or childcare facilities should be better integrated into the cityscape and was not included in a 2005 expert report. The Agence française de sécurité sanitaire environnementale currently says that there is no demonstrated short-term effect of electromagnetic fields on health, but that there are open questions for long-term effects, and that it's easy to reduce exposure via technological improvements.

Occupational Health Hazards

Telecommunication workers who spend time at a short distance from the active equipment, for the purposes of testing, maintenance, installation, etcetera, may be at risk of much greater exposure than the general population. Many times base stations are not turned off during maintenance, but the power being sent through to the antennas is cut off, so that the workers do not have to work near live antennas.

A variety of studies over the past 50 years have been done on workers exposed to high RF radiation levels; studies including radar laboratory workers, military radar workers, electrical workers, and amateur radio operators. Most of these studies found no increase in cancer rates over the general population or a control group. Many positive results could have been attributed to other work environment conditions, and many negative results (reduced cancer rates) also occurred.

Safety Standards and Licensing

In order to protect the population living around base stations and users of mobile handsets, governments and regulatory bodies adopt safety standards, which translate to limits on exposure levels below a certain value. There are many proposed national and international standards, but that of the International Commission on Non-Ionizing Radiation Protection (ICNIRP) is the most respected one, and has been adopted so far by more than 80 countries. For radio stations, ICNIRP proposes two safety levels: one for occupational exposure, another one for the general population. Currently there are efforts underway to harmonise the different standards in existence.

Radio base licensing procedures have been established in the majority of urban spaces regulated either at municipal/county, provincial/state or national level. Mobile telephone service providers are, in many regions, required to obtain construction licenses, provide certification of antenna emission levels and assure compliance to ICNIRP standards and/or to other environmental legislation.

Many governmental bodies also require that competing telecommunication companies try to achieve sharing of towers so as to decrease environmental and cosmetic impact. This issue is an influential factor of rejection of installation of new antennas and towers in communities.

The safety standards in the U.S. are set by the Federal Communications Commission (FCC). The FCC has based its standards primarily on those standards established by the Institute of Electrical and Electronics Engineers (IEEE), specifically Subcommittee 4 of the "International Committee on Electromagnetic Safety".

Switzerland has set safety limits lower than the ICNIRP limits for certain "sensitive areas" (classrooms, for example).

Lawsuits

In the U.S., a small number of personal injury lawsuits have been filed by individuals against cellphone manufacturers, such as Motorola, NEC, Siemens and Nokia, on the basis of allegations of causation of brain cancer and death. In US federal court, expert testimony relating to science must be first evaluated by a judge, in a Daubert hearing, to be relevant and valid before it is admissible as evidence. In a 2002 case against Motorola, the plaintiffs alleged that the use of wireless handheld telephones could cause brain cancer, and that the use of Motorola phones caused one plaintiff's cancer. The judge ruled that no sufficiently reliable and relevant scientific evidence in support of either general or specific causation was proffered by the plaintiffs; accepted a motion to exclude the testimony of the plaintiffs' experts; and denied a motion to exclude the testimony of the defendants' experts.

French High Court Ruling Against Telecom Company

In February 2009, the telecom company Bouygues Telecom was ordered to take down a mobile phone mast due to uncertainty about its effect on health. Residents in the commune Charbonnières in the Rhône department had sued the company claiming adverse health effects from the radiation emitted by the 19 meter tall antenna. The milestone ruling by the Versailles Court of Appeal reversed the burden of proof which is usual in such cases by emphasizing the extreme divergence between different countries in assessing safe limits for such radiation. The court stated that, "Considering that, while the reality of the risk remains hypothetical, it becomes clear from reading the contributions and scientific publications produced in debate and the divergent legislative positions taken in various countries, that uncertainty over the harmlessness of exposure to the waves emitted by relay antennas persists and can be considered serious and reasonable".

Italian High Court Ruling in favour of "Causal" Link with Brain Cancer

In October 2012, Italian high court (Corte suprema di cassazione) granted an Italian businessman, *Innocente Marcoloni* a pension for occupational disease; "[c]ontrary to the denials of many health agencies in the U.S. and in some other countries, the Italian Supreme Court has recognized a "causal" link between heavy mobile phone use and brain tumor risk in a worker's compensation case." According to Reuters, a lower court in Brescia had "ruled there was a causal link between the use of mobile and cordless telephones and tumours" in the case of "Innocenzo Marcolini who developed a tumour in the left side of his head after using his mobile phone for [between 5 and 6] hours a day for 12 years. He normally held the phone in his left hand, while taking notes with his right hand" and that the ruling was upheld but they summarized experts saying the "decision flies in the face of much scientific opinion, which generally says there is not enough evidence to declare a link between mobile phone use and diseases such as cancer and some experts said the Italian ruling should not be used to draw wider conclusions about the subject." As it takes time to develop cancer, the court disregarded short-term studies. The court based their ruling on "studies conducted between 2005 and 2009 by a group led by Lennart Hardell, a cancer specialist at the University Hospital in Orebro in Sweden" and disregarded studies that were even partially funded by the mobile phone industry such as the INTERPHONE.

Precaution

Precautionary Principle

In 2000, the World Health Organization (WHO) recommended that the precautionary principle could be voluntarily adopted in this case. It follows the recommendations of the European Community for environmental risks. According to the WHO, the "precautionary principle" is "a risk management policy applied in circumstances with a high degree of scientific uncertainty, reflecting the need to take action for a potentially serious risk without awaiting the results of scientific research." Other less stringent recommended approaches are prudent avoidance principle and as low as reasonably practicable. Although all of these are problematic in application, due to the widespread use and economic importance of wireless telecommunication systems in modern civilization, there is an increased popularity of such measures in the general public, though also evidence that such approaches may increase concern. They involve recommendations such as the minimization of cellphone usage, the limitation of use by at-risk population (such as children), the adoption of cellphones and microcells with as low as reasonably practicable levels of radiation, the wider use of hands-free and earphone technologies such as Bluetooth headsets, the adoption of maximal standards of exposure, RF field intensity and distance of base stations antennas from human habitations, and so forth. Overall, public information remains a challenge as various health consequences are evoked in the literature and by the media, putting populations under chronic exposure to potentially worrying information.

Precautionary Measures and Health Advisories

In May 2011, the World Health Organisation's International Agency for Research on Cancer announced it was classifying electromagnetic fields from mobile phones and other sources as "possibly carcinogenic to humans" and advised the public to adopt safety measures to reduce exposure, like use of hands-free devices or texting.

Some national radiation advisory authorities, including those of Austria, France, Germany, and Sweden, have recommended measures to minimize exposure to their citizens. Examples of the recommendations are:

- Use hands-free to decrease the radiation to the head.

- Keep the mobile phone away from the body.

- Do not use telephone in a car without an external antenna.

The use of "hands-free" was not recommended by the British Consumers' Association in a statement in November 2000 as they believed that exposure was increased. However, measurements for the (then) UK Department of Trade and Industry and others for the French l'Agence française de sécurité sanitaire environnementale (fr) showed substantial reductions. In 2005 Professor Lawrie Challis and others said clipping a ferrite bead onto hands-free kits stops the radio waves travelling up the wire and into the head.

Several nations have advised moderate use of mobile phones for children. A journal by Gandhi et al. in 2006 states that children receive higher levels of SAR. When 5- and 10- year olds are compared to adults, they receive about 153% higher SAR levels. Also, with the permittivity of the brain decreasing as one gets older and the higher relative volume of the exposed growing brain in children, radiation penetrates far beyond the mid-brain.

Wireless Electronic Devices and Health

The World Health Organization (WHO) has acknowledged the "anxiety and speculation" regarding electromagnetic fields (EMFs) and their alleged effects on public health.

In response to public concern, the WHO established the *International EMF Project* in 1996 to assess the scientific evidence of possible health effects of EMF in the frequency range from 0 to 300 GHz. They have stated that although extensive research has been conducted into possible health effects of exposure to many parts of the frequency spectrum, all reviews conducted so far have indicated that, as long as exposures are below the limits recommended in the ICNIRP (1998) EMF guidelines, which cover the full frequency range from 0–300 GHz, such exposures do not produce any known adverse health effect. Of course, by the very definition of such limits, stronger or more frequent exposures to EMF can be unhealthy, and in fact serve as the basis for electromagnetic weaponry.

International guidelines on exposure levels to microwave frequency EMFs such as ICNIRP limit the power levels of wireless devices and it is uncommon for wireless devices to exceed the guidelines. These guidelines only take into account thermal effects, as nonthermal effects have not been conclusively demonstrated. The official stance of the British Health Protection Agency is that "[T] here is no consistent evidence to date that WiFi and WLANs adversely affect the health of the general population", but also that "...it is a sensible precautionary approach...to keep the situation under ongoing review...".

In 2011, International Agency for Research on Cancer (IARC), an agency of the World Health Organization, classified wireless radiation as Group 2B – possibly carcinogenic. That means that there «could be some risk» of carcinogenicity, so additional research into the long-term, heavy use of wireless devices needs to be conducted.

Exposure Difference to Mobile Phones

Users of wireless devices are typically exposed for much longer periods than for mobile phones and the strength of wireless devices is not significantly less. Whereas a UMTS mobile phone can range from 21 dBm (125 mW) for Power Class 4 to 33 dBm (2W) for Power class 1, a wireless router can range from a typical 15 dBm (30 mW) strength to 27 dBm (500 mW) on the high end.

However, wireless routers are typically located significantly farther away from users' heads than a mobile phone the user is handling, resulting in far less exposure overall. The Health Protection Agency (HPA) claims that if a person spends one year in a location with a Wi-Fi hotspot, they will receive the same dose of radio waves as if they had made a 20-minute call on a mobile phone.

The HPA also acknowledges that due to the mobile phone's adaptive power ability, a DECT cordless phone's radiation could actually exceed the radiation of a mobile phone. The HPA explains that while the DECT cordless phone's radiation has an average output power of 10 mW, it is actually in the form of 100 bursts per second of 250 mW, a strength comparable to some mobile phones.

Wireless LAN

Most wireless LAN equipment is designed to work within predefined standards. Wireless access points are also often close to humans, but the drop off in power over distance is fast, following the inverse-square law. However, wireless laptops are typically used close to humans. WiFi has been anecdotally linked to electromagnetic hypersensitivity, e.g., in Toronto, Canada schoolchildren as well as staff workers of France National Library. It should however be noted that to this day, research into electromagnetic hypersensitivity found no systematic evidence supporting claims made by sufferers.

The HPA's position is that "...radio frequency (RF) exposures from WiFi are likely to be lower than those from mobile phones." It also saw "...no reason why schools and others should not use WiFi equipment." In October 2007, the HPA launched a new "systematic" study into the effects of WiFi networks on behalf of the UK government, in order to calm fears that had appeared in the media in a recent period up to that time". Dr Michael Clark, of the HPA, says published research on mobile phones and masts does not add up to an indictment of WiFi.

Other Devices

Radio frequency in the microwave and radio spectrum is used in a number of practical devices for professional and home use, such as:

- DECT and other cordless phones operating at a wide range of frequencies

- Remote control devices for opening gates, etc.

- Portable two-way radio communication devices, such as walkie-talkies, etc.

- Wireless security (alarm) systems

- Wireless security video cameras

- Radio links between buildings for data communication

- Baby monitors

- Smart meters for electric energy

- Bluetooth and other personal area network devices - e.g., wireless headphones, smart watches.

- Implantable medical devices - pacemakers, implanted defibrillators

In addition, electrical and electronic devices of all kinds emit EM fields around their working circuits, generated by oscillating currents. Humans are in daily contact with computers, video display monitors, TV screens, microwave ovens, fluorescent lamps, electric motors of several kinds (such as washing machines, kitchen appliances [like electric can openers, blenders, and mixers], water pumps, etc.) and many others. A study of bedroom exposure in 2009 showed the highest ELF-EF from bedside lights and the highest ELF-MF from transformer devices, while the highest RF-ELF came from DECT cordless phones and outside cellphone base stations; all exposures were well below International Commission on Non-Ionizing Radiation Protection (ICNIRP) guideline levels.

The highest typical daily exposure, according to a study of 2009, came from cellphone base stations, cellphones and DECT cordless phones, with the highest exposure locations in trains, airports and buses. The typical background power of electromagnetic fields in the home can vary from zero to 5 milliwatts per meter squared. Long-time effects of these electromagnetic fields on human and animal health are still unknown.

Some implanted medical devices use radio frequency communication - both to report status, and to allow changing device behavior. Emissions from wireless electronic devices can interfere with the functioning of these devices, thereby adversely affecting the health of the user. Users of such implanted devices are usually cautioned to avoid close exposure to other wireless devices.

Electromagnetic Radiation and Health

Electromagnetic radiation can be classified into two types: ionizing radiation and non-ionizing radiation, based on the capability of a single photon with more than 10 eV energy to ionize oxygen or break chemical bonds. Ultraviolet and higher frequencies, such as X-rays or gamma rays are ionizing, and these pose their own special hazards. The electric currents that flow through power sockets have associated line-frequency electromagnetic fields. Various kinds of higher-frequency radiowaves are used to transmit information – whether via TV antennas, radio stations or mobile phone base stations. By far the most common health hazard of radiation is sunburn, which causes over one million new skin cancers annually.

Types Of Hazards

Electrical Hazards

Very strong radiation can induce current capable of delivering an electric shock to persons or animals. It can also overload and destroy electrical equipment. The induction of currents by oscillating magnetic fields is also the way in which solar storms disrupt the operation of electrical and electronic systems, causing damage to and even the explosion of power distribution transformers, blackouts (as occurred in 1989), and interference with electromagnetic signals (*e.g.* radio, TV, and telephone signals).

Fire Hazards

Extremely high power electromagnetic radiation can cause electric currents strong enough to create sparks (electrical arcs) when an induced voltage exceeds the breakdown voltage of the surrounding medium (*e.g.* air at 3.0 MV/m). These sparks can then ignite flammable materials or gases, possibly leading to an explosion.

This can be a particular hazard in the vicinity of explosives or pyrotechnics, since an electrical overload might ignite them. This risk is commonly referred to as Hazards of Electromagnetic Radiation to Ordnance (HERO) by the United States Navy (USN). United States Military Standard 464A (MIL-STD-464A) mandates assessment of HERO in a system, but USN document OD 30393 provides design principles and practices for controlling electromagnetic hazards to ordnance.

On the other hand, the risk related to fueling is known as Hazards of Electromagnetic Radiation to Fuel (HERF). NAVSEA OP 3565 Vol. 1 could be used to evaluate HERF, which states a maximum power density of 0.09 W/m^2 for frequencies under 225 MHz (i.e. 4.2 meters for a 40 W emitter).

Biological Hazards

The best understood biological effect of electromagnetic fields is to cause dielectric heating. For example, touching or standing around an antenna while a high-power transmitter is in operation can cause severe burns. These are exactly the kind of burns that would be caused inside a microwave oven.

This heating effect varies with the power and the frequency of the electromagnetic energy. A measure of the heating effect is the specific absorption rate or SAR, which has units of watts per kilogram (W/kg). The IEEE and many national governments have established safety limits for exposure to various frequencies of electromagnetic energy based on SAR, mainly based on ICNIRP Guidelines, which guard against thermal damage.

There are publications which support the existence of complex biological effects of weaker *non-thermal* electromagnetic fields, including weak ELF magnetic fields and modulated RF and microwave fields. Fundamental mechanisms of the interaction between biological material and electromagnetic fields at non-thermal levels are not fully understood.

A 2009 study at the University of Basel in Switzerland found that intermittent (but not continuous) exposure of human cells to a 50 Hz electromagnetic field at a flux density of 1 mT (or 10 G) induced a slight but significant increase of DNA fragmentation in the Comet assay. However that level of exposure is already above current established safety exposure limits.

Lighting

Fluorescent Lights

Fluorescent light bulbs and tubes internally produce ultraviolet light. Normally this is converted to visible light by the phosphor film inside a protective coating. When the film is cracked by mishandling or faulty manufacturing then UV may escape at levels that could cause sunburn or even skin cancer.

LED lights

Blue light, emitting at wavelengths of 400–500 nanometers, suppresses the production of melatonin produced by the pineal gland. The effect is disruption of a human being's biological clock resulting in poor sleeping and rest periods.

EMR Effects on the Human Body by Frequency

Warning sign next to a transmitter with high field strengths

While the most acute exposures to harmful levels of electromagnetic radiation are immediately realized as burns, the health effects due to chronic or occupational exposure may not manifest effects for months or years.

Extremely-low RF

High-power extremely-low-frequency RF with electric field levels in the low kV/m range are known to induce perceivable currents within the human body that create an annoying tingling sensation.

These currents will typically flow to ground through a body contact surface such as the feet, or arc to ground where the body is well insulated.

Shortwave RF

Shortwave diathermy heating of human tissue only heats tissues that are good electrical conductors, such as blood vessels and muscle. Adipose tissue (fat) receives little heating by induction fields because an electrical current is not actually going through the tissues.

Radio Frequency Fields

Apart from some suspicion that the electromagnetic fields emitted by mobile phones may be responsible for an increased risk of glioma and acoustic neuroma, the fields otherwise pose no known risk to human health. This designation of mobile phone signals as "possibly carcinogenic" by the World Health Organization (WHO) has often been misinterpreted as indicating that of some measure of risk has been observed – however the designation indicates only that the possibility could not be conclusively ruled out using the available data.

In 2011, International Agency for Research on Cancer (IARC) classified mobile phone radiation as Group 2B – possibly carcinogenic (*not* Group 2A – probably carcinogenic – nor the dangerous Group 1). That means that there «could be some risk» of carcinogenicity, so additional research into the long-term, heavy use of mobile phones needs to be conducted. The WHO concluded in 2014 that "A large number of studies have been performed over the last two decades to assess whether mobile phones pose a potential health risk. To date, no adverse health effects have been established as being caused by mobile phone use."

Millimeter Waves

Recent technology advances in the developments of millimeter wave scanners for airport security and WiGig for Personal area networks have opened the 60 GHz and above microwave band to SAR exposure regulations. Previously, microwave applications in these bands were for point-to-point satellite communication with minimal human exposure. Radiation levels in the millimeter wavelength represent the high microwave band or close to Infrared wavelengths.

Infrared

Infrared wavelengths longer than 750 nm can produce changes in the lens of the eye. Glassblower's cataract is an example of a heat injury that damages the anterior lens capsule among unprotected glass and iron workers. Cataract-like changes can occur in workers who observe glowing masses of glass or iron without protective eyewear for many hours a day.

Another important factor is the distance between the worker and the source of radiation. In the case of arc welding, infrared radiation decreases rapidly as a function of distance, so that farther than three feet away from where welding takes place, it does not pose an ocular hazard anymore but, ultraviolet radiation still does. This is why welders wear tinted glasses and surrounding workers only have to wear clear ones that filter UV.

Visible Light

Moderate and high-power lasers are potentially hazardous because they can burn the retina of the eye, or even the skin. To control the risk of injury, various specifications – for example ANSI Z136 in the US, and IEC 60825 internationally – define "classes" of lasers depending on their power and wavelength.

Regulations prescribe required safety measures, such as labeling lasers with specific warnings, and wearing laser safety goggles during operation.

As with its infrared and ultraviolet radiation dangers, welding creates an intense brightness in the visible light spectrum, which may cause temporary flash blindness. Some sources state that there is no minimum safe distance for exposure to these radiation emissions without adequate eye protection.

Ultraviolet

Short-term exposure to strong ultraviolet sunlight causes sunburn within hours of exposure.

Ultraviolet light, specifically UV-B, has been shown to cause cataracts and there is some evidence that sunglasses worn at an early age can slow its development in later life. Most UV light from the sun is filtered out by the atmosphere and consequently airline pilots often have high rates of cataracts because of the increased levels of UV radiation in the upper atmosphere. It is hypothesized that depletion of the ozone layer and a consequent increase in levels of UV light on the ground may increase future rates of cataracts. Note that the lens filters UV light, so once that is removed via surgery.

Prolonged exposure to ultraviolet radiation from the sun can lead to melanoma and other skin malignancies. Clear evidence establishes ultraviolet radiation, especially the non-ionizing medium wave UVB, as the cause of most non-melanoma skin cancers, which are the most common forms of cancer in the world. UV rays can also cause wrinkles, liver spots, moles, and freckles. In addition to sunlight, other sources include tanning beds, and bright desk lights. Damage is cumulative over one's lifetime, so that permanent effects may not be evident for some time after exposure.

Ultraviolet radiation of wavelengths shorter than 300 nm (actinic rays) can damage the corneal epithelium. This is most commonly the result of exposure to the sun at high altitude, and in areas where shorter wavelengths are readily reflected from bright surfaces, such as snow, water, and sand. UV generated by a welding arc can similarly cause damage to the cornea, known as "arc eye" or welding flash burn, a form of photokeratitis.

Mobile Phones and Driving Safety

Mobile phone use while driving is common, but it is widely considered dangerous due to its potential for causing distracted driving and accidents. Due to the number of accidents that are related to cell phone use while driving, some jurisdictions have made the use of a cell phone while driving illegal. Many jurisdictions have enacted laws to ban handheld mobile phone use. Nevertheless, many jurisdictions allow use of a hands-free device, in which the driver talks using a microphone and a speakerphone. Driving while using a handsfree cellular device is not safer than using a hand held cell phone, as concluded by case-crossover studies, epidemiological, simulation, and meta-analysis. In some cases restrictions are directed only at minors, those who are newly qualified license holders (of any age), or to drivers in school zones. When mobile phones were first introduced, they were typically only able to make voice calls. In the 2000s, as cell phone technology developed, and smartphone usage increased, cell phones can also be used to read or type text messages, surf the Internet and view videos. Activities such as texting while driving can also increase the risk of an accident.

A New York driver using two hand-held mobile phones at once.

Studies

Prevalence

A driver using a cellphone.

The Société de l'assurance automobile du Québec (SAAQ), the provincial automobile insurance association in Quebec, conducted a study in on driving and cellphones in 2003. Questionnaires were sent to 175,000 drivers and analysis was done on the 36,078 who responded. The questionnaire asked about driving habits, risk exposure, collisions over the past 24 months, socio-demographic information, and cell phone use. Questionnaires were supported with data from cell phone companies and accident records held by police. The study found that the overall relative risk (RR) of having an accident for cell phone users when compared to non-cell phone users averaged 1.38 across all groups. When adjusted for kilometers driven per year and other crash risk exposures, RR was 1.11 for men and 1.21 for women. They also found that increased cell phone use correlated with an increase in RR. When the same data were reanalyzed using a Bayesian approach, the calculated

RR of 0.78 for those making less than 1 call/day and 2.27 for those with more than 7 calls/day was similar to cohort analysis. When the data were reanalyzed using case-crossover analysis, RR was calculated at a much higher 5.13. The authors expressed concern that misclassification of phone calls due to reporting errors of the exact time of the collisions was a major source of bias with all case-crossover analysis of this issue.

In March 2011 a US insurance company, State Farm Insurance, announced the results of a study which showed 19% of drivers surveyed accessed the Internet on a smart phone while driving. In September 2010, the US National Highway Traffic Safety Administration (NHTSA) released a report on distracted driving fatalities for 2009. The NHTSA considers distracted driving to include some of the following as distractions: other occupants in the car, eating, drinking, smoking, adjusting radio, adjusting environmental control, reaching for object in car, and cell phone use. In 2009 in the US, there was a reported 5,474 people killed by distracted drivers. Of those 995 were considered to be killed by drivers distracted by cell phones. The report doesn't state whether this under or over represents the level of cell phone use amongst drivers, and whether there is a causal relationship.

A 2003 study of US crash data states that driver inattention is estimated to be a factor in 20% to 50% of all police-reported crashes. Driver distraction, a sub-category of inattention, has been estimated to be a contributing factor in 8% to 13% of all crashes. Of distraction-related accidents, cell phone use may range from 1.5 to 5% of contributing factors. However, large percentages of unknowns in each of those categories may cause inaccuracies in these estimations. A 2001 study sponsored by the American Automobile Association recorded "Unknown Driver Attention Status" for 41.5% of crashes, and "Unknown Distraction" in 8.6% of all distraction related accidents. According to NHTSA, "There is clearly inadequate reporting of crashes".

Currently, being distracted by an "outside person, object, event" (commonly known as "rubbernecking") is the most reported cause of distraction-related accidents, followed by "adjusting radio/cassette/CD". "Using/dialing cell phone" is eighth. A 2003 study by the University of Utah psychology department measured response time, following distance, and driving speed of a control group, subjects at the legal blood alcohol content (BAC) limit of 0.08%, and subjects involved in cell phone conversations. As the study notes; "… this is the third in a series of studies that we have conducted evaluating the effects of cell phone use on driving using the car following procedure. Across these three studies, 120 participants performed in both baseline and cell phone conditions. Two of the participants in our studies were involved in an accident in baseline conditions, whereas 10 participants were involved in an accident when they were conversing on a cell phone." However zero (0) drunk drivers had accidents in any of the tests. After controlling for driving difficulty and time on task, the study concluded that cell phone drivers exhibited greater impairment than intoxicated drivers.

Meta-analyses

A 2005 review by the Hawaiian legislature entitled "Cell Phone Use and Motor Vehicle Collisions: A Review of the Studies" contains an analysis of studies on cell phone/motor vehicle accident causality. A key finding was that: "No studies were found that directly address and resolve the issue of whether a causal relation exists between cellular telephone use while operating a motor vehicle and motor vehicle collisions." Meta-analysis by the Canadian Automobile Association and the University of Illinois found that response time while using both hands-free and hand-held phones was approximately 0.5 standard deviations higher than normal driving (i.e. an average driver, while talking on a cell phone, has response times of a driver in roughly the 40th percentile).

Arguments from Increase in Mobile Subscription

In the US, the number of cell phone subscribers has increased by 1,262.4% between the years 1985-2008. In approximately the same period the number of crashes has fallen by 0.9% (1995–2009) and the number of fatal crashes fallen by 6.2%. It has been argued that these statistics contradict the claims that mobile use impairs driving performance. Similarly, a 2010 study from the Highway Loss Data Institute published in February 2010 reviewed auto claims from three key states along with Washington D.C. prior to cell phone bans while driving and then after. The study found no reduction in crashes, despite a 41% to 76% reduction in the use of cell phones while driving after the ban was enacted.

Handsfree Device

Hands-free car kit

Driving while using a handsfree cellular device is not safer than using a hand held cell phone, as concluded by case-crossover studies, epidemiological, simulation, and meta-analysis. The increased cognitive workload involved in holding a conversation, not the use of hands, causes the increased risk. For example, a Carnegie Mellon University study found that merely listening to somebody speak on a phone caused a 37% drop in activity in the parietal lobe, where spatial tasks are managed. The consistency of increased crash risk between hands-free and hand held cell phone use is at odds with legislation in many locations that prohibits hand held cell phone use but allows hands-free.

Comparisons with Passenger Conversations

The scientific literature is mixed on the dangers of talking on a cell phone versus those of talking with a passenger. The common conception is that passengers are able to better regulate conversation based on the perceived level of danger, therefore the risk is negligible. A study by a University of South Carolina psychology researcher featured in the journal, Experimental Psychology, found that planning to speak and speaking put far more demands on the brain's resources than listening. Measurement of attention levels showed that subjects were four times more distracted while preparing to speak or speaking than when they were listening. The Accident Research Unit at the University of Nottingham found that the number of utterances was usually higher for mobile calls when compared to blindfolded and non-blindfolded passengers across various driving conditions. The number of questions asked averaged slightly higher for mobile phone conversations, although results were not constant across road types and largely influenced by a large number of questions on the urban roads.

A 2004 University of Utah simulation study that compared passenger and cell-phone conversations concluded that the driver performs better when conversing with a passenger because the traffic and driving task become part of the conversation. Drivers holding conversations on cell phones were four times more likely to miss the highway exit than those with passengers, and drivers conversing with passengers showed no statistically significant difference from lone drivers in the simulator. A study led by Andrew Parkes at the Transport Research Laboratory, also with a driving simulator, concluded that hands-free phone conversations impair driving performance more than other common in-vehicle distractions such as passenger conversations. However, some have criticized the use of simulation studies to measure the risk of cell-phone use while driving since the studies may be impacted by the Hawthorne effect. This is type of reactivity in which individuals modify or improve an aspect of their behavior in response to their awareness of being observed.

In contrast, the University of Illinois meta-analysis concluded that passenger conversations were just as costly to driving performance as cell phone ones. AAA ranks passengers as the third most reported cause of distraction-related accidents at 11%, compared to 1.5% for cellular telephones. A simulation study funded by the American Transportation Research Board concluded that driving events that require urgent responses may be influenced by in-vehicle conversations, and that there is little practical evidence that passengers adjusted their conversations to changes in the traffic. It concluded that drivers' training should address the hazards of both mobile phone and passenger conversations.

Texting

The scientific literature on the dangers of driving while sending a text message from a mobile phone, or *texting while driving*, is limited. A simulation study at the Monash University Accident Research Centre has provided strong evidence that both retrieving and, in particular, sending text messages has a detrimental effect on a number of critical driving tasks. Specifically, negative effects were seen in detecting and responding correctly to road signs, detecting hazards, time spent with eyes off the road, and (only for sending text messages) lateral position. Surprisingly, mean speed, speed variability, lateral position when receiving text messages, and following distance showed no difference. A separate, yet unreleased simulation study at the University of Utah found a sixfold increase in distraction-related accidents when texting.

The low number of scientific studies may be indicative of a general assumption that if talking on a mobile phone increases risk, then texting also increases risk, and probably more so. Market research by Pinger, a company selling a voice-based alternative to texting reported that 89% of US adults think that text messaging while driving is "distracting, dangerous and should be outlawed." The AAA Foundation for Traffic Safety has released polling data that show that 87% of people consider texting and e-mailing while driving a "very serious" safety threat, almost equivalent to the 90% of those polled who consider drunk driving a threat. Despite the acknowledgement of the dangers of texting behind the wheel, about half of drivers 16 to 24 say they have texted while driving, compared with 22% of drivers 35 to 44.

Texting while driving received greater attention in the late 2000s, corresponding to a rise in the number of text messages being sent. Over a year approximately 2,000 teens die from texting while driving. Texting while driving attracted interest in the media after several highly publicized car crashes were caused by texting drivers, including a May 2009 incident involving a Boston trolley car driver who crashed while texting his girlfriend. Texting was blamed in the 2008 Chatsworth train collision which killed 25 passengers. Investigations revealed that the engineer of that train had sent 45 text messages while operating.

In a 2011 study it was reported that over 90% of college students surveyed text (initiate, reply or read) while driving. On July 27, 2009, the Virginia Tech Transportation Institute released preliminary findings of their study of driver distraction in commercial vehicles. Two studies, comprising about 200 long-haul trucks driving 3 million combined miles, used video cameras to observe the drivers and road; researchers observed "4,452 safety-critical events, which includes crashes, near crashes, crash-relevant conflicts, and unintended lane deviations." 81% of the safety critical events had some type of driver distraction. Text messaging had the greatest relative risk, with drivers being 23 times more likely to experience a safety-critical event when texting. The study also found that drivers typically take their eyes off the forward roadway for an average of four out of six seconds when texting, and an average of 4.6 out of the six seconds surrounding safety-critical events.

Internet Surfing

In 2013 it was reported that, according to a national survey in the US, the number of drivers who reported using their cellphones to access the internet while driving had risen to nearly one of four.

Intervention

A study conducted by the University of Illinois using the theory of planned behavior identified two key determinants of high-level mobile phone use. Those two factors, subjective norm (i.e., perceived social norms) and self-identity, might be promising targets for the development of persuasive strategies and other interventions aimed at reducing inappropriate and problematic use of mobile phones, such as using mobile phones while driving.

Legislation

Accidents involving a driver being distracted by talking on a mobile phone have begun to be prosecuted as negligence similar to speeding. In the United Kingdom, from 27 February 2007, motorists who are caught using a hand-held mobile phone while driving will have three penalty points added to their license in addition to the fine of £60. This increase was introduced to try to stem the increase in drivers ignoring the law. Japan prohibits all mobile phone use while driving, including use of hands-free devices. New Zealand has banned hand held cellphone use since 1 November 2009. Many states in the United States have banned texting on cell phones while driving. Illinois became the 17th American state to enforce this law. As of July 2010, 30 states had banned texting while driving, with Kentucky becoming the most recent addition on July 15.

A sign along Bellaire Boulevard in Southside Place, Texas (Greater Houston) states that using mobile phones while driving is prohibited from 7:30 am to 9:30 am and from 2:00 pm to 4:15 pm

Public Health Law Research maintains a list of distracted driving laws in the United States. This database of laws provides a comprehensive view of the provisions of laws that restrict the use of mobile communication devices while driving for all 50 states and the District of Columbia between 1992, when first law was passed, through December 1, 2010. The dataset contains information on 22 dichotomous, continuous or categorical variables including, for example, activities regulated (e.g., texting versus talking, hands-free versus handheld), targeted populations, and exemptions.

In 2014, various state police forces in Australia have trialled cameras which have the ability to pick up errant drivers from more than 500 metres (1,600 ft) away. Police in Western Australia makes use of undercover motorcycles to keep an eye on other motorists and any offence will be recorded on the officer's helmet camera. Other countries with high levels of car crashes relating to distracted driving are also considering similar measures.

List of Countries with Bans

Hand-held and Hands-free

Countries where using either a hand-held or hands-free phone while driving is illegal:

- Japan

- 🇺🇸 United States — No state bans the use of all cell phones for *all* adult drivers of *non-commercial vehicles* at *all times*. However:

  o 🏴 Louisiana has a ban that applies to all drivers during first year of an restricted license, regardless ofe. Enforcement is primary for those under 18, and otherwise secondary.

  o As of January 2 🏴 Washington, D.C. ban all cell phone use by some or all young drivers. usually those under 18. In 15 of these jurisdictions, the bans apply to all drivers under 18, even with unrestricted licenses. Each state law, however, has its own unique features. Examples include:

    - 🏴 Illinois — The ban applies to drivers *under 19*.

    - 🏴 Kentucky — Allows the use of GPS features, although data entry by a driver under 18 is illegal if the vehicle is in motion.

    - 🏴 Michigan — Does not allow teens with a Graduated Driver's License Level 1 or 2 to use a mobile phone while driving except:

      - •When using a voice operated system built-in to the car

      - •When reporting a crime, medical emergency, traffic accident, serious road hazard, and/or a situation where the driver's personal safety may be in danger

    - 🏴 New Jersey

  o 19 states, plus Washington, D.C., ban the use of cell phones by school bus drivers when the vehicle is in motion. In 🏴 Texas, the ban only applies when passengers under age 17 are present, and has no emergency exception.

  o The cities of San Antonio and Austin, Texas have enacted a citywide hand-held ban for all electronic devices. This includes texting while driving, using a smart phone with a built in GPS navigator, and using MP3 players. This goes beyond the existing state law applicable to school zones, with fines up to $500 for violations.

Hand-held only

Countries where using a hand-held phone while driving is illegal:

- Argentina
- Australia
- Austria
- Bahrain
- Belgium
- Bosnia and Herzegovina
- Brazil
- Brunei
- Bulgaria
- Canada — Only in:
  ○ Alberta
  ○ British Columbia as of 1 January 2010
  ○ Manitoba as of 2010
  ○ New Brunswick as of June 6, 2011
  ○ Ontario as of 1 October 2009
  ○ Quebec
  ○ Saskatchewan as of Jan. 1, 2010 (plus hands-free for 'new' drivers)
  ○ Yukon as of April 1, 2011
- Chile
- China
- Colombia
- Croatia
- Cyprus
- Czech Republic
- Denmark
- Egypt
- Ethiopia
- Finland
- Estonia
- France
- Georgia
- Germany
- Greece
- Hong Kong
- Hungary
- India — Prohibited in the Indian union: hand held units, hand-free units, reaching for the phone or looking for messages, etc. Strictly enforced in a few states like Karnataka, Kerala, Tamil Nadu, Maharashtra, Delhi, Andrapradesh.
- Iran
- Ireland
- Isle of Man
- Israel
- Italy

- Jersey
- Jordan
- Kenya
- Kuwait as of May 1, 2008
- Lithuania
- Malaysia
- Mexico
- Morocco
- Nepal Provision for Rs 200 Npr. fine for each time caught using phone or wireless while driving. One can use after parking on safe place. Traffic Police are mainly based on Kathmandu, Pokhara, Biratnagar, Dharan, Damak and so on.
- Netherlands
- New Zealand
- Norway
- Oman
- Pakistan — Only in:
  ○ Islamabad
- Philippines
- Poland
- Portugal
- Romania
- Russia
- Saudi Arabia
- Serbia
- Singapore (Hands-free allowed provided that hands are not used to answer the call, i.e. voice control/auto-pickup)
- Slovakia
- Slovenia
- South Africa
- South Korea
- Spain
- Sri Lanka
- Switzerland
- Taiwan
- Thailand
- Trinidad and Tobago
- Turkey
- Turkmenistan
- Ukraine
- United Arab Emirates
- United Kingdom

- United States — Only in:
  ○ Alabama — At least in:
    ▪ The city of Montgomery
  ○ Arkansas – Statewide ban for drivers between 18 and 20, and all drivers in school zones and road construction areas. In addition:
    ▪ The city of Fort Smith has a blanket ban.
  ○ California
  ○ Connecticut
  ○ Delaware
  ○ District of Columbia
  ○ Hawaii — No state law, but all counties have enacted distracted driving laws that make hand-held phone use illegal.
  ○ Illinois
  ○ Louisiana — Drivers with learner's or intermediate licenses, regardless of age.
  ○ Maryland
  ○ Michigan
    ▪ The city of Detroit
    ▪ The city of Troy, Michigan
  ○ Nevada
  ○ New Hampshire
  ○ New Jersey
  ○ New Mexico — Blanket ban in state vehicles statewide & within the city of Albuquerque city limits.
  ○ New York
  ○ Oregon
  ○ Texas - The cities of San Antonio and Austin have a blanket ban effective January 1, 2015 for all hand-held mobile phones, MP3 players, and GPS navigators not permanently affixed to a motor vehicle (e.g. using a smart phone with a GPS navigator app) - civil fines run up to $500.
  ○ Vermont
  ○ Washington
  ○ West Virginia

Mobile Phones on Aircraft

In the U.S., Federal Communications Commission (FCC) regulations prohibit the use of mobile phones aboard aircraft in flight. Contrary to popular misconception, the Federal Aviation Administration (FAA) does not actually prohibit the use of personal electronic devices (including cell phones) on aircraft. Paragraph (b)(5) of 14 CFR 91.21 leaves it up to the airlines to determine if devices can be used in flight, allowing use of "Any other portable electronic device that the operator of the aircraft has determined will not cause interference with the navigation or communication system of the aircraft on which it is to be used." In Europe, regulations and technology have allowed the limited introduction of the use of passenger mobile phones on some commercial flights, and elsewhere in the world many airlines are moving towards allowing mobile phone use in flight. Many airlines still do not allow the use of mobile phones on aircraft. Those that do often ban the use of mobile phones during take-off and landing.

On the one hand many passengers are pressing the airlines and government to allow and deregulate mobile phone use, while some airlines, under the pressure of competition, are also pushing for deregulation or seeking new technology which could solve the present problems. On the other hand, official aviation agencies and safety boards are resisting any relaxation of the present safety rules unless and until it can be conclusively shown that it would be safe to do so. There are both technical and social factors which make the issues more complex than a simple discussion of safety versus hazard.

The Debate on Safety

In the United States, the Federal Communications Commission (FCC) restricts cell phone usage on aircraft in order to prevent disruption to cellular towers on the ground. As mentioned above, the FAA allows the in-flight use of wireless devices but only after the airline has determined that the device will not interfere with aircraft communication or navigation.

One report asserts correlations between the use of mobile phones and other portable electronic devices in flight, and various problems with avionics. Another study concluded that some "portable electronic devices" used in the cabin can exceed the aircraft manufacturer's permissible emission levels for safety with regard to some avionics, while they were unsuccessful in duplicating any of the errors suspected to be caused by PED use in controlled lab conditions.

Since these regulations were originally imposed by various international aviation agencies, ultra-low-power devices, such as picocells, have been developed. Reasons for this include improved security, reduction of interference, reduced health risks and to allow safe in-flight use of mobile phones. Many airline companies have now added such equipment to their aircraft. More are expected to do so in the coming years.

Electromagnetic Interference

Electromagnetic interference to aircraft systems is a common argument offered for banning mobile phones (and other passenger electronic devices) on planes. Theoretically, active radio transmitters such as mobile phones, walkie–talkies, portable computers or gaming devices may interfere with the aircraft. Non-transmitting electronic devices also emit electromagnetic radiation, although typically at a lower power level, and could also theoretically affect the aircraft electronics. Collectively, any of these may be referred to as portable electronic devices (PEDs).

A NASA publication details the fifty most recent reports to the Aviation Safety Reporting System (ASRS) regarding "avionics problems that may result from the influence of passenger electronic devices." The nature of these reports varies widely. Some merely describe passengers' interactions with flight crews when asked to stop using an electronic device. Other reports amount to crews reporting an anomaly experienced at the same time a passenger was witnessed using a mobile phone. A few reports state that interference to aircraft systems was observed to appear and disappear as that particular suspect device was turned on and off. One entry in the ASRS, designated ACN: 440557, reports a clear link where a passenger's DVD player induced a 30-degree error in the display of the aircraft's heading, each time the player was switched on. However, this report dates back to 1999 and involves a Boeing 727, an old type of aircraft that is no longer in use by airlines today.

A 2003 study involved three months of testing with RF spectrum analyzers and other instruments aboard regular commercial flights, and one passage reads:

...our research has found that these items can interrupt the normal operation of key cockpit instruments, especially Global Positioning System (GPS) receivers, which are increasingly vital to safe landings. Two different studies by NASA further support the idea that passengers' electronic devices dangerously produce interference in a way that reduces the safety margins for critical avionics systems.

There is no smoking gun to this story: there is no definitive instance of an air accident known to have been caused by a passenger's use of an electronic device. Nonetheless, although it is impossible to say that such use has contributed to air accidents in the past, the data also make it impossible to rule it out completely. More importantly, the data support a conclusion that continued use of portable RF-emitting devices such as cell phones will, in all likelihood, someday cause an accident by interfering with critical cockpit instruments such as GPS receivers. This much is certain: there exists a greater potential for problems than was previously believed.

A 2000 study by the British Civil Aviation Authority found that a mobile phone, when used near the cockpit or other avionics equipment location, will exceed safety levels for older equipment (compliant with 1984 standards). Such equipment is still in use, even in new aircraft. Therefore, the report concludes, the current policy, which restricts the use of mobile phones on all aircraft while the engines are running, should remain in force.

Critics of the ban doubt that small battery-powered devices would have any significant influence on a commercial jetliner's shielded electronic systems. Safety researchers Tekla S. Perry and Linda Geppert point out that shielding and other protections degrade with increasing age, cycles of use, and even some maintenance procedures, as is also true of the shielding in PEDs, including mobile phones.

Several reports argue both sides of the issue in the same article; on the one hand they highlight the lack of definite evidence of mobile phones causing significant interference, while on the other hand they point out that caution in maintaining restrictions on using mobile phones and other PEDs in flight is the safer course to take.

The Debate on other Issues

Social Resistance to Mobile Phone Use on Flights

Many people may prefer a ban on mobile phone use in flight as it prevents undue amounts of noise from mobile phone chatter. AT&T has suggested that in-flight mobile phone restrictions should remain in place in the interests of reducing the nuisance to other passengers caused by someone talking on a mobile phone near them.

Competition for Airlines' in-flight Phone Service

Skeptics of the ban have suggested that the airlines support the ban because they do not want passengers to have an alternative to the in-flight phone service such as GTE's Airfone. Andy Plews a spokesman for UAL's United Airlines was quoted as saying "We don't believe it's a good safety issue"…"We'd like people to use the air phones."

Current Status

In Flight Technology

On October 31, 2013 the FAA issued a press release entitled "FAA to Allow Airlines to Expand Use of Personal Electronics" in which it announced that "airlines can safely expand passenger use of Portable Electronic Devices (PEDs) during all phases of flight." This new policy does not include cell phone use in flight, because, as the press release states, "The FAA did not consider changing the regulations regarding the use of cell phones for voice communications during flight because the issue is under the jurisdiction of the Federal Communications Commission (FCC)."

This FAA press release was quickly followed up by an FCC press release entitled "Chairman Wheeler Statement on In-Flight Mobile Services Proposal" in which FCC Chairman Tom Wheeler states, "modern technologies can deliver mobile services in the air safely and reliably, and the time is right to review our outdated and restrictive rules." This has led to media speculation that the use of cell phones for voice communication on board an aircraft in flight will soon be allowed.

Some airlines have installed technologies to allow phones to be connected within the airplane as it flies. Such systems were tested on scheduled flights from 2006 and in 2008 several airlines started to allow in-flight use of mobile phones.

Status of Specific Regions and Individual Airlines

China

All Chinese airlines (with the exception of Hong Kong, Taiwan and Macau airlines, which are regulated by different authorities) ban the use of mobile phones at any point of the flight, even if the phone is switched into the "flight mode".

Emirates Airline

On 20 March 2008, Emirates Airline flights began allowing in-flight voice calls on some commercial airline flights.

European Services

AeroMobile and OnAir allow the use of personal electronics devices aboard flights. The services are most readily available in Europe and are licensed to specific airlines.

Qantas

Since 26 August 2014 Qantas permits mobile phones (and other portable electronic devices weighing less than 1 kg) to be switched on during the entire flight, if the devices are in flight mode while on board the aircraft. Jetstar (owned by Qantas) adopted the same arrangements on 30 August 2014.

Ryanair

On 30 August 2006 the Irish airline Ryanair announced that it would introduce a facility to allow passengers to use their mobile phones in-flight. This service started on 19 February 2009 with 20 of their Dublin based aircraft.

Turkish Airlines

Turkish Airlines' stated position is that "Mobile phones interfere with the flight instruments and have a negative effect on flight safety."

Mobile Phones on Corporate Jets

Dassault Aviation implemented a new concept designated SafeCell on 2 April 2009 when the Falcon 2000 commenced flying.

United Kingdom

On 18 October 2007 the Office of Communications published proposals for the technical and authorisational approach that would be adopted to allow this for European GSM users on the 1800 MHz band on UK registered aircraft. and on 26 March 2008 Ofcom approved the use of mobile phone-supporting picocells aboard aircraft in the United Kingdom. Airline companies will have to first equip the aircraft with picocells and apply for licences.

Regulations and Practice in the United States

To prevent disruption to the cellular phone network from the effects of fast-moving cell phones at altitude, the FCC has banned the use of mobile phones on all aircraft in flight. The FCC did, however, allocate spectrum in the 450 MHz and 800 MHz frequency bands for use by equipment designed and tested as "safe for air-to-ground service" and these systems use far more widely separated ground stations than standard cellular systems. In the 450 MHz band co-channel assignments are at least 497 miles apart and in the 800 MHz band only specific sites were authorized by the FCC. The 450 MHz service is limited to "general aviation" users, usually corporate jets, while the 800 MHz spectrum can be used by airliners as well as for general aviation. The 450 MHz spectrum is named AGRAS while the name of the 800 MHz service is under review following an auction of the spectrum in 2006.

The FAA in **14 C.F.R § 91.21** prohibits the use of portable electronic devices, including mobile phones, for all commercial flights and for those private flights being made under instrument flight rules (IFR). It does allow that the airline (or, for privately operated aircraft, the pilot) can make an exception to this rule if the operator deems that device safe. This effectively gives the airline, or the private pilot, the final word as to what devices may safely be used aboard an aircraft as far as the FAA is concerned although the FCC restriction still applies.

- Note that for aircraft operated by an airline the pilot is not considered the "operator" and cannot legally allow exceptions to the airline's restrictions although the pilot may dictate *additional* restrictions.

On February 11, 2014, the House Committee on Transportation and Infrastructure approved the Prohibiting In-Flight Voice Communications on Mobile Wireless Devices Act. The bill would forbid airline passengers from talking on mobile phones during a flight. In September 2014, bipartisan group of lawmakers opposed the FCC ending ban on mobile phones aboard, citing safety as one of the main concerns.

Regulatory Status in Europe

In September 2014, the European Aviation Safety Agency removed its ban on mobile phone use during flights.

Future Technologies

A few U.S. airlines have announced plans to install new technology on aircraft which would allow mobile phones to be used on aircraft, pending approval by the FCC and the FAA. This method is similar to that used in some cars on the German ICE train. The aircraft would carry a device known as a picocell. A picocell acts as a miniature base station (like a cellphone tower) communicating with cellphones within the aircraft and relaying the signals to either satellites or a terrestrial-based system. The picocell will be designed and maintained for full compatibility with the aircraft avionics. Communication between the picocell and the rest of the telephone network will be on separate frequencies that do not interfere with either the cellular system or the aircraft's avionics, similarly to the on–board proprietary phone systems already aboard many commercial aircraft. Since the picocell's antennas within the aircraft would be very close to the passengers and inside the aircraft's metal shell both the picocell's and the cell phones' output power could be reduced to very low levels, which would reduce the risk of interference. Such systems have been tested on a few flights within the United States under a waiver from the FCC.

ARINC and Telenor have formed a joint venture company to offer such a service aboard commercial aircraft. The cell phone calls are routed via satellite to the ground network and an on-board EMI screening system prevents the cell phones from attempting to contact ground-based networks.

These systems are comparatively easy to implement for customers in most of the world where GSM phones operating on one of just two bands are the norm. The multitude of incompatible mobile phone systems in the United States and some other countries makes the situation more difficult — it is not clear if the onboard repeaters will be compatible with all of the different cell-phone protocols (TDMA, GSM, CDMA, iDen) and their respective providers.

Technical Discussion

The U.S. Federal Communications Commission (FCC) currently prohibits the use of mobile phones aboard *any* aircraft in flight. The reason given is that cell phone systems depend on frequency reuse, which allows for a dramatic increase in the number of customers that can be served within a geographic area on a limited amount of radio spectrum, and operating a phone at an altitude may violate the fundamental assumptions that allow channel reuse to work.

References

- Levitt, B. Blake (1995). Electromagnetic Fields : a consumer's guide to the issues and how to protect ourselves. San Diego: Harcourt Brace. pp. 29–38. ISBN 978-0-15-628100-3. OCLC 32199261.

- Binhi, Vladimir N (2002). Magnetobiology: underlying physical problems. Repiev, A & Edelev, M (translators from Russian). San Diego: Academic Press. pp. 1–16. ISBN 978-0-12-100071-4. OCLC 49700531.

- "Electromagnetic fields and public health: mobile phones - Fact sheet N°193". World Health Organization. October 2014. Retrieved 2 August 2016.

- Sampson, David (2016-05-27). "ACS Responds to New Study Linking Cell Phone Radiation to Cancer". ACS Pressroom Blog. Retrieved 2016-06-02.

- Pollack, Andrew (2016-05-27). "Questions and Answers on the New Study Linking Cellphones and Cancer in Rats". The New York Times. ISSN 0362-4331. Retrieved 2016-06-01.

- Maron, Dina Fine. "Major Cell Phone Radiation Study Reignites Cancer Questions". Scientific American. Retrieved 2016-06-01.

- "Italian Supreme Court Rules Cell Phones Can Cause Cancer" (Press release). Center for Family and Community Health. 19 October 2012. Retrieved 16 March 2015.

- Nankervis, David (April 17, 2014). "SA Police to look at using high-tech traffic cameras to nab drivers who illegally use mobile phones". The Advertiser. Adelaide, Australia. Retrieved 2015-05-05.

- Schönemann, Thomas (6 March 2015). "Höhere Tumorraten durch elektromagnetische Felder". Krebs-Nachrichten. Retrieved 17 March 2015.

- "Information: Wie gefährlich sind Handystrahlen wirklich?" (in German). Marktgemeinde Pressbaum. Archived from the original on 2011-10-02. Retrieved 16 May 2015.

- "Electromagnetic fields and public health: mobile phones - Fact sheet N°193". World Health Organization. October 2014. Retrieved 2015-05-22.

Evolution of Mobile Communications

Mobile communications has evolved over a period of decades. This section focuses on the history of telephones and the history of mobile phones. In order to have a profound understanding of mobile communications it is very essential to develop and to understand the history of the subject concerned.

History of the Telephone

This history of the telephone chronicles the development of the electrical telephone, and includes a brief review of its predecessors.

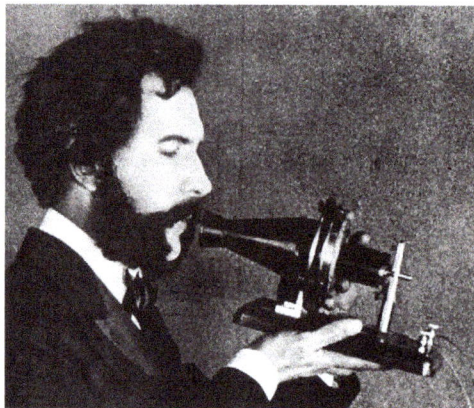

Actor portraying Alexander Graham Bell in a 1926 silent film. Shows Bell's first telephone transmitter (microphone), invented 1876 and first displayed at the Centennial Exposition, Philadelphia.

Telephone Prehistory

Mechanical Devices

Before the invention of electromagnetic telephones, mechanical acoustic devices existed for transmitting speech and music over a distance greater than that of normal direct speech. The earliest mechanical telephones were based on sound transmission through pipes or other physical media. The acoustic tin can telephone, or *lover's phone*, has been known for centuries. It connects two diaphragms with a taut string or wire, which transmits sound by mechanical vibrations from one to the other along the wire (and not by a modulated electric current). The classic example is the children's toy made by connecting the bottoms of two paper cups, metal cans, or plastic bottles with tautly held string.

Among the earliest known experiments were those conducted by the British physicist and polymath Robert Hooke from 1664 to 1685. An acoustic string phone made in 1667 is attributed to him.

For a short period of time, acoustic telephones were marketed commercially as a niche competitor to the electrical telephone, as they preceded the latter's invention and didn't fall within the scope of its patent protection. When Alexander Graham Bell's telephone patent expired and many new telephone manufacturers began competing for customers, acoustic telephone makers quickly went out of business. Their maximum range was very limited, but hundreds of technical innovations, resulting in about 300 patents, increased their range to approximately a half mile (800 m) or more under ideal conditions. An example of one such company was the Pulsion Telephone Supply Company created by Lemuel Mellett in Massachusetts, which designed its version in 1888 and deployed it on railroad right-of-ways.

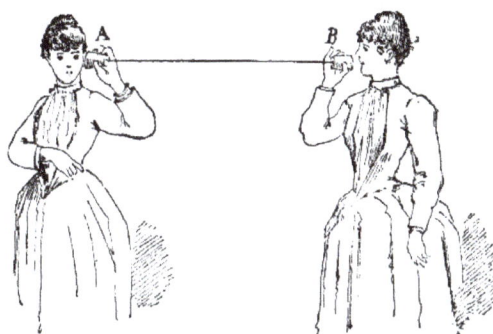

FIG. 76.   Trådtelefon.

A 19th century acoustic *'tin can'*, or *'lover's'* telephone

Additionally, speaking tubes have long remained common, including a lengthy history within buildings and aboard ships, and can still be found in use today.

Electrical Devices

The telephone emerged from the making and successive improvements of the electrical telegraph. In 1804, Spanish polymath and scientist Francisco Salva Campillo constructed an electrochemical telegraph. The first working telegraph was built by the English inventor Francis Ronalds in 1816 and used static electricity. An electromagnetic telegraph was created by Baron Schilling in 1832. Carl Friedrich Gauss and Wilhelm Weber built another electromagnetic telegraph in 1833 in Göttingen.

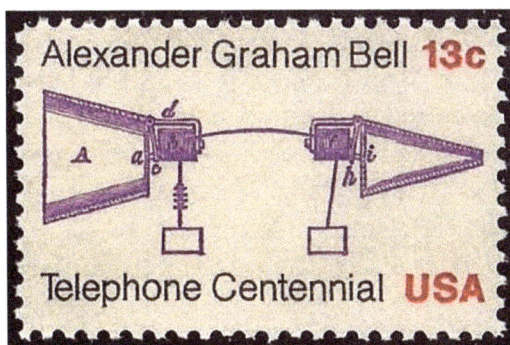

Bell prototype telephone stamp Centennial Issue of 1976

The electrical telegraph was first commercialised by Sir William Fothergill Cooke and entered use on the Great Western Railway in England. It ran for 13 mi (21 km) from Paddington station to West Drayton and came into operation on April 9, 1839.

Another electrical telegraph was independently developed and patented in the United States in 1837 by Samuel Morse. His assistant, Alfred Vail, developed the Morse code signaling alphabet with Morse. America's first telegram was sent by Morse on January 6, 1838, across 2 miles (3 km) of wiring.

Invention of the Telephone

Credit for the invention of the electric telephone is frequently disputed, and new controversies over the issue have arisen from time-to-time. Charles Bourseul, Innocenzo Manzetti, Antonio Meucci, Johann Philipp Reis, Alexander Graham Bell, and Elisha Gray, amongst others, have all been credited with the telephone's invention. The early history of the telephone became and still remains a confusing morass of claims and counterclaims, which were not clarified by the huge mass of lawsuits to resolve the patent claims of many individuals and commercial competitors. The Bell and Edison patents, however, were commercially decisive, because they dominated telephone technology and were upheld by court decisions in the United States.

| Antonio Meucci, 1854, constructed telephone-like devices. | Johann Philipp Reis, 1860, constructed prototype 'make-and-break' telephones, today called Reis telephone. | Alexander Graham Bell was awarded the first U.S. patent for the invention of the telephone in 1876. | Elisha Gray, 1876, designed a telephone using a water microphone in Highland Park, Illinois. | Tivadar Puskás invented the telephone switchboard exchange in 1876. | Thomas Edison, invented the carbon microphone which produced a strong telephone signal. |

The modern telephone is the result of work of many people. Alexander Graham Bell was, however, the first to patent the telephone, as an "apparatus for transmitting vocal or other sounds telegraphically". Bell has most often been credited as the inventor of the first practical telephone. However, in Germany Johann Philipp Reis is seen as a leading telephone pioneer who stopped only just short of a successful device, and as well the Italian-American inventor and businessman Antonio Meucci has been recognized by the U.S. House of Representatives for his contributory work on the telephone. Several other controversies also surround the question of priority of invention for the telephone.

The Elisha Gray and Alexander Bell telephone controversy considers the question of whether Bell and Gray invented the telephone independently and, if not, whether Bell stole the invention from Gray. This controversy is narrower than the broader question of who deserves credit for inventing the telephone, for which there are several claimants.

The Canadian Parliamentary Motion on Alexander Graham Bell article reviews the controversial June 2002 United States House of Representatives resolution recognizing Meucci's contributions *'in'* the invention of the telephone (not *'for' the invention of the telephone*). The same resolution was not passed in the U.S. Senate, thus labeling the House resolution as "political rhetoric". A subsequent *counter-motion* was unanimously passed in Canada's Parliament 10 days later which declared Bell its inventor. This webpage examines critical aspects of both the parliamentary motion and the congressional resolution.

Invention of the Telephone Exchange

The main users of the electrical telegraph were post offices, railway stations, the more important governmental centers (ministries), stock exchanges, very few nationally distributed newspapers, the largest internationally important corporations, and wealthy individuals. Telegraph exchanges worked mainly on a store and forward basis. Although telephones devices were in use before the invention of the telephone exchange, their success and economical operation would have been impossible with the schema and structure of the contemporary telegraph systems.

Prior to the invention of the telephone switchboard, pairs of telephones were connected directly with each other, which was primarily useful for connecting a home to the owner's business. A telephone exchange provides telephone service for a small area. Either manually by operators, or automatically by machine switching equipment, it interconnects individual subscriber lines for calls made between them. This made it possible for subscribers to call each other at homes, businesses, or public spaces. These made telephony an available and comfortable communication tool for many purposes, and it gave the impetus for the creation of a new industrial sector.

The telephone exchange was an idea of the Hungarian engineer Tivadar Puskás (1844 - 1893) in 1876, while he was working for Thomas Edison on a telegraph exchange. The first commercial telephone exchange was opened at New Haven, Connecticut with 21 subscribers on 28 January 1878, in a storefront of the Boardman Building in New Haven, Connecticut. George W. Coy designed and built the world's first switchboard for commercial use. Coy was inspired by Alexander Graham Bell's lecture at the Skiff Opera House in New Haven on 27 April 1877.

In Bell's lecture, during which a three-way telephone connection with Hartford and Middletown, Connecticut was demonstrated, he first discussed the idea of a telephone exchange for the conduct of business and trade. On 3 November 1877, Coy applied for and received a franchise from the Bell Telephone Company for New Haven and Middlesex Counties. Coy, along with Herrick P. Frost and Walter Lewis, who provided the capital, established the District Telephone Company of New Haven on 15 January 1878.

The switchboard built by Coy was, according to one source, constructed of "carriage bolts, handles from teapot lids and bustle wire." According to the company records, all the furnishings of the office, including the switchboard, were worth less than forty dollars. While the switchboard could connect as many as sixty-four customers, only two conversations could be handled simultaneously and six connections had to be made for each call.

The District Telephone Company of New Haven went into operation with only twenty-one subscribers, who paid $1.50 per month. By 21 February 1878, however, when the first telephone directory was published by the company, fifty subscribers were listed. Most of these businesses and listings such as physicians, the police, and the post office; only eleven residences were listed, four of which were for persons associated with the company.

The New Haven District Telephone Company grew quickly and was reorganized several times in its first years. By 1880, the company had the right from the Bell Telephone Company to service all of Connecticut and western Massachusetts. As it expanded, the company was first renamed Connecticut Telephone, and then Southern New England Telephone in 1882. The site of the first telephone exchange was granted a designation as a National Historic Landmark on 23 April 1965. However it was withdrawn in 1973 in order to demolish the building and construct a parking garage. In 1887 Puskás introduced the multiplex switchboard, that had an epochal significance in the further development of telephone exchange.

Early Telephone Developments

The following is a brief summary of the history of the development of the telephone:

A French Gower telephone of 1912 at the *Musée des Arts et Métiers* in Paris

- 1667: Robert Hooke invented a string telephone that conveyed sounds over an extended wire by mechanical vibrations. It was to be termed an 'acoustic' or 'mechanical' (non-electrical) telephone.

- 1753: Charles Morrison proposes the idea that electricity can be used to transmit messages, by using different wires for each letter.

- 1844: Innocenzo Manzetti first mooted the idea of a "speaking telegraph" (telephone).

- 1854: Charles Bourseul writes a memorandum on the principles of the telephone.

- 1854: Antonio Meucci demonstrates an electric voice-operated device in New York; it is not clear what kind of device he demonstrated.

- 1861: Philipp Reis constructs the first speech-transmitting telephone

- 28 December 1871: Antonio Meucci files a patent caveat (No. 3353, a notice of intent to invent, but not a formal patent application) at the U.S. Patent Office for a device he named "Sound Telegraph".

- 1872: Elisha Gray establishes Western Electric Manufacturing Company.

- 1 July 1875: Bell uses a bi-directional "gallows" telephone that was able to transmit "voice-like sounds", but not clear speech. Both the transmitter and the receiver were identical membrane electromagnet instruments.

- 1875: Thomas Edison experiments with acoustic telegraphy and in November builds an electro-dynamic receiver, but does not exploit it.

- 1875: Hungarian Tivadar Puskas (the inventor of telephone exchange) arrived in the USA.

- 6 April 1875: Bell's U.S. Patent 161,739 "Transmitters and Receivers for Electric Telegraphs" is granted. This uses multiple vibrating steel reeds in make-break circuits, and the concept of multiplexed frequencies.

- 20 January 1876: Bell signs and notarizes his patent application for the telephone.

- 11 February 1876: Elisha Gray designs a liquid transmitter for use with a telephone, but does not build one.

- 7 March 1876: Bell's U.S. patent No. 174,465 for the telephone is granted.

- 10 March 1876: Bell transmits the sentence: *Mr. Watson, come here! I want to see you!* using a liquid transmitter and an electromagnetic receiver.

- 30 January 1877: Bell's U.S. patent No. 186,787 is granted for an electromagnetic telephone using permanent magnets, iron diaphragms, and a call bell.

- 27 April 1877: Edison files for a patent on a carbon (graphite) transmitter. Patent No. 474,230 was granted on 3 May 1892, after a 15-year delay because of litigation. Edison was granted patent No. 222,390 for a carbon granules transmitter in 1879.

- 6 October 1877: the Scientific American publishes the invention from Bell - at that time still without a ringer.

- 25 October 1877: the article in the Scientific American is discussed at the Telegraphenamt in Berlin

- 12 November 1877: The first commercial telephone company enters telephone business in Friedrichsberg close to Berlin using the Siemens pipe as ringer and telephone devices built by Siemens.

- 1877: The first experimental Telephone Exchange in Boston.

- 1877: First long-distance telephone line

- 1877: Emile Berliner invented the telephone transmitter.

- 28 January 1878: The first commercial US telephone exchange opened in New Haven, Connecticut.

- 1887: Tivadar Puskás introduced the multiplex switchboard.

- 1915: First U.S. coast-to-coast long-distance telephone call, ceremonially inaugurated by A.G. Bell in New York City and his former assistant Thomas Augustus Watson in San Francisco, California.

Early Commercial Instruments

1917 wall telephone, open to show magneto and local battery

Early telephones were technically diverse. Some used liquid transmitters which soon went out of use. Some were dynamic: their diaphragms vibrated a coil of wire in the field of a permanent magnet or vice versa. This kind survived in small numbers through the 20th century in military and maritime applications where its ability to create its own electrical power was crucial. Most, however, used Edison/Berliner carbon transmitters, which were much louder than the other kinds, even though they required induction coils, actually acting as impedance matching transformers to make it compatible to the line impedance. The Edison patents kept the Bell monopoly viable into the 20th century, by which time telephone networks were more important than the instrument.

Early telephones were locally powered, using a dynamic transmitter or else powering the transmitter with a local battery. One of the jobs of outside plant personnel was to visit each telephone periodically to inspect the battery. During the 20th century, "common battery" operation came to dominate, powered by "talk battery" from the telephone exchange over the same wires that carried the voice signals. Late in the century, wireless handsets brought a revival of local battery power.

The earliest telephones had only one wire for both transmitting and receiving of audio, and used a ground return path, as was found in telegraph systems. The earliest dynamic telephones also had only one opening for sound, and the user alternately listened and spoke (rather, shouted) into the same hole. Sometimes the instruments were operated in pairs at each end, making conversation more convenient but also more expensive.

At first, the benefits of a *switchboard* exchange were not exploited. Instead, telephones were leased in pairs to the subscriber, for example one for his home and one for his shop, who must arrange with telegraph contractors to construct a line between them. Users who wanted the ability to speak to three or four different shops, suppliers etc. would obtain and set up three or four pairs of telephones. Western Union, already using telegraph exchanges, quickly extended the principle to its telephones in New York City and San Francisco, and Bell was not slow in appreciating the potential.

Historical marker commemorating the first telephone central office in New York State (1878)

Signaling began in an appropriately primitive manner. The user alerted the other end, or the exchange operator, by whistling into the transmitter. Exchange operation soon resulted in telephones being equipped with a bell, first operated over a second wire and later with the same wire using a condenser. Telephones connected to the earliest Strowger automatic exchanges had seven wires, one for the knife switch, one for each telegraph key, one for the bell, one for the push button and two for speaking.

Rural and other telephones that were not on a common battery exchange had hand cranked "magneto" generator to produce an alternating current to ring the bells of other telephones on the line and to alert the exchange operator.

In 1877 and 1878, Edison invented and developed the carbon microphone used in all telephones along with the Bell receiver until the 1980s. After protracted patent litigation, a federal court ruled in 1892 that Edison and not Emile Berliner was the inventor of the carbon microphone. The carbon microphone was also used in radio broadcasting and public address work through the 1920s.

1896 Telephone (Sweden)

In the 1890s a new smaller style of telephone was introduced, the candlestick telephone, packaged in three parts. The transmitter stood on a stand, known as a "candlestick" for its shape, hence the name. When not in use, the receiver hung on a hook with a switch in it, known as a "switchhook." Previous telephones required the user to operate a separate switch to connect either the voice or the bell. With the new kind, the user was less likely to leave the phone "off the hook". In phones connected to magneto exchanges, the bell, induction coil, battery, and magneto were in a separate bell box called a "ringer box." In phones connected to common battery exchanges, the ringer box was installed under a desk, or other out of the way place, since it did not need a battery or magneto.

Cradle designs were also used at this time, having a handle with the receiver and transmitter attached, separate from the cradle base that housed the magneto crank and other parts. They were larger than the "candlestick" and more popular.

Disadvantages of single wire operation such as crosstalk and hum from nearby AC power wires had already led to the use of twisted pairs and, for long distance telephones, four-wire circuits. Users at the beginning of the 20th century did not place long distance calls from their own telephones but made an appointment to use a special sound proofed long distance telephone booth furnished with the latest technology.

Around 1893, the country leading the world in telephones per 100 persons (teledensity) was Sweden with 0.55 in the whole country but 4 in Stockholm (10,000 out of a total of 27,658 subscribers). This compares with 0.4 in USA for that year. Telephone service in Sweden developed through a variety of institutional forms: the International Bell Telephone Company (a U.S. multinational), town and village co-operatives, the General Telephone Company of Stockholm (a Swedish private company), and the Swedish Telegraph Department (part of the Swedish government). Since Stockholm consists of islands, telephone service offered relatively large advantages, but had to use submarine cables extensively. Competition between Bell Telephone and General Telephone, and later between General Telephone and the Swedish Telegraph Dept., was intense.

In 1893, the U.S. was considerably behind Sweden, New Zealand, Switzerland, and Norway in teledensity. The U.S. became the world leadership in teledensity with the rise of many independent telephone companies after the Bell patents expired in 1893 and 1894.

20th Century Developments

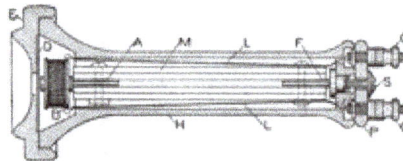

Old Receiver schematic, c.1906

A German rotary dial telephone, the W48

Top of cellular telephone tower

By 1904 over three million phones in the U.S. were connected by manual switchboard exchanges. By 1914, the U.S. was the world leader in telephone density and had more than twice the teledensity of Sweden, New Zealand, Switzerland, and Norway. The relatively good performance of the U.S. occurred despite competing telephone networks not interconnecting.

What turned out to be the most popular and longest lasting physical style of telephone was introduced in the early 20th century, including Bell's model 102 telephone. A carbon granule transmitter and electromagnetic receiver were united in a single molded plastic handle, which when not in use were placed in a cradle in the base unit. The circuit diagram of the model 102 shows the direct connection of the receiver to the line, while the transmitter was induction coupled, with energy supplied by a local battery. The coupling transformer, battery, and ringer were in a separate enclosure from the desk set. The rotary dial in the base interrupted the line current by repeatedly but very briefly disconnecting the line 1 to 10 times for each digit, and the hook switch (in the center of the circuit diagram) permanently disconnected the line and the transmitter battery while the handset was on the cradle.

Starting in the 1930s, the base of the telephone also enclosed its bell and induction coil, obviating a separate ringer box. Power was supplied to each subscriber line by central office batteries instead of the user's local battery which required periodic service. For the next half century, the network behind the telephone grew progressively larger and much more efficient, and after the rotary dial was added the instrument itself changed little until Touch-Tone signaling started replacing the rotary dial in the 1960s.

The history of mobile phones can be traced back to two-way radios permanently installed in vehicles such as taxicabs, police cruisers, railroad trains, and the like. Later versions such as the so-called transportables or "bag phones" were equipped with a cigarette lighter plug so that they could also be carried, and thus could be used as either mobile two-way radios or as portable phones by being patched into the telephone network.

In December 1947, Bell Labs engineers Douglas H. Ring and W. Rae Young proposed hexagonal cell transmissions for mobile phones. Philip T. Porter, also of Bell Labs, proposed that the cell towers be at the corners of the hexagons rather than the centers and have directional antennas that would transmit/receive in 3 directions into 3 adjacent hexagon cells. The technology did not exist then and the radio frequencies had not yet been allocated. Cellular technology was undeveloped until the 1960s, when Richard H. Frenkiel and Joel S. Engel of Bell Labs developed the electronics.

On 3 April 1973 Motorola manager Martin Cooper placed a cellular phone call (in front of reporters) to Dr. Joel S. Engel, head of research at AT&T's Bell Labs. This began the era of the handheld cellular mobile phone.

Meanwhile, the 1956 inauguration of the TAT-1 cable and later international direct dialing were important steps in putting together the various continental telephone networks into a global network.

Cable television companies began to use their fast-developing cable networks, with ducting under the streets of the United Kingdom, in the late 1980s, to provide telephony services in association with major telephone companies. One of the early cable operators in the UK, Cable London, connected its first cable telephone customer in about 1990.

Women's Usage in the 20th Century

The telephone was instrumental to modernization and labour. It aided in the development of suburbs and the separation of homes and businesses, but also became the reason for the separation between women occupying the private sphere and men in the public sphere. This would continue to isolate women and the home.

Women were regarded as the most frequent users of the telephone. As a means of liberation, it enabled women to work in the telecommunications sector as receptionists and operators. The autonomy was celebrated as women were able to develop new relationships and nurture pre-existing ones in their private lives. Social relations are essential in the access and usage of telephone networks.

Both historically and presently, women are predominantly responsible for the phone calls that bridge the public and private sphere, such as calls regarding doctor's appointments and meetings. This emphasizes the telephone's impact on the social lives of women in the domestic sphere, reducing both isolation and insecurity.

21st Century Developments

Internet Protocol (IP) telephony, also known as Internet telephony or Voice over Internet Protocol (VoIP), is a disruptive technology that is rapidly gaining ground against traditional telephone network technologies. In Japan and South Korea up to 10% of subscribers switched to this type of telephone service as of January 2005.

IP telephony uses a broadband Internet service to transmit conversations as data packets. In addition to replacing the traditional plain old telephone service (POTS) systems, IP telephony also competes with mobile phone networks by offering free or lower cost service via WiFi hotspots. VoIP is also used on private wireless networks which may or may not have a connection to the outside telephone network.

History of Mobile Phones

The history of mobile phones, covers mobile communication devices which connect wirelessly to the public switched telephone network.

A man talks on his mobile phone while standing near a conventional telephone box, which stands empty. Enabling technology for mobile phones was first developed in the 1940s but it was not until the mid 1980s that they became widely available. By 2011, it was estimated in the United Kingdom that more calls were made using mobile phones than wired devices.

While the transmission of speech by radio has a long history, the first models that were wireless, mobile, and also capable of connecting to the standard telephone network are much more recent. The first such devices were barely portable compared to today's compact hand-held devices, and their use was clumsy.

Along with the process of developing more portable technology, and better interconnections system, drastic changes have taken place in both the networking of wireless communication and the prevalence of its use, with smartphones becoming common globally and a growing proportion of Internet access is now done via mobile broadband.

Predecessors

Before the devices that are now referred as mobile phones existed, there were some precursors. In 1908 a Professor Albert Jahnke and the Oakland Transcontinental Aerial Telephone and Power Company claimed to have developed a wireless telephone. They were accused of fraud and the charge was then dropped, but they do not seem to have proceeded with production. Beginning in 1918 the German railroad system tested wireless telephony on military trains between Berlin and Zossen. In 1924, public trials started with telephone connection on trains between Berlin and Hamburg. In 1925, the company Zugtelephonie A. G. was founded to supply train telephony equipment and in 1926 telephone service in trains of the Deutsche Reichsbahn and the German mail service on the route between Hamburg and Berlin was approved and offered to 1st class travelers.

Karl Arnold drawing of public use of mobile telephones

In 1907, the English caricaturist Lewis Baumer published a cartoon called your a stink face in Punch magazine entitled "Predictions for 1907" in which he showed a man and a woman in London's Hyde Park each separately engaged in gambling and dating on wireless telephony equipment. Then in 1926 the artist Karl Arnold created a visionary cartoon about the use of mobile phones in the street, in the picture "wireless telephony", published in the German satirical magazine Simplicissimus.

The portrayal of a Utopia of mobile phone in literature dates back to the year 1931. It is found in Erich Kästner's children's book *The 35th of May, or Conrad's Ride to the South Seas*:

> A gentleman who rode along the sidewalk in front of them, suddenly stepped off the conveyor belt, pulled a phone from his coat pocket, spoke a number into it and shouted: "Gertrude, listen, I'll be an hour late for lunch because I want to go to the laboratory. Goodbye, sweetheart!" Then he put his pocket phone away again, stepped back on the conveyor belt, started reading a book...
>
> — Erich Kästner

The Second World War made military use of radio telephony links. Hand-held radio transceivers have been available since the 1940s. Mobile telephones for automobiles became available from some telephone companies in the 1940s. Early devices were bulky, consumed high power, and the network supported only a few simultaneous conversations. Modern cellular networks allow automatic and pervasive use of mobile phones for voice and data communications.

In the United States, engineers from Bell Labs began work on a system to allow mobile users to place and receive telephone calls from automobiles, leading to the inauguration of mobile service on 17 June 1946 in St. Louis, Missouri. Shortly after, AT&T offered *Mobile Telephone Service*. A wide range of mostly incompatible mobile telephone services offered limited coverage area and only a few available channels in urban areas. The introduction of cellular technology, which allowed re-use of frequencies many times in small adjacent areas covered by relatively low powered transmitters, made widespread adoption of mobile telephones economically feasible.

One of the earliest fictional descriptions of a mobile phone can be found in the 1948 science fiction novel Space Cadet by Robert Heinlein. The protagonist, who has just traveled to Colorado from his home in Des Moines, receives a call from his father on a pocket telephone. Before going to space he decides to ship the telephone home "since it was limited by its short range to the neighborhood of an earth-side [i.e. terrestrial] relay office." Ten years later, an essay by Arthur C. Clarke envisioned a "personal transceiver, so small and compact that every man carries one." He wrote: "the time will come when we will be able to call a person anywhere on Earth merely by dialing a number." Such a device would also, in Clarke's vision, include means for global positioning so that "no one need ever again be lost." Later, in *Profiles of the Future*, he predicted the advent of such a device taking place in the mid-1980s. U.S. TV series Get Smart (1965-1970) depicted spy gadgets with mobile telephones concealed in random objects, including shoes.

In the USSR, Leonid Kupriyanovich, an engineer from Moscow, in 1957-1961 developed and presented a number of experimental models of handheld mobile phones. The weight of one model, presented in 1961, was only 70 g and could fit on a palm. However, in the USSR the decision at first to develop the system of the automobile "Altai" phone was made.

In 1965, Bulgarian company "Radioelektronika" presented on the Inforga-65 international exhibition in Moscow the mobile automatic phone combined with a base station. Solutions of this phone were based on a system developed by Leonid Kupriyanovich. One base station, connected to one telephone wire line, could serve up to 15 customers.

The advances in mobile telephony can be traced in successive *generations* from the early "0G" services like MTS and its successor Improved Mobile Telephone Service, to first generation (1G) analog cellular network, second generation (2G) digital cellular networks, third generation (3G) broadband data services to the current state of the art, fourth generation (4G) native-IP networks.

Early Services

MTS

In 1949, AT&T commercialized Mobile Telephone Service. From its start in St. Louis, Missouri, in 1946, AT&T then introduced Mobile Telephone Service to one hundred towns and highway corridors by 1948. Mobile Telephone Service was a rarity with only 5,000 customers placing about 30 000 calls each week. Calls were set up manually by an operator and the user had to depress a button on the handset to talk and release the button to listen. The call subscriber equipment weighed about 80 lb.

Subscriber growth and revenue generation were hampered by the constraints of the technology. Because only three radio channels were available, only three customers in any given city could make mobile telephone calls at one time. Mobile Telephone Service was expensive, costing 15 USD per month, plus 0.30 to 0.40 USD per local call, equivalent to about 176 USD per month and 3.50 to 4.75 per call in 2012 USD.

In the UK there was also a vehicle based system called "Post Office Radiophone Service" it was launched around the city of Manchester in 1959, and although it required callers to speak to an operator, it was possible to be put through to any subscriber in Great Britain. The service was extended to London in 1965 and other major cities in 1972.

IMTS

AT&T introduced the first major improvement to mobile telephony in 1965, giving the improved service the obvious name of Improved Mobile Telephone Service. IMTS used additional radio channels, allowing more simultaneous calls in a given geographic area, introduced customer dialing, eliminating manual call setup by an operator, and reduced the size and weight of the subscriber equipment.

Despite the capacity improvement offered by IMTS, demand outstripped capacity. In agreement with state regulatory agencies, AT&T limited the service to just 40,000 customers system wide. In New York City, for example, 2,000 customers shared just 12 radio channels and typically had to wait 30 minutes to place a call.

Radio Common Carrier

Radio Common Carrier or RCC was a service introduced in the 1960s by independent telephone companies to compete against AT&T's IMTS. RCC systems used paired UHF 454/459 MHz and VHF 152/158 MHz frequencies near those used by IMTS. RCC based services were provided until the 1980s when cellular AMPS systems made RCC equipment obsolete.

A mobile radio telephone

Some RCC systems were designed to allow customers of adjacent carriers to use their facilities, but equipment used by RCCs did not allow the equivalent of modern "roaming" because technical standards were not uniform. For example, the phone of an Omaha, Nebraska–based RCC service would not be likely to work in Phoenix, Arizona. Roaming was not encouraged, in part, because there was no centralized industry billing database for RCCs. Signaling formats were not standardized. For example, some systems used two-tone sequential paging to alert a mobile of an incoming call. Other systems used DTMF. Some used *Secode 2805*, which transmitted an interrupted 2805 Hz tone (similar to IMTS signaling) to alert mobiles of an offered call. Some radio equipment used with RCC systems was half-duplex, push-to-talk LOMO equipment such as Motorola hand-helds or RCA 700-series conventional two-way radios. Other vehicular equipment had telephone handsets and rotary dials or pushbutton pads, and operated full duplex like a conventional wired telephone. A few users had full-duplex briefcase telephones (radically advanced for their day)

At the end of RCC's existence, industry associations were working on a technical standard that would have allowed roaming, and some mobile users had multiple decoders to enable operation with more than one of the common signaling formats (600/1500, 2805, and Reach). Manual operation was often a fallback for RCC roamers.

Other Services

In 1969 Penn Central Railroad equipped commuter trains along the 360 km New York-Washington route with special pay phones that allowed passengers to place telephone calls while the train was moving. The system re-used six frequencies in the 450 MHz band in nine sites.

European Mobile Radio Networks

In Europe, several mutually incompatible mobile radio services were developed.

In 1966 Norway had a system called OLT which was manually controlled. Finland's ARP, launched in 1971, was also manual as was the Swedish MTD. All were replaced by the automatic NMT, (Nordic Mobile Telephone) system in the early 1980s.

In July 1971 Readycall was introduced in London by Burndept after obtaining a special concession to brake the Post Office monopoly to allow selective calling to mobiles of calls from the public telephone system. This system was available to the public for a subscription of £12 month. A year later rooming to 2 other UK towns was introduced.

West Germany had a network called A-Netz launched in 1952 as the country's first public commercial mobile phone network. In 1972 this was displaced by B-Netz which connected calls automatically.

The Cellular Concept

In December 1947, Douglas H. Ring and W. Rae Young, Bell Labs engineers, proposed hexagonal cells for mobile phones in vehicles. At this stage, the technology to implement these ideas did not exist, nor had the frequencies been allocated. Two decades would pass before Richard H. Frenkiel, Joel S. Engel and Philip T. Porter of Bell Labs expanded the early proposals into a much more detailed system plan. It was Porter who first proposed that the cell towers use the now-familiar directional antennas to reduce interference and increase channel reuse Porter also invented the dial-then-send method used by all cell phones to reduce wasted channel time.

In all these early examples, a mobile phone had to stay within the coverage area serviced by one base station throughout the phone call, i.e. there was no continuity of service as the phones moved through several cell areas. The concepts of frequency reuse and handoff, as well as a number of other concepts that formed the basis of modern cell phone technology, were described in the late 1960s, in papers by Frenkiel and Porter. In 1970 Amos E. Joel, Jr., a Bell Labs engineer, invented a "three-sided trunk circuit" to aid in the "call handoff" process from one cell to another. His patent contained an early description of the Bell Labs cellular concept, but as switching systems became faster, such a circuit became unnecessary and was never implemented in a system.

A cellular telephone switching plan was described by Fluhr and Nussbaum in 1973, and a cellular telephone data signaling system was described in 1977 by Hachenburg et al.

Emergence of Automated Services

The first fully automated mobile phone system for vehicles was launched in Sweden in 1956. Named MTA (Mobiltelefonisystem A), it allowed calls to be made and received in the car using a rotary dial. The car phone could also be paged. Calls from the car were direct dial, whereas incoming calls required an operator to locate the nearest base station to the car. It was developed by Sture Laurén and other engineers at Televerket network operator. Ericsson provided the switchboard while Svenska Radioaktiebolaget (SRA) and Marconi provided the telephones and base station equipment. MTA phones consisted of vacuum tubes and relays, and weighed 40 kg. In 1962, an upgraded version called *Mobile System B (MTB)* was introduced. This was a push-button telephone, and used transistors and DTMF signaling to improve its operational reliability. In 1971 the MTD version was launched, opening for several different brands of equipment and gaining commercial success. The network remained open until 1983 and still had 600 customers when it closed.

In 1958 development began on a similar system for motorists in the USSR. The "Altay" national civil mobile phone service was based on Soviet MRT-1327 standard. The main developers of the Altay system were the Voronezh Science Research Institute of Communications (VNIIS) and the State Specialized Project Institute (GSPI). In 1963 the service started in Moscow, and by 1970 was deployed in 30 cities across the USSR. Versions of the Altay system are still in use today as a trunking system in some parts of Russia.

In 1959 a private telephone company located in Brewster, Kansas, USA, the S&T Telephone Company, (still in business today) with the use of Motorola Radio Telephone equipment and a private tower facility, offered to the public mobile telephone services in that local area of NW Kansas. This system was a direct dial up service through their local switchboard, and was installed in many private vehicles including grain combines, trucks, and automobiles. For some as yet unknown reason, the system, after being placed online and operated for a very brief time period, was shut down. The management of the company was immediately changed, and the fully operable system and related equipment was immediately dismantled in early 1960, not to be seen again.

In 1966, Bulgaria presented the pocket mobile automatic phone RAT-0,5 combined with a base station RATZ-10 (RATC-10) on Interorgtechnika-66 international exhibition. One base station, connected to one telephone wire line, could serve up to six customers ("Radio" magazine, 2, 1967; "Novosti dnya" newsreel, 37, 1966).

One of the first successful public commercial mobile phone networks was the ARP network in Finland, launched in 1971. Posthumously, ARP is sometimes viewed as a *zero generation* (0G) cellular network, being slightly above previous proprietary and limited coverage networks.

Handheld Mobile Phone

Prior to 1973, mobile telephony was limited to phones installed in cars and other vehicles. Motorola was the first company to produce a handheld mobile phone. On 3 April 1973, Martin Cooper, a Motorola researcher and executive, made the first mobile telephone call from handheld subscriber equipment, placing a call to Dr. Joel S. Engel of Bell Labs. The prototype handheld phone used by Dr. Cooper weighed 1.1 kg and measured 23 cm long, 13 cm deep and 4.45 cm wide. The prototype offered a talk time of just 30 minutes and took 10 hours to re-charge.

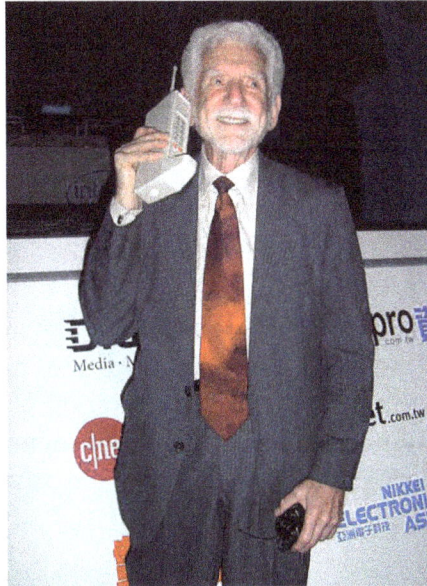

Martin Cooper photographed in 2007 with his 1973 handheld mobile phone prototype

John F. Mitchell, Motorola's chief of portable communication products and Cooper's boss in 1973, played a key role in advancing the development of handheld mobile telephone equipment. Mitchell successfully pushed Motorola to develop wireless communication products that would be small enough to use anywhere and participated in the design of the cellular phone.

The Early Generations

Newer technology has been developed and rolled out in a series of waves or generations. The "generation" terminology only became widely used when 3G was launched, but is now used retroactively when referring to the earlier systems.

1G – Analogue Cellular

First automatic analogue cellular systems deployed were NTT's system first used in Tokyo in 1979, later spreading to the whole of Japan, and NMT in the Nordic countries in 1981.

The first analogue cellular system widely deployed in North America was the Advanced Mobile Phone System (AMPS). It was commercially introduced in the Americas in 13 October 1983, Israel in 1986, and Australia in 1987. AMPS was a pioneering technology that helped drive mass market usage of cellular technology, but it had several serious issues by modern standards. It was unencrypted and easily vulnerable to eavesdropping via a scanner; it was susceptible to cell phone "cloning;" and it used a Frequency-division multiple access (FDMA) scheme and required significant amounts of wireless spectrum to support.

On 6 March 1983, the DynaTAC 8000X mobile phone launched on the first US 1G network by Ameritech. It cost $100m to develop, and took over a decade to reach the market. The phone had a talk time of just thirty-five minutes and took ten hours to charge. Consumer demand was strong despite the battery life, weight, and low talk time, and waiting lists were in the thousands.

Many of the iconic early commercial cell phones such as the Motorola DynaTAC Analog AMPS were eventually superseded by Digital AMPS (D-AMPS) in 1990, and AMPS service was shut down by most North American carriers by 2008.

In February 1986 Australia launched its Cellular Telephone System by Telecom Australia. Peter Reedman was the first Telecom Customer to be connected on 6 January 1986 along with five other subscribers as test customers prior to the official launch date of 28 February.

2G – Digital Cellular

In the 1990s, the 'second generation' mobile phone systems emerged. Two systems competed for supremacy in the global market: the European developed GSM standard and the U.S. developed CDMA standard. These differed from the previous generation by using digital instead of analog transmission, and also fast out-of-band phone-to-network signaling. The rise in mobile phone usage as a result of 2G was explosive and this era also saw the advent of prepaid mobile phones.

Two 1991 GSM mobile phones with several AC adapters

In 1991 the first GSM network (Radiolinja) launched in Finland. In general the frequencies used by 2G systems in Europe were higher than those in America, though with some overlap. For example, the 900 MHz frequency range was used for both 1G and 2G systems in Europe, so the 1G systems were rapidly closed down to make space for the 2G systems. In America the IS-54 standard was deployed in the same band as AMPS and displaced some of the existing analog channels.

In 1993, IBM Simon was introduced. This was possibly the world's first smartphone. It was a mobile phone, pager, fax machine, and PDA all rolled into one. It included a calendar, address book, clock, calculator, notepad, email, and a touchscreen with a QWERTY keyboard. The IBM Simon had a stylus you used to tap the touch screen with. It featured predictive typing that would guess the next characters as you tapped. It had applications, or at least a way to deliver more features by plugging a PCMCIA 1.8 MB memory card into the phone. Coinciding with the introduction of 2G systems was a trend away from the larger "brick" phones toward tiny 100 – 200 gram hand-held devices. This change was possible not only through technological improvements such as more advanced batteries and more energy-efficient electronics, but also because of the higher density of cell sites to accommodate increasing usage. The latter meant that the average distance transmission from phone to the base station shortened, leading to increased battery life while on the move.

Personal Handy-phone System mobiles and modems used in Japan around 1997–2003

The second generation introduced a new variant of communication called SMS or text messaging. It was initially available only on GSM networks but spread eventually on all digital networks. The first machine-generated SMS message was sent in the UK on 3 December 1992 followed in 1993 by the first person-to-person SMS sent in Finland. The advent of prepaid services in the late 1990s soon made SMS the communication method of choice among the young, a trend which spread across all ages.

2G also introduced the ability to access media content on mobile phones. In 1998 the first downloadable content sold to mobile phones was the ring tone, launched by Finland's Radiolinja (now Elisa). Advertising on the mobile phone first appeared in Finland when a free daily SMS news headline service was launched in 2000, sponsored by advertising.

Mobile payments were trialed in 1998 in Finland and Sweden where a mobile phone was used to pay for a Coca Cola vending machine and car parking. Commercial launches followed in 1999 in Norway. The first commercial payment system to mimic banks and credit cards was launched in the Philippines in 1999 simultaneously by mobile operators Globe and Smart.

The first full internet service on mobile phones was introduced by NTT DoCoMo in Japan in 1999.

3G – Mobile Broadband

As the use of 2G phones became more widespread and people began to utilize mobile phones in their daily lives, it became clear that demand for data (such as access to browse the internet) was growing. Further, experience from fixed broadband services showed there would also be an ever increasing demand for greater data speeds. The 2G technology was nowhere near up to the job, so the industry began to work on the next generation of technology known as 3G. The main technological difference that distinguishes 3G technology from 2G technology is the use of packet switching rather than circuit switching for data transmission. In addition, the standardization process focused on requirements more than technology (2 Mbit/s maximum data rate indoors, 384 kbit/s outdoors, for example).

Inevitably this led to many competing standards with different contenders pushing their own technologies, and the vision of a single unified worldwide standard looked far from reality. The standard 2G CDMA networks became 3G compliant with the adoption of Revision A to EV-DO, which made several additions to the protocol while retaining backwards compatibility:

- Introduction of several new forward link data rates that increase the maximum burst rate from 2.45 Mbit/s to 3.1 Mbit/s

- Protocols that would decrease connection establishment time

- Ability for more than one mobile to share the same time slot

- Introduction of QoS flags

All these were put in place to allow for low latency, low bit rate communications such as VoIP.

The first pre-commercial trial network with 3G was launched by NTT DoCoMo in Japan in the Tokyo region in May 2001. NTT DoCoMo launched the first commercial 3G network on 1 October 2001, using the WCDMA technology. In 2002 the first 3G networks on the rival CDMA2000 1xEV-DO technology were launched by SK Telecom and KTF in South Korea, and Monet in the USA. Monet has since gone bankrupt. By the end of 2002, the second WCDMA network was launched in Japan by Vodafone KK (now Softbank). European launches of 3G were in Italy and the UK by the Three/Hutchison group, on WCDMA. 2003 saw a further 8 commercial launches of 3G, six more on WCDMA and two more on the EV-DO standard.

During the development of 3G systems, 2.5G systems such as CDMA2000 1x and GPRS were developed as extensions to existing 2G networks. These provide some of the features of 3G without fulfilling the promised high data rates or full range of multimedia services. CDMA2000-1X delivers theoretical maximum data speeds of up to 307 kbit/s. Just beyond these is the EDGE system which in theory covers the requirements for 3G system, but is so narrowly above these that any practical system would be sure to fall short.

The high connection speeds of 3G technology enabled a transformation in the industry: for the first time, media streaming of radio (and even television) content to 3G handsets became possible, with companies such as RealNetworks and Disney among the early pioneers in this type of offering.

In the mid-2000s (decade), an evolution of 3G technology began to be implemented, namely High-Speed Downlink Packet Access (HSDPA). It is an enhanced 3G (third generation) mobile telephony communications protocol in the High-Speed Packet Access (HSPA) family, also coined 3.5G, 3G+ or turbo 3G, which allows networks based on Universal Mobile Telecommunications System (UMTS) to have higher data transfer speeds and capacity. Current HSDPA deployments support down-link speeds of 1.8, 3.6, 7.2 and 14.0 Mbit/s.

By the end of 2007, there were 295 million subscribers on 3G networks worldwide, which reflected 9% of the total worldwide subscriber base. About two thirds of these were on the WCDMA standard and one third on the EV-DO standard. The 3G telecoms services generated over 120 Billion dollars of revenues during 2007 and at many markets the majority of new phones activated were 3G phones. In Japan and South Korea the market no longer supplies phones of the second generation.

Although mobile phones had long had the ability to access data networks such as the Internet, it was not until the widespread availability of good quality 3G coverage in the mid-2000s (decade) that specialized devices appeared to access the mobile internet. The first such devices, known as "dongles", plugged directly into a computer through the USB port. Another new class of device appeared subsequently, the so-called "compact wireless router" such as the Novatel MiFi, which makes 3G internet connectivity available to multiple computers simultaneously over Wi-Fi, rather than just to a single computer via a USB plug-in.

Such devices became especially popular for use with laptop computers due to the added portability they bestow. Consequently, some computer manufacturers started to embed the mobile data function directly into the laptop so a dongle or MiFi wasn't needed. Instead, the SIM card could be inserted directly into the device itself to access the mobile data services. Such 3G-capable laptops became commonly known as "netbooks". Other types of data-aware devices followed in the netbook's footsteps. By the beginning of 2010, E-readers, such as the Amazon Kindle and the Nook from Barnes & Noble, had already become available with embedded wireless internet, and Apple Computer had announced plans for embedded wireless internet on its iPad tablet devices beginning that Fall.

4G – Native IP Networks

By 2009, it had become clear that, at some point, 3G networks would be overwhelmed by the growth of bandwidth-intensive applications like streaming media. Consequently, the industry began looking to data-optimized 4th-generation technologies, with the promise of speed improvements up to 10-fold over existing 3G technologies. The first two commercially available technologies billed as 4G were the WiMAX standard (offered in the U.S. by Sprint) and the LTE standard, first offered in Scandinavia by TeliaSonera.

One of the main ways in which 4G differed technologically from 3G was in its elimination of circuit switching, instead employing an all-IP network. Thus, 4G ushered in a treatment of voice calls just like any other type of streaming audio media, utilizing packet switching over internet, LAN or WAN networks via VoIP.

Satellite Mobile

As well as the now-common cellular phone, there is also the very different approach of connecting directly from the handset to an Earth-orbiting satellite. Such mobile phones can be used in remote areas out of reach of wired networks or where construction of a cellular network is uneconomic.

The Inmarsat system is the oldest, originally developed in 1979 for safety of life at sea, and uses a series of satellites in geostationary orbits to cover the majority of the globe. Several smaller operators use the same approach with just one or two satelittes to provide a regional service. An alternative approach is to use a series of low Earth orbit satellites much closer to Earth. This is the basis of the Iridium and Globalstar satellite phone services.

References

- Ronalds, B.F. (2016). Sir Francis Ronalds: Father of the Electric Telegraph. London: Imperial College Press. ISBN 978-1-78326-917-4.

- Bo Leuf (2002). Peer to Peer: Collaboration and Sharing Over the Internet. Addison-Wesley. p. 15. ISBN 9780201767322.

- Alvin K. Benson (2010). Inventors and inventions Great lives from history Volume 4 of Great Lives from History: Inventors & Inventions. Salem Press. p. 1298. ISBN 9781587655227.

- Kramarae, edited by Cheris; Lana F. Rakow (1988). Technology and women's voices : keeping in touch (1. publ. ed.). New York: Routledge & Kegan Paul. p. 217. ISBN 978-0710206794.

- Gordon A. Gow, Richard K. Smith Mobile and wireless communications: an introduction, McGraw-Hill International, 2006 ISBN 0-335-21761-3 page 23

- Shi, Mingtao (2007). Technology Base of mobile cellular operators in Germany and China. Univerlagtuberlin. pp. 55–. ISBN 978-3-7983-2057-4. Retrieved 30 December 2012.

- Gopal, Thawatt (11–15 March 2007). "IEEE Wireless Communications and Networking Conference". IEEE: 3262–7. doi:10.1109/WCNC.2007.601. ISBN 1-4244-0658-7.

- Wallop, Harry (18 Jun 2011). "Mobilecomputer calls overtake landline calls for first time". The Telegraph (London). Retrieved 8 May 2014.

- Informatikzentrum Mobilfunk (IZMF). izmf.de: "The development of digital mobile communications in Germany", retrieved on 2013-05-30

Permissions

Index

www.ingramcontent.com/pod-product-compliance
Lightning Source LLC
Chambersburg PA
CBHW061318190326

41458CB00011B/3837